Interface Technology for Computer-Controlled Manufacturing Processes

MANUFACTURING ENGINEERING AND MATERIALS PROCESSING

A Series of Reference Books and Textbooks

SERIES EDITORS

Geoffrey Boothroyd

Department of Mechanical Engineering
University of Massachusetts
Amherst, Massachusetts

George E. Dieter

Dean, College of Engineering
University of Maryland
College Park, Maryland

1. Computers in Manufacturing, *U. Rembold, M. Seth, and J. S. Weinstein*
2. Cold Rolling of Steel, *William L. Roberts*
3. Strengthening of Ceramics: Treatments, Tests, and Design Applications, *Henry P. Kirchner*
4. Metal Forming: The Application of Limit Analysis, *Betzalel Avitzur*
5. Improving Productivity by Classification, Coding, and Data Base Standardization: The Key to Maximizing CAD/CAM and Group Technology, *William F. Hyde*
6. Automatic Assembly, *Geoffrey Boothroyd, Corrado Poli, and Laurence E. Murch*
7. Manufacturing Engineering Processes, *Leo Alting*
8. Modern Ceramic Engineering: Properties, Processing, and Use in Design, *David W. Richerson*
9. Interface Technology for Computer-Controlled Manufacturing Processes, *Ulrich Rembold, Karl Armbruster, and Wolfgang Ühlzmann*

OTHER VOLUMES IN PREPARATION

Interface Technology for Computer-Controlled Manufacturing Processes

ULRICH REMBOLD
Institut für Informatik III
Universität Karlsruhe
Karlsruhe, Federal Republic of Germany

KARL ARMBRUSTER
Protec GmbH, Karlsruhe
Karlsruhe, Federal Republic of Germany

WOLFGANG ÜLZMANN
Fachhochschule Wedel
Wedel, Federal Republic of Germany

MARCEL DEKKER, INC. New York and Basel

Library of Congress Cataloging in Publication Data

Rembold, Ulrich
Interface technology for computer-controlled manufacturing processes.

(Manufacturing engineering and materials processing; 9)

Includes bibliographies and index.
1. Manufacturing processes--Data processing.
2. Computer interfaces. I. Armbruster, Karl, [date]
II. Ülzmann, Wolfgang, [date]. III. Title.
IV. Series.
TS183.R45 1983 670.42'7 82-22166
ISBN 0-8247-1836-4

MARCEL DEKKER, INC.
270 Madison Avenue, New York, New York 10016

Current printing (last digit):
10 9 8 7 6 5 4 3 2 1

PRINTED IN THE UNITED STATES OF AMERICA

PREFACE

The interfacing of control computers with manufacturing processes is a very special art that requires thorough knowledge of computer architecture, software, measuring equipment, and control devices. The complex subject of interfacing is seldom taught at universities, and is, in general, not well understood by the engineers and scientists responsible for the implementation of control computer systems using microcomputers and minicomputers. Usually, the application engineer must struggle to design a functional interface and to write the software to operate it. Often, different design solutions are found by engineers with different backgrounds, resulting in expensive and complex systems. Typically electrical engineers try to solve a problem predominantly with hardware components because they may not be familiar with software, whereas programmers try to obtain a pure software solution since they do not know the hardware. In reality, the best solution is a properly balanced software/hardware tradeoff.

In many installations designing and programming the interface and connecting the computer to the process requires 30 to 50 percent of the total development effort. Thus an attempt is made here to treat interface technology from a practical point of view. The authors have spent many years in both industry and universities implementing minicomputers and microcomputers for a wide spectrum of applications. A thorough discussion of the hardware and software aspects of interfacing and the interactions of these two technologies is presented. The book will be of value to the student concerned with the use of computers for process control and instrumentation. It also will aid application engineers in manufacturing and processing industries to conceive and design interfaces for computer control of machines and instrumentation. Because online laboratory computers employ the same interface technology, this material will help researchers and scientists to build and use more advanced laboratory instrumentation.

To understand this book the reader should know the fundamentals of assembly programming, have some knowledge of digital electronics, and be familiar with the basic electrical principles of sensors and control elements.

First the book gives an overview of the historical development of interfacing practices used in process control technology. It is necessary to understand the principles of conventional control elements since in many computer installations an attempt is being made to update the control system of an old plant. However, the reader will find that better solutions can be obtained when computers are installed in new plants together with modern measuring and control devices.

The basic operation of minicomputers and microcomputers is briefly explained and several commonly used sensors are discussed, although it is not the purpose of the book to explain sensor technology in detail.

Attention is given to the architecture of data channels and to communication principles. Since interface technology has been changing along with the rapid development of integrated circuits, some future development trends are pointed out.

The individual functions of the interface are explained. Since the interface must communicate with the computer and the process, it often has to be able to receive and send control signals and information to and from many different components that have different signal requirements. Because the interrupt structure of the computer has a pronounced influence on the design of the interface, it is shown how different interrupt structures of currently available computers work.

A description is given of the most important hardware components incorporated into the design of interfaces. It is shown how to generate and count pulses and how these pulses are used to form and interpret control signals. To obtain proper sensor signals from the process and to send control information to control devices it is necessary to convert digital information to analog information, or vice versa, and in many cases to amplify these signals. The components that process this information are discussed in detail. It is further shown how controllers, stepping motors, and special-purpose devices are interfaced and operated.

Principal aspects of interface software are explained. It is shown how the design of an interface can be improved by the proper selection of hardware and software technology. A discussion is given of the art of writing a driver for a peripheral and of how the driver communicates with the operating system. Since the use of properly formed signals often eliminates interference problems, light is shed on different grounding and wiring procedures.

One chapter is devoted to special interface bus systems. These systems were designed to make standard buses available to the user and to simplify the implementation of control tasks. Unfortunately, most of these buses are expensive and are only used where a high payoff can be expected from a computer installation or where error-free communication is of extreme importance. Particular attention is given to the CAMAC, IEC, Ethernet, and PDV bus systems.

In the last chapter the most important I/O modules for microcomputers are explained. Large-scale integration makes it possible to accommodate many control functions of an interface on a single chip. Very advanced I/O modules even contain a microprocessor that performs preprocessing and control operations normally done by the main processor.

This book is based on an interface course for students of computer science at the University of Karlsruhe. There were several people who made it possible to write the book; they deserve our sincere thanks. The numerous drawings and graphs were made by Miss Maria Hornung. Miss Cornelia Xanke and Mrs. Cynthia Klohr typed the manuscript and the tables. Miss Mariann Ogilvie helped proofread the manuscript.

Ulrich Rembold
Karl Armbruster
Wolfgang Ülzmann

CONTENTS

1

THE AUTOMATION OF PROCESSES

1.1 TECHNICAL PROCESSES

The word *process* has many different meanings in the English language. For example, there are cognitive processes, technical processes, growing processes, and so on. In this book we define a process as follows: A *technical process* is the transformation or transportation of matter, energy, and/or information. The state of a technical process can be measured and controlled with the help of technical devices.

Technical processes can be very simple or complex. Because in manufacturing many different subprocesses are combined to form one single entity, it is impossible to list all technical processes. It is also very difficult to give uniform guidelines to show how individual subprocesses can be combined in a complete entity. When identical products are manufactured at different locations, the manufacturing process in each case will consist of a different structure, since the number of products made, fabrication practices at the plant, labor market, raw material supply, modes of transportation, and so on, will vary.

Processes can be divided into three categories: continuous processes, discrete processes, and hybrid processes.

1.1.1 Continuous Process

The continuous processes are, in general, deterministic processes in which a continuous stream of material or energy is transported or manufactured (Fig. 1.1) [1]. Typical processes of this type are as follows:

Continuous steel production
Chemical processes
Continuous material flow
Energy production
Petrochemical processes

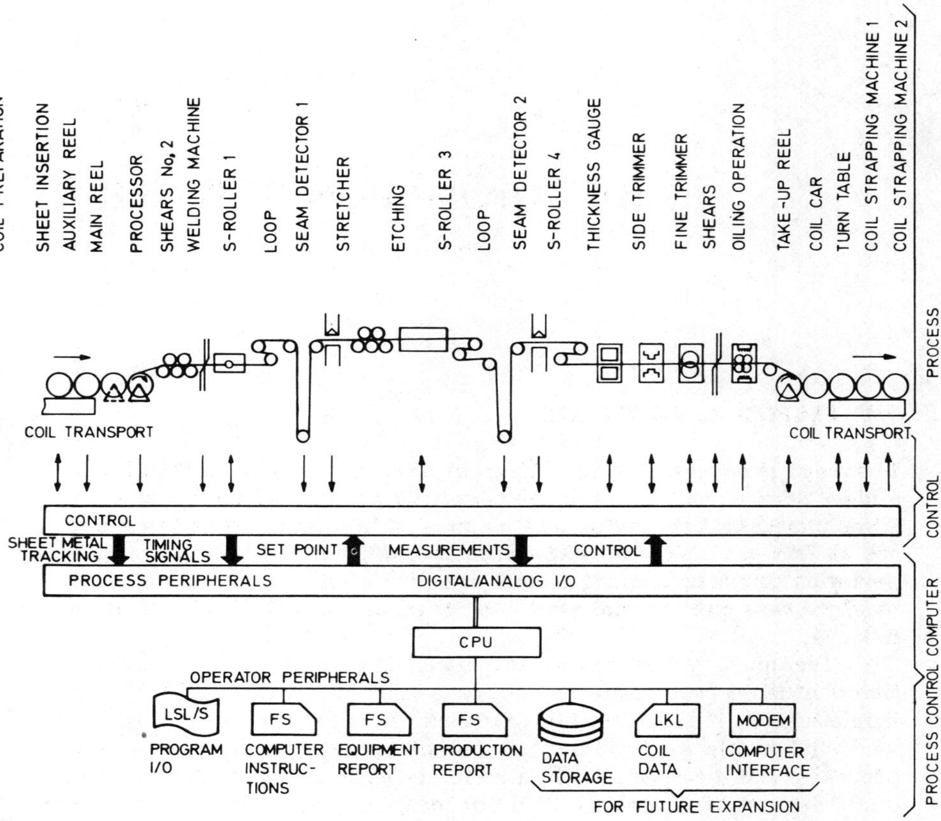

FIG. 1.1 Continuous metal surface treatment process. LSL/S: paper tape reader and punch; FS: I/O peripheral; LKL: card reader. (Courtesy of Brown, Boveri, and Cie, Aktiengesellschaft, Mannheim, Federal Republic of Germany.)

1.1.2 Discrete Process

The discrete processes are usually of a stochastic nature. They are used for the production of discrete products (Fig. 1.2) [2]. They include

Single and mass production
Single and line assembly
Discrete material movement
Material storage and warehousing
Information processing

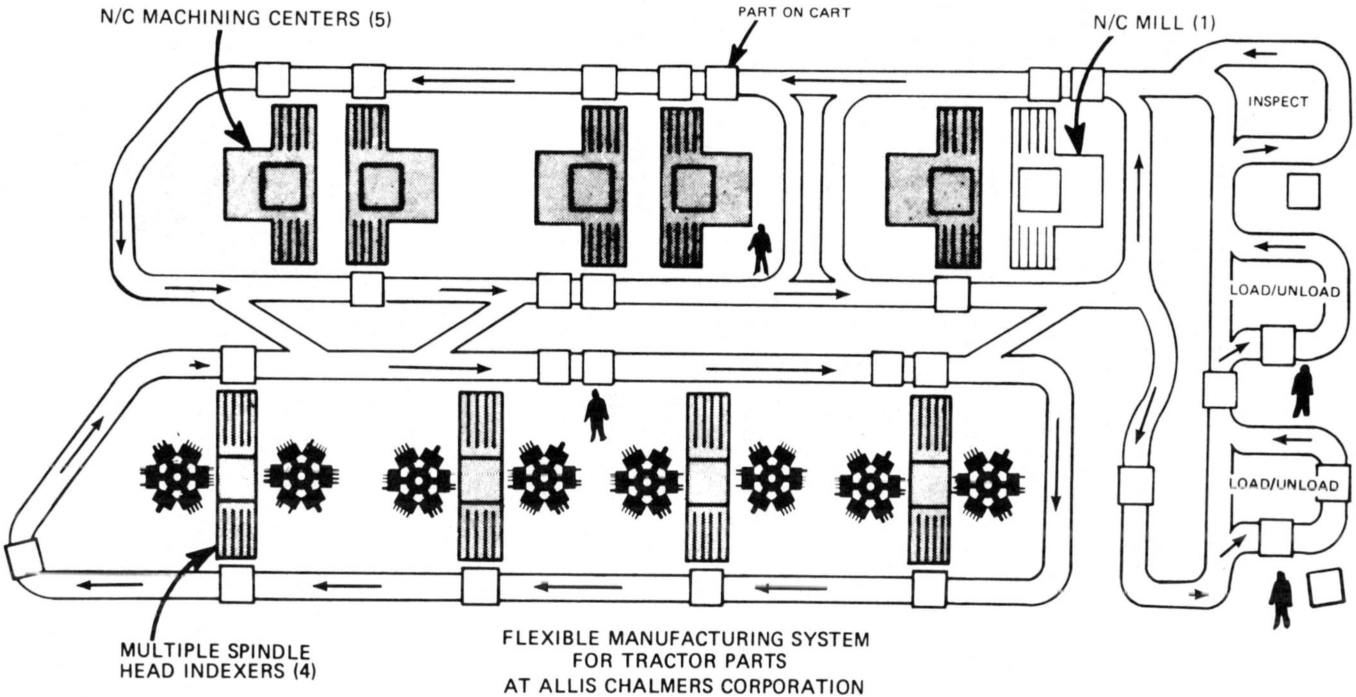

FIG. 1.2 Physical arrangement of a typical FMS. (Courtesy of Kearney & Trecker Corp., Milwaukee, Wisconsin.)

1.1.3 Hybrid Process

The majority of all manufacturing processes are of a hybrid nature. For example, materials for a continuous chemical process are delivered by individual trucks, processed by a continuous manufacturing process, and the product is shipped in individual containers. Similarly, in discrete automobile production many continuous processes are employed, for example, for pretreating of coiled steel.

There are various control systems and control strategies used for the different types of processes. In the continuous process measurement data can be sent to the computer and evaluated at periodic intervals. After comparing the measured value with the set-point value, the controller can be adjusted at periodic intervals. This method requires neither a very sophisticated hardware interface nor a complex operating system. The computer can be operated in polling mode to supervise the process controller. In discrete processes, on the other hand, the computer receives an interrupt signal and processes an event immediately after its occurrence. Otherwise, current information may get lost if too much additional information immediately succeeds it. In this case the computer must have interrupt capabilities with different levels of priority. The interfaces connecting the computer to the processes and the operating system which allows interrupt processing are by far more complex. Formerly, these two completely different types of process control were a selection criteria for hardware and the operating system. Today, most computer systems are of universal design; thus they are applicable for continuous and discrete processes. In reality, the former strict division is not practical because most factories employ hybrid processes.

1.2 THE TASK OF THE CONTROL COMPUTER

1.2.1 Interfacing with the Process

Control computers have been successfully employed for the last 15 years for controlling and optimizing technical processes. They are widely accepted in areas where conventional control methods for leading processes do not render satisfactory results. The control computer allows central control, thereby making possible better supervision of the entire process. By combining the hardware and software technology it is possible to conceive flexible manufacturing systems adaptable to changing market conditions and new, improved manufacturing methods. A control system supervised by a control computer, in general, is by one degree of magnitude more complex than a conventional system. Therefore, such a system must be conceived and designed by well-trained personnel acquainted with the process and the hardware software technology. It is often not possible for one person to have extensive knowledge of all these various technologies. For this reason a team of experts whose members cooperate well must be formed.

Typical functions presently performed by control computers are data acquisition, functional supervision of machines, quality control, and direct control of manufacturing equipment. Today there is hardly any industry which has not successfully experimented with process control computer technology. For this reason it is almost impossible to present control computer technology as a single entity. However, many elements of process control are similar and can be combined as basic building blocks for complex control systems.

Process control with the help of the computer has not developed overnight. It arose from conventional control technologies. Many well-known measuring and control elements are used in process control by computer and form the backbone of a complete control system. In most cases it is not possible to connect these basic control elements directly to the computer; however, interface circuits must be designed to transfer process signals. In almost all applications the effort of designing, building, and installing these circuits between the computer and the process can be very substantial. Often the interfacing task consumes between 30 to 50 percent of the installation cost of the computer system. In the following section a brief review of the historic development of analog controllers will be given. The reader should keep in mind that with the addition of a computer and with increasing complexity of the control task, the interface also increases substantially in complexity.

A similar development took place for digital control systems. This, however, is not shown in detail. In this discussion we only wish to get the point across that the development took place in different stages from the simple to the complex computer-controlled system.

1.2.2 Manual Control

Originally, all processes were controlled manually. The control system consisted of a control element, an instrument, and an operator (Fig. 1.3). With this method it was necessary that the operator be familiar with the process and that the control process not be too complex to be handled by a human operator. After reading the instrument, a decision had to be made on whether it was necessary to correct the process. With the help of experience gathered over a long period of time, the new position of the control element was determined and corrected accordingly. In general, this correction had to be made several times in succession to obtain the correct operation conditions. The human operator was the basic controller of the feedback loop.

1.2.3 Manual Control Assisted by Automatic Recorder

By connecting an automatic recorder to the control instrument, a considerable improvement in the control task was reached (Fig. 1.4).

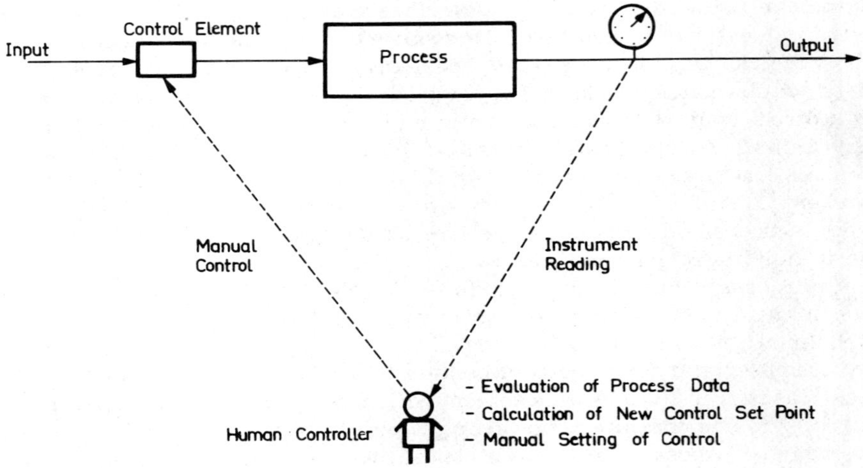

FIG. 1.3 Manual control of a process.

Thus it was possible for the operator to perform an online trend analysis and to observe in advance the possible occurrences of out-of-limit conditions. If necessary, an immediate correction could be made. Likewise, it was possible to collect historic data which allowed the study of the behavior of the process and of the control element.

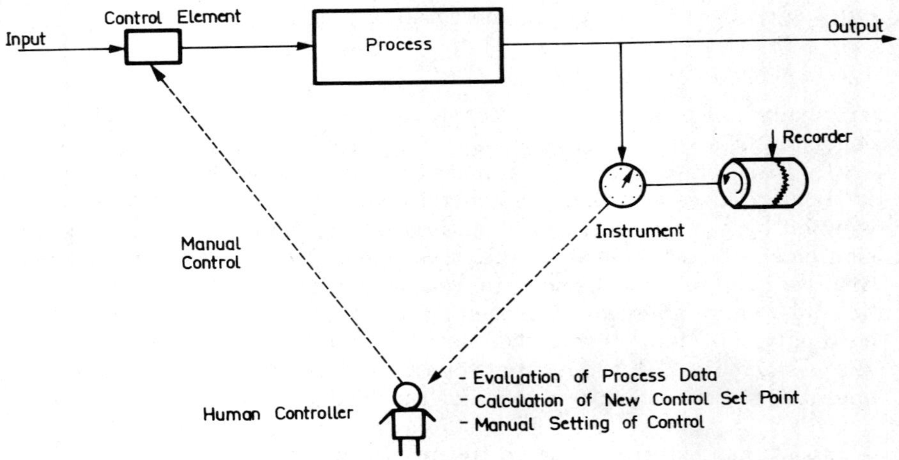

FIG. 1.4 Manual control of a process with the help of an automatic recorder.

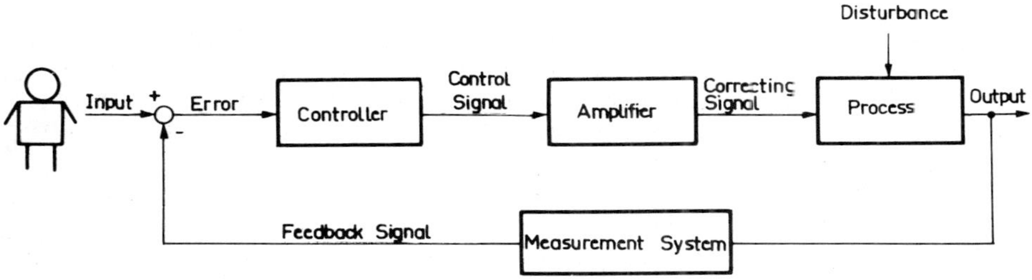

FIG. 1.5 Local automatic control.

1.2.4 Local Automatic Control

The automatic controller removed the operator from the feedback loop. Now only the set point of the process had to be entered by hand. The controller compared the actual operating point with the desired set point and if necessary automatically introduced a correction (Fig. 1.5). In general, one individual controller was installed to supervise one given process variable.

1.2.5 Central Automatic Control

In complex processes it became cumbersome to supervise all these individual controllers and to introduce the necessary corrections. With the increase in the number of control variables, central control rooms were installed (Fig. 1.6) from which the individual controllers could

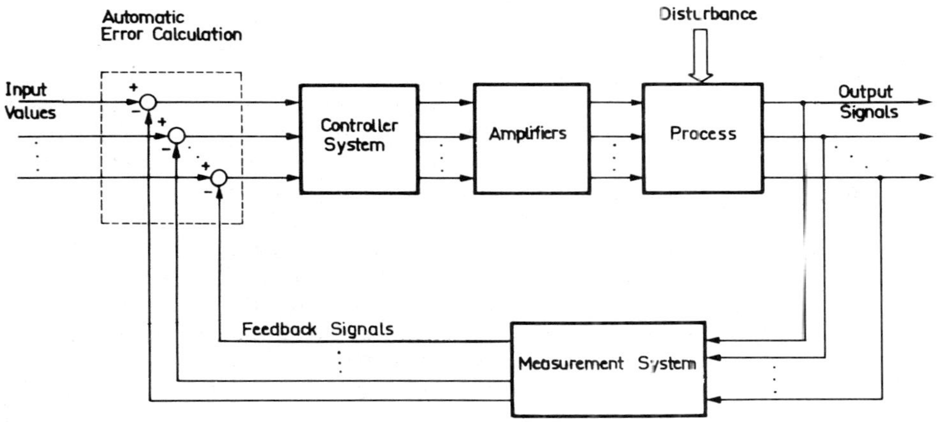

FIG. 1.6 Control system with multiple input and output signals.

be operated. With this arrangement it was possible to obtain an adequate overview of the state of the entire process and to observe deviations from the desired operating conditions. The central control room, however, quickly became unmanageable since it was often necessary to supervise hundreds of control variables with the help of a larger staff of operators. The control panels were improved by introducing graphic displays, which allowed supervision of important process elements and control variables.

The central automatic control system still had many disadvantages:

Achieving process optimization was difficult.
Recording process variables was cumbersome.
Because of the great number of corrections necessary for improvement of the product and the number of people needed for this job, central control strategies were difficult to develop.
The reaction time to critical process conditions was too long.
Because of dependence on the operator, the quality of the product and the efficiency of the process varied considerably.

1.2.6 Computer Assisted Control

With the development of the control computer, industry suddenly had an instrument which eliminated many of the disadvantages of conventional control technologies. The particular advantage of the computer is its ability to perform logic decisions, to process control tasks at high speed, to store process data, and to perform its own control decisions. Interfacing of measuring instruments and controllers with the computer and the process can be achieved by many different methods. The programming and hardware efforts must be adapted to the particular functions of the installations and the desired degree of automation. Computers are being adapted to existing installations and are also integrated into new manufacturing processes. Generally, it is a very difficult task to add on a computer to existing process equipment. Usually, this equipment was not conceived for process control by computer, and a simple adaptation of the controllers is not possible. In a brand new installation the total control system can be developed according to a well-balanced hardware/software approach. If a new process is planned, without the use of the computer, often it is an advantage to select instruments (e.g., IEC-bus compatible, Chap. 9) and controllers which can later easily be adapted to a computer. When a new technology is selected there are many considerations to be made, for example, cost, function of controllers, complexity of control, management philosophy, safety. At present there are basically three different tasks which can be assigned to the computer. These will be discussed briefly.

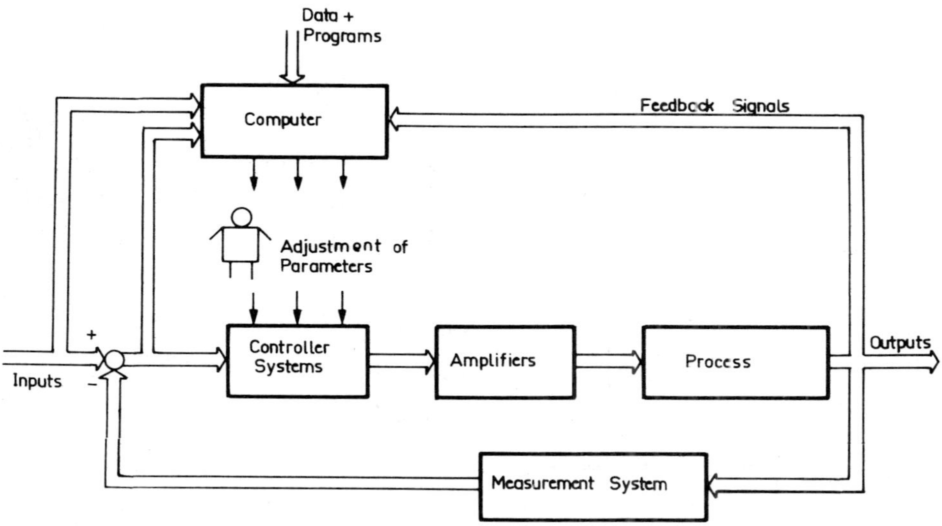

FIG. 1.7 Adaptive control by supervising computer (computer data acquisition).

Adaptive Control by Supervising Computer

This type of control system is used to collect and process factory operating data in discrete as well as in continuous processes. In general, the computer is easy to program; however, the necessary interface technology as a whole may be extremely complex and expensive. Operating data is automatically taken, but the feedback control for the process is handled by personnel (Fig. 1.7). The task of the computer is to read, process, and store data as well as to calculate new operating points and to communicate these to the operator. Thus the computer is connected to the process indirectly. This application is often called factory data acquisition.

Control of the Process

Until now process control by computer was mainly used for continuous production. It was not very common in discrete processes. The computer can be connected to the process according to various strategies. The reliability of the installation is an important factor for the selection of a strategy. In direct process control the computer is provided with a control algorithm, able to operate the process according to a preset static or alterable dynamic strategy. All the abovementioned functions of a data acquisition system can be incorporated in the same system.

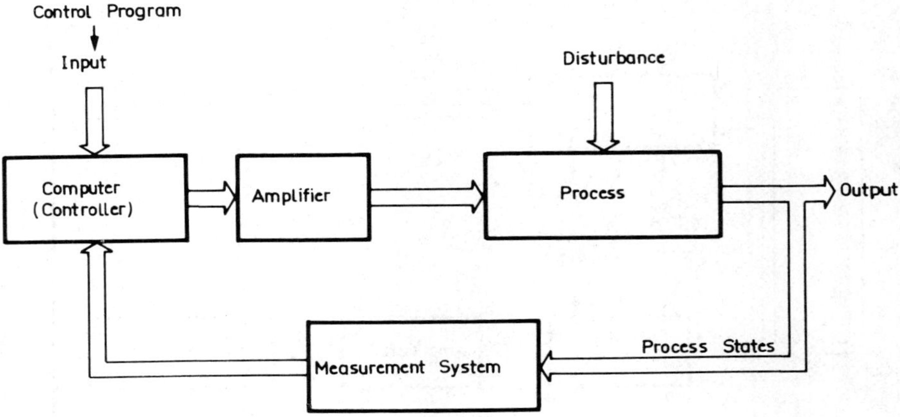

FIG. 1.8 Direct digital control.

Direct Digital Control

In this arrangement the computer itself performs the control function of the controller and gives instructions directly to the control element (Fig. 1.8). Many different control elements can be operated by the same computer in a time-sharing mode. This method, however, is not often used for continuous processes because the reliability of the system can be low. If the computer fails, the entire system is inoperable. The method does offer some advantages in digital control. Here too, there is the need for high reliability, and therefore programmable controllers are often introduced between the computer and the process.

Feedforward Control

With the feedforward control method the computer observes the performances of the process and it sends the necessary information on set-point adjustments to the controller (Fig. 1.9). Many controllers, in this case, can be supervised in a timesharing mode. The advantage of this arrangement is that upon failure of the computer, the process can still be controlled by the conventional controllers.

1.2.7 Optimizing the Process

With the process control computer it is possible to control the process in such a manner that throughput is optimized, manufacturing costs of the product are minimized, pollution is minimized, etc. For this purpose it is necessary to build a mathematical model from which an optimal control strategy can be determined during operation of the instal-

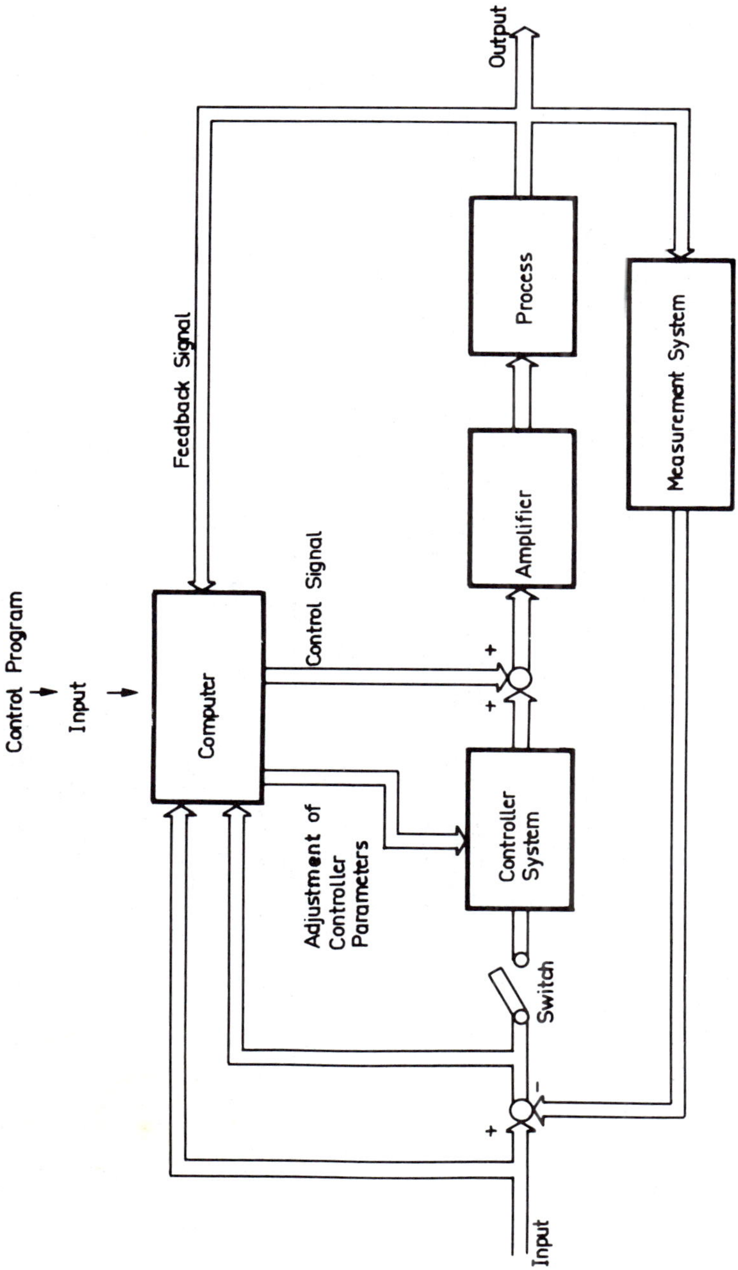

FIG. 1.9 Feedforward control.

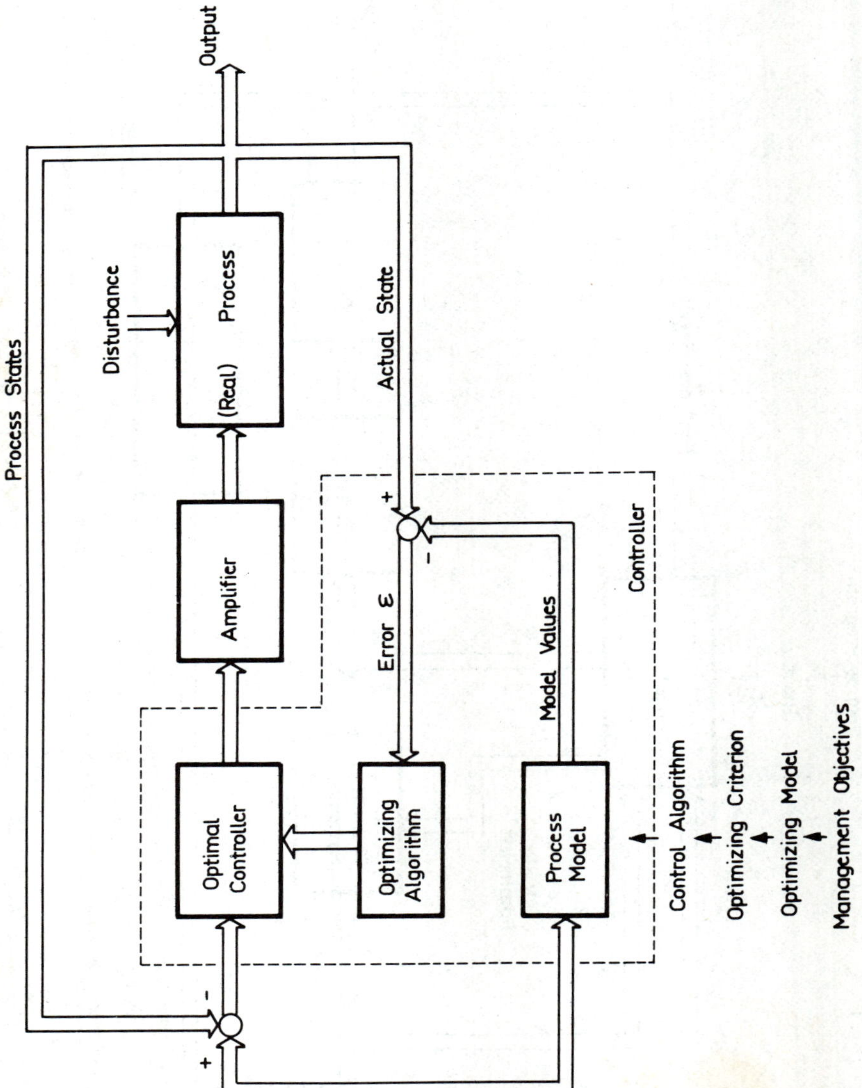

FIG. 1.10 Process optimization using parallel model.

lation. Unfortunately, most process optimizing models become so large that they cannot be maintained on the control computer itself. They must then be run and maintained on a supervisory computer system which calculates optimizing strategies. The results are sent to the process computer, via a data communication link, for further usage. For process optimizing three different methods are used. Each of them imposes different requirements for interfacing.

Process Optimization Using a Parallel Model

In this concept the computer contains a mathematical model of the process, which determines the optimal control strategy for a given operating condition and a predefined goal of management (Fig. 1.10).

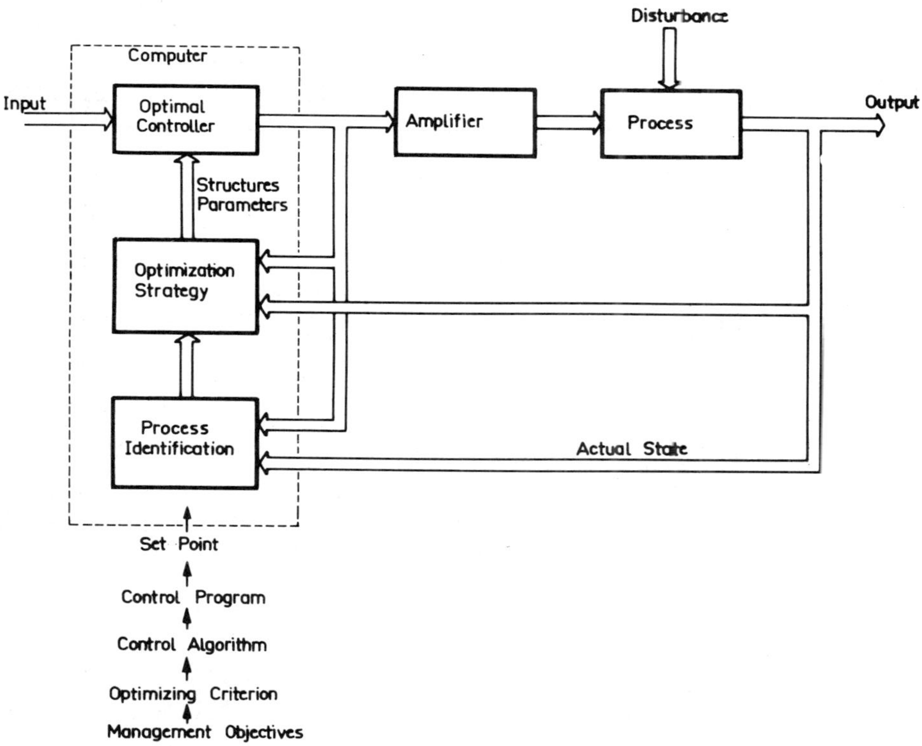

FIG. 1.11 Process optimization with feedback control.

Process Optimization with Feedback Control

For feedback control it is not necessary to have a mathematical model on hand to describe the process. The optimal control strategy is determined online with the aid of experimental search methods (Fig. 1.11).

Process Optimization with Feedback and Parallel Model

If material parameters and process variables are difficult to describe, the process can be controlled with a combined feedforward and feedback method (Fig. 1.12).

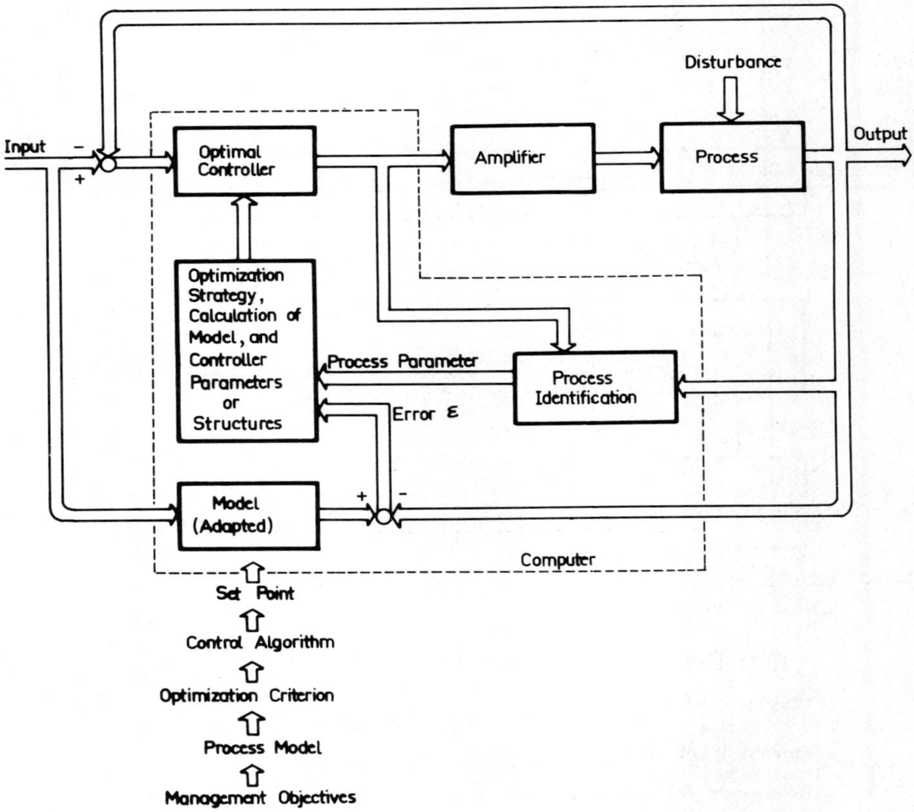

FIG. 1.12 Process optimization.

1.2.8 The Importance of Interface Technology

For the implementation of an entire control computer installation, many development phases must be completed. The success of an installation depends on the careful management of the individual phases and the smooth transition from one stage to the next. All phases are of equal importance. If an error occurs in one phase, it is difficult to eliminate it from the next one. The development cycle of an installation can be divided into the following phases:

System analysis
Planning of the process control computer installation
Test runs with pilot installation and control strategies
Programming
Interfacing of the computer to the process
Start-up of the installation
Operating the installation

This book is only concerned with the interface technology. However, it is difficult to define where this phase starts and where it ends. Upon conception of the control computer installation, one must already plan for the lowest cost interface strategy. The task of the software and hardware must be defined during design of the interface. For example, individual pulses from a turbine flow meter can be counted with the aid of the computer. To evaluate the pulses the computer calls up a program which continuously updates the results and stores these in a data buffer for later use. If there are several flow meters in use (Fig. 1.13), each instrument must have its own digital input line and its own software counter. In order not to lose a pulse, the polling sequence of the computer must be shorter than the minimal time span between the occurrence of two pulses.

It can easily be seen that with a high number of pulses and the use of several flow meters the computer is quite busy counting pulses. Thus it has little time for other assignments because the software is

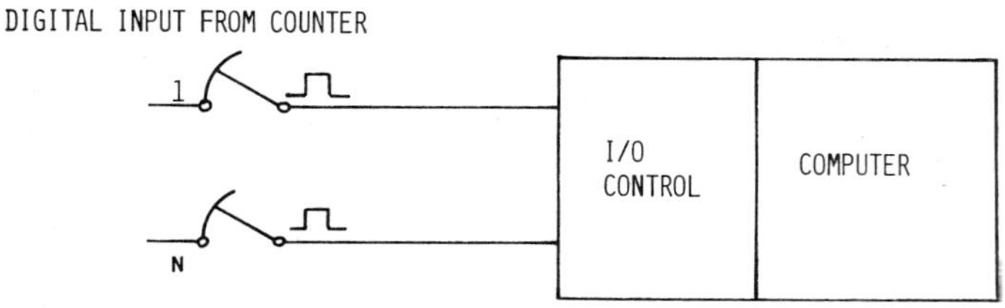

FIG. 1.13 Direct input of digital counting signals.

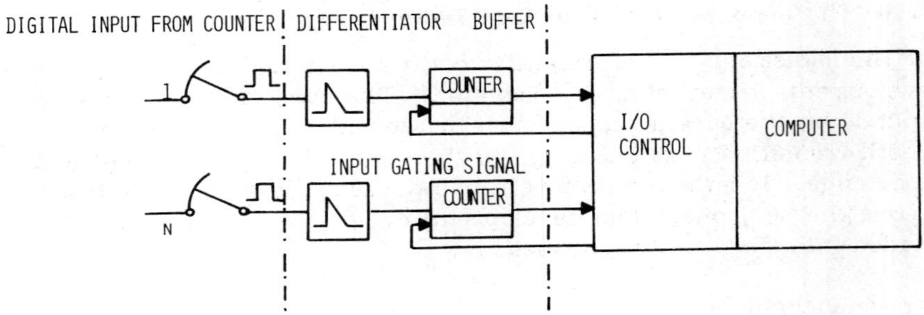

FIG. 1.14 Input of digital counting signals via input buffer.

mainly involved with counting. Such an installation can be improved with the addition of a data buffer in the interface. For better recognition the pulses are differentiated and continuously entered in the interface and stored in its data buffer. At predetermined intervals, the computer interrogates the data buffer of the interface. In this scheme there are still individual data lines running to the computer, but now only the sums of the counters are entered into the computer memory at periodic intervals (Fig. 1.14). Here the internal software counters are substituted for by external hardware counters. Now the computer is relieved of routine work, and it has more time for important assignments.

If signals are to be transferred over great distances, this installation can be further improved by using a multiplexer (Fig. 1.15). When addressing a digital input line, the computer sends the address of the desired flow meter to the multiplexer, which in turn conveys the contents of the corresponding data buffer to the computer. This

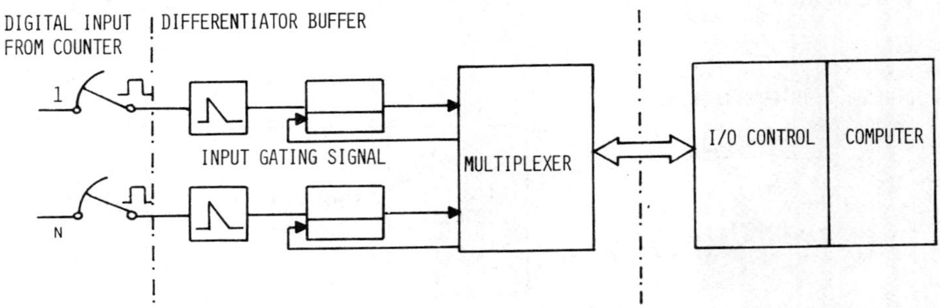

FIG. 1.15 Input of digital counting signals via input buffer and multiplexer.

scheme has its merits when the total cost of individual signal lines between the peripherals and computer exceeds the cost of a multiplexer. In this case the running time of the program is also significantly lower.

A further improvement in software running time can be obtained by providing the multiplexer with hardware polling capabilities. At predetermined cycle times the multiplexer polls the counters and the counting information is transferred from the data buffer of the peripheral to the computer. Here the computer does not have to send individual addresses of flow meters to the multiplexer.

1.2.9 Task of the Interface Technology

The task of a simplified interface will be described with the help of an example. Figure 1.16 shows a manufacturing installation controlled by a computer. Each machine tool has digital and analog control loops. To connect these loops with the computer, an interface must be designed. To keep the signal-line cost low, the information between the computer and interface is transferred in serial form. In this installation it is also desired to display the quantity of manufactured parts and to report possible problems with the machine tools on an alphanumeric display located at the interface. Thus the interface has to perform the following tasks.

Decode the control signals of the computer and the machine tool.
Decode addresses.
Transform serial information into parallel information and vice versa.
Operate the alphanumeric display.
Transform signal levels between the machine tool and the interface.
Analog-to-digital (A/D) and digital-to-analog (D/A) conversion of signals.
Multiplex signals of various machine tools.
Store information in a temporary buffer.
Count pulses.

A detailed analysis of this problem would result in many additional subfunctions. It is necessary to transform the software instructions of the computer into signals understandable to the interface hardware. The computer itself must understand the flow of signals delivered by the interface hardware and transfer it into software information.

1.2.10 Functions of the Interface

The preceding example shows that the interface may have to perform a multitude of different functions. The conception and design of the different interface functions must be done by the development engineer of the computer system, who must be familiar with the process peripherals, computer, and software. In the following chapter an attempt

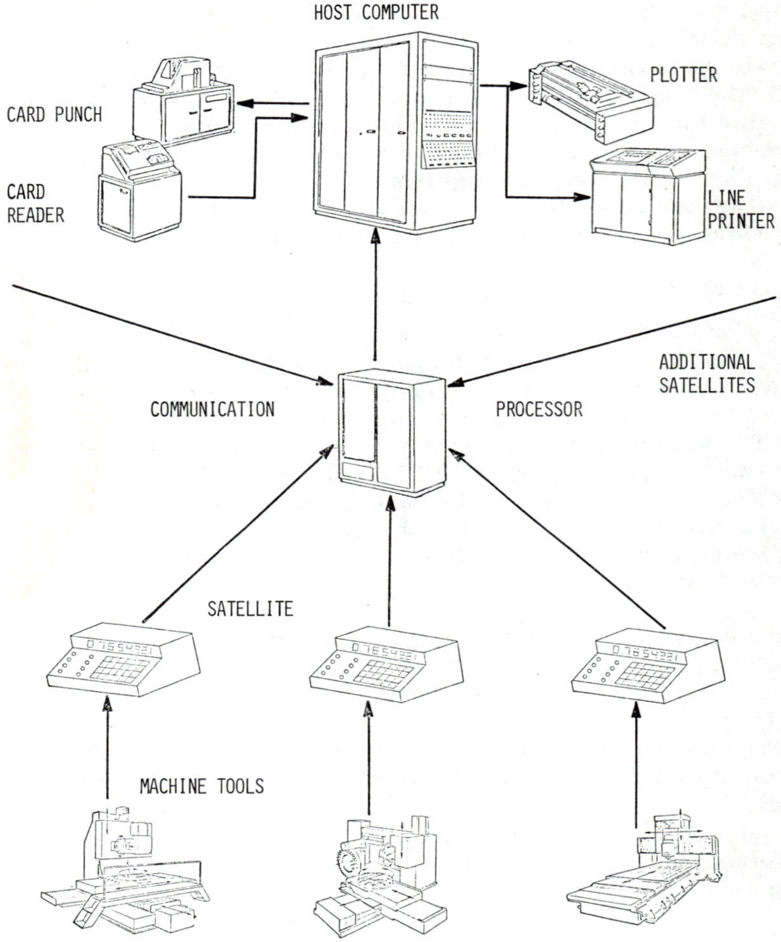

FIG. 1.16 Hierarchical data acquisition system controlling machine tools.

is made to convey this knowledge. Much of the material has been derived from practical experience. In an introductory survey some fundamental functions of the interface are discussed.

Transfer of Control Signals

Data exchange between the process peripherals and the computer takes place according to predescribed protocols to assure that information arrives at the right time and in the correct sequence.

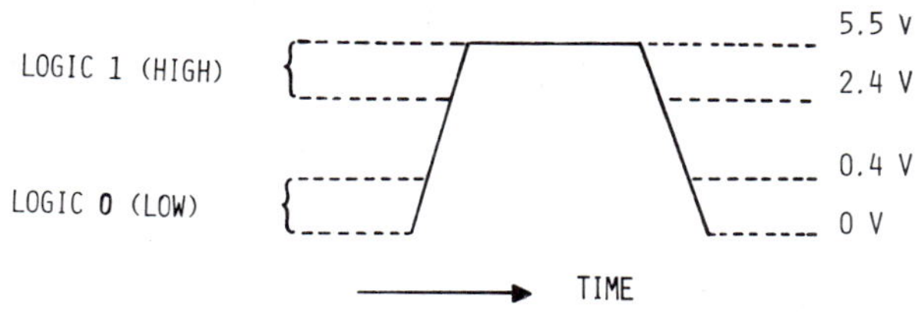

FIG. 1.17 Voltage levels used with TTL logic.

Therefore, it is mandatory that the interface be capable of performing the following functions (Secs. 5.4 and 6.2):

Generate and recognize square waves.
Synchronize the operation of the computer with that of the peripheral.
Request data and make data available.
Supervise the computer/peripheral data communication.

The length of the required pulses depends on the architecture of the computer, the design of the peripheral equipment, time requirements of control functions, and many other variables (see Sec. 4.3). The voltage level of the control pulse is determined by the integrated circuit technology used for the computer and interface. For transistor-transistor logic (TTL) the signal level is shown in Fig. 1.17. Interfaces mostly use circuits based on this technology.

Data Transfer

Data can be transferred in various modes, for example, parallel or serial. It is necessary that the interface contain logic circuits capable of performing the required data transfer (see Chaps. 4 and 5). The mode of transfer can be selected arbitrarily or can depend on the computer or peripheral equipment employed. For example, a teletype will transfer characters and numbers in a 7-bit serial mode (Fig. 1.18).

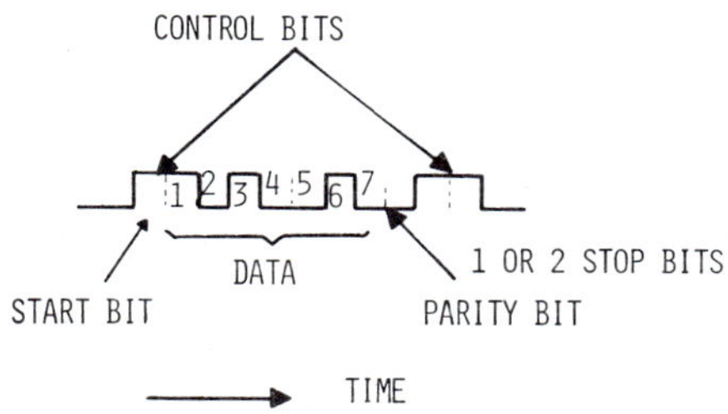

FIG. 1.18 Transmission of the letter R by a teletype.

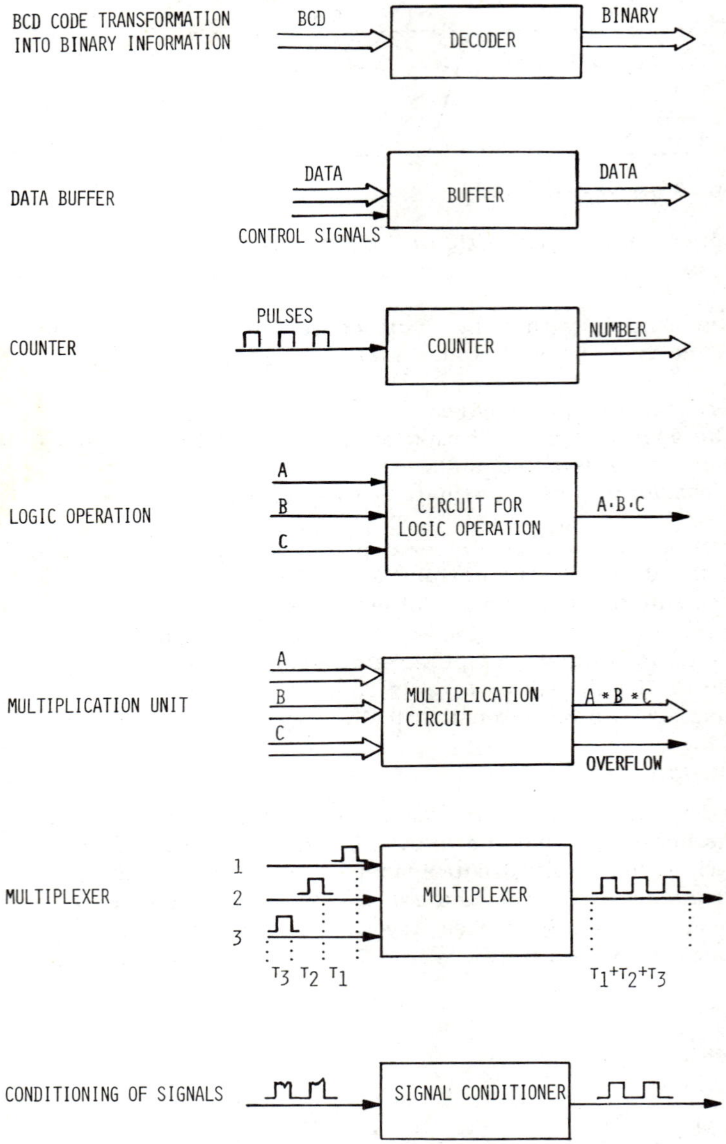

FIG. 1.19 Typical tasks of the interface.

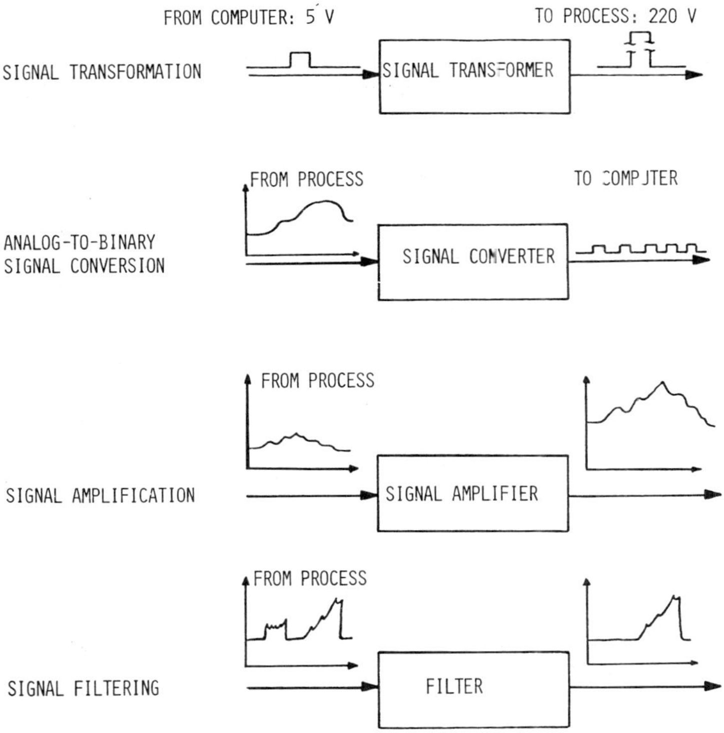

FIG. 1.19 *(continued)*.

However, for the complete transfer of an alphanumeric character also, a parity bit and three additional control bits are needed. Thus data transfer for this device is done with 11 bits.

Coding and Decoding

In many cases the interface must perform code transformation in order to be able to understand the information and to make it available in proper form for further processing (Fig. 1.19; see also Secs. 5.6 and 7.1). Code transformation is necessary, for example, to operate digital displays or octal or decimal counters, for address decoding or to energize mechanical printing mechanism of the printer, as well as for many other purposes.

Buffering of Data

Since the computer in almost every application is much faster than the process peripheral, synchronous data transfer with a peripheral is impossible or extremely difficult (see Sec. 6.2). Therefore, a data buffer has to be provided in the interface, to which information can be entered or from which it can be removed (Fig. 1.19; see also Sec. 5.4.2). This buffer may consist of simple flip-flops or complex storage devices. For example, during data output from the computer, the information must be buffered in the interface until the peripheral equipment is able to use it. The process of transferring the information is initiated by a control pulse, then acknowledged, and finally terminated.

Counter

Counters in interface circuits have a multitude of tasks to perform (see Sec. 5.6). They may supervise proper time-sequential processing of control signals, count data to be transferred, or operate as counters for process signals (Fig. 1.19). In Fig. 1.14 a counter is shown that adds up pulses transferred by a transducer.

Logic Operations

Process control requires many logic operations. For example, if a machine tool has been provided with three different safety mechanisms and an alarm signal is to be sent to the computer only if all three mechanisms fail, then an AND gate can be used to transfer the information (Fig. 1.19). Logic operations are also necessary to perform many of the control functions of the interface.

Arithmetic Registers

A computer can be relieved of many of its routine tasks if an interface performs arithmetic operations and then transfers only the result (Fig. 1.19). Such arithmetic operations are often required to process instrument readings.

Signal Multiplexing

One of the important functions of the interface is to switch process signals. For example, to reduce cable cost, signals of different transducers are often sent over one cable with the help of a multiplexer (Fig. 1.19; see also Sec. 5.6).

Shaping of Signals

When signals are transferred, they are often distorted in phase and amplitude. Therefore, regeneration of pulses is often needed. This assures that information can be further processed (Fig. 1.19).

For shaping signals a Schmitt trigger can be used (Sec. 5.4.3). Quite often this circuit is also of advantage when specific information must be filtered out from analog signals.

Signal Level Transformation

In most cases, when a computer is connected to the process, a signal-level transformation must be performed (Fig. 1.19). This is necessary because the control element of the process generally operates at 24, 115, or 220 volts or at another potential and the input/output signals of the computer must actuate, for example, TTL circuits of maximum 5 V (Sec. 8.1). Such signal-level transformers often can be purchased as ready-made components and generally are not designed by the user.

Signal Conversion

In many applications measuring signals are generated in analog form. Likewise, many of the control elements of the process are actuated by analog signals. Since the control computer is only capable of processing digital information, the analog signals entering the computer must be digitized. In the same manner, control signals coming from the computer must be transformed into analog signals (Sec. 5.3). These tasks will be performed by A/D and D/A converters (Fig. 1.19).

Signal Amplification

The majority of signals coming from analog transducers must be amplified. They are too weak to be transferred over long distances. There are many different signal amplifiers suitable for this task (Secs. 3.5 and 5.2). Recently operational amplifiers have been accepted widely (Fig. 1.19).

Signal Filtering and Elimination of Ground Potentials

In factories it is often impossible to obtain a clear signal because noise is superimposed over the process signal. There are many different filter circuits available for elimination of noise (Fig. 1.19). Ground potential generated by grounding different system components at different ground terminals can be eliminated by a variety of different methods (Sec. 8.3).

Communication Processor

In many well-conceived technical applications the interface must function as a communication processor. This unit may reduce the workload of the main computer considerably. The task can be very diversified and in general is tailored toward the individual needs of the process. Typical functions of this computer are as follows:

Controlling the data communication process
Multiplexing data channels
Data concentration
Parity check
Code transformation
Data analysis
Data storage

1.3 CHAIN OF CONTROL ELEMENTS OF THE COMPUTER INTERFACE

To connect a computer to a process, a chain of control elements is needed (Fig. 1.20). Each control element in this chain has assigned to it a special task. It is important that the tasks be understood correctly.

1.3.1 Input/Output Control Unit

The input/output control unit connects the central processing unit with the input/output channels. Generally, this control unit can be pur-

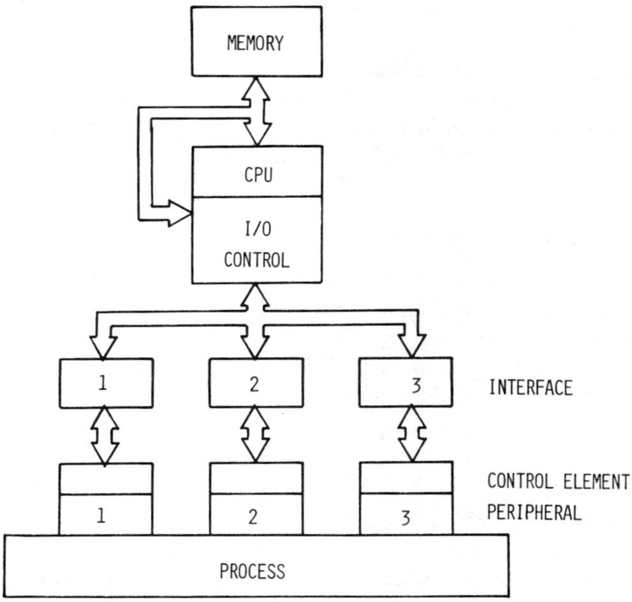

FIG. 1.20 Chain of control elements in a computer system.

chased from the computer manufacturer. Since it is very complex it is seldom redesigned or altered by the user to adapt it to a specific process (Chap. 5). For example, in a computer with a universal bus, the input/output control unit is responsible for the following functions:

1. Input/output of control signals (Sec. 6.2.3)
   a. Synchronization of the computer with the peripheral
   b. Output of function signals to the peripheral
   c. Reception of status signals (busy/ready)
   d. Output of busy signals for the bus
   e. Output of control signals to indicate the direction of data transfer (Sec. 6.4)
2. Interrupt signals (Secs. 6.2.3 and 6.3.3)
   a. Initiation of an interrupt
   b. Recognition of interrupt signals
   c. Polling of interrupt request
3. Transfer of addresses (Sec. 6.2.3)
   a. Output of a peripheral's address
4. Data transfer
   a. Input/output of data (Secs. 6.2.3 and 6.3.3)
   b. Input of parity information from the peripheral

1.3.2 Interface

Many basic functions of the interface have already been described in the previous sections. The interface has a dual task since it forms the link between the computer and the process. On the computer side it must react to the instructions issued by the computer and perform the correct data transfer (Chap. 6). Here the following functions must be carried out:

1. Input/output of control signal
   a. Synchronization of the computer with the peripheral
   b. Transfer of function signals for the peripheral
   c. Output of status signals
   d. Output of busy signals for the bus
   e. Output of direction signals to indicate the data flow on the bus
2. Interrupt signals
   a. Output of interrupt requests
3. Transfer of addresses
   a. Input/output of data
   b. Transfer of parity errors from the peripheral
4. Data buffering
   a. Input/output buffering of process data (Sec. 5.4 and Chap. 6)

On the peripheral side of the interface, control information must be given to the controller of the peripheral itself (Chap. 8). In addition, the correct function of the peripheral must be supervised. For example, with a teletype it is necessary to convert the signals coming from the computer into the proper code for the electromechanical drive so that it is possible to print the desired characters. The interface must carry out the following data exchange with the control element of the peripherals (Chaps. 3 and 8):

1. Check the status of the peripheral.
2. Initiate the operation of the peripheral.
3. Sense and report difficulties with the peripheral.
4. Transform the computer data into control data for the process.
5. Transform process data into control data for the computer.

1.3.3 The Control Element of the Peripheral

This element must be able to perform the required control of the peripheral to pass on control and measurement parameters, and to recognize and report difficulties. To it the following tasks are assigned:

1. Control the operation of the peripheral.
2. Report the status of the peripheral.
3. Control the peripheral's functions.
4. Transmit errors.
5. Transfer measurement data.
6. Transfer control data.

BIBLIOGRAPHY

N. N. KT's World of Manufacturing Systems, Kearney & Trecker Corporation, Milwaukee, 1980.

Sass, D., and W. Uhl, Prozessrechner für Bandbearbeitungsanlagen, BBC Nachrichten, Jahrgang 58, 1976, Heft 1.

2

THE ARCHITECTURE OF CONTROL COMPUTER SYSTEMS

2.1 THE BASIC CONCEPT OF A CONTROL COMPUTER

Control computers have an architecture similar to that of the general digital computer, whose structure originates from ideas of von Neumann and co-workers [1]. In this detailed description dating back to 1946 von Neumann outlined the basic architecture and organization of the computer of the future. In the process control computer the internal functional units of the so-called von Neumann's machine are supplemented by an interrupt unit and special input/output units for the process peripherals. These additions are an essential criterion for the evaluation of the power and efficiency of a process control computer.

Figure 2.1 shows a simplified block diagram of a control computer. In accordance with von Neumann's concept the computer can be divided into the following units: arithmetic unit, control unit, internal working registers, input/output unit, interrupt unit, and central memory. The arithmetic unit, the working or internal memory, and the control unit form the central processing unit (CPU). In microprocessors these units are integrated into one chip. This chip can be assembled to a complete control computer by supplementing it with a memory and an input/output unit. There are also "one-chip microcomputers" available where only one chip contains a complete control computer.

The computer designed by von Neumann had only peripherals for operator communication such as the console keyboard, card reader, printer, etc. This type of equipment is called *data peripheral*. At a later date the von Neumann computer was greatly improved by connecting mass memory to it. The control computer in addition has further elements which allow it to communicate with a process and to control it. They are called *process peripherals*. The interaction of the described functional units of the control computer can be seen in Fig. 2.1, which shows the control and data paths.

In the following sections of this chapter data transfer within the computer is described in more detail. The arithmetic unit in Fig. 2.1

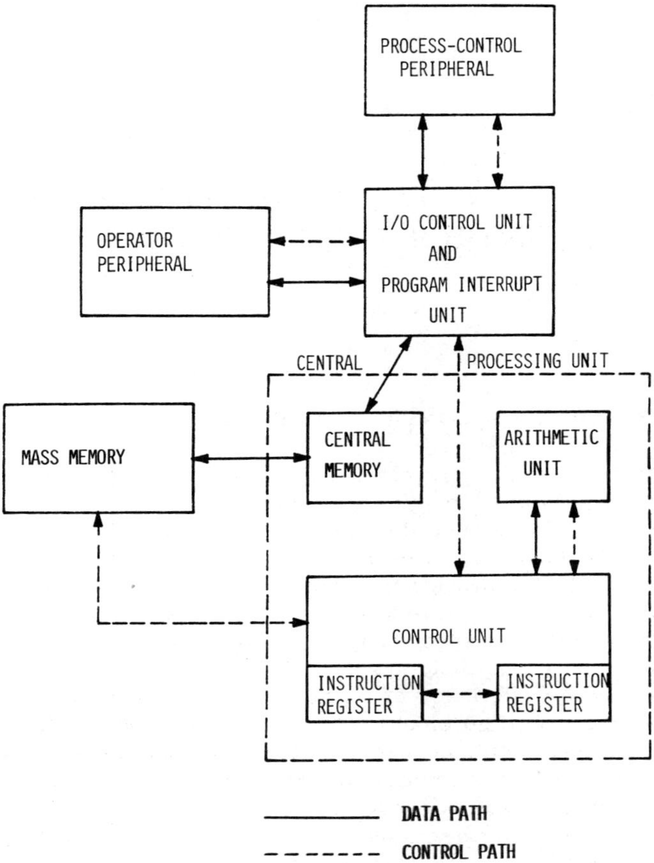

FIG. 2.1 Simplified block diagram of a process control computer.

consists essentially of one or more accumulators and several registers. The control unit controls the data transfer and the execution of arithmetic operations. The individual steps of these operations in general are subdivided into elementary operations and are controlled by machine instructions. Both data and instructions are stored in the central memory, whose design and technology is described in Sec. 2.7. The program interrupt unit responds to interrupts or alarms triggered by the peripheral equipment. This ability to respond in real time to an external request characterizes the control computer. The program interrupt unit, also called *interrupt logic*, is covered in detail in Sec. 2.4.

The design of the communication paths over which instructions and data are transferred, as well as the organization of the central memory and the registers, are criteria for the efficiency of a control computer. The influence of the data format and the internal data paths on the performance of the computer are described in Sec. 2.2 and 2.4. The traditional architecture of the register is listed in Sec. 2.3.

The register concept of a central processing unit of a typical control computer will now be explained with the help of a modern microcomputer. The following three operations can be distinguished when the computer executes a program:

Input
Execution
Output

During an input operation data or instructions are transferred from the central memory or peripheral equipment to the central processing unit. The instruction has two components: the instruction field, containing the instruction code, and the address field, containing the address of the operand.

When, for example, a load instruction is executed, the control unit, initiated by the instruction code, selects the address in memory given by the address field of the instruction and transfers the selected data to a specified working register. Analogously, data from the central processing unit is transferred to a register of a peripheral equipment.

For the execution phase the operation code of the actual instruction is decoded by the control unit, which then initiates the desired operations. In many new computers the instruction is divided into small elementary steps, so-called machine cycles or microoperations. In the control unit these machine cycles can be combined to one machine instruction by hardware or by software. In the first case the hardware has been rigidly designed to perform the microoperations which are necessary to control the execution of the instruction. In the second method the instructions are stored as a variable microprogram in an read-only memory. With microprogramming it is possible to custom design instructions for individual tasks. Because of this flexibility the method of microprogramming is being used more and more in modern process control computers.

2.2 INSTRUCTIONS

Computers used for the control of processes must be capable of handling the following information:

OPERATION FIELD	MODIFICATION FIELD	OPERAND FIELD (ADDRESS)

FIG. 2.2 Instruction format.

Digital signals
Analog signals
Alphanumeric characters and special characters

Data formats with a length of 8 to 32 bits in general are sufficient to perform the task of a control computer. In commercial computers longer formats are often used. Normally, modern microcomputers use a word length of 8 to 16 bits. One word may consist of two separately controlled bytes. In some slice processors as well as electronic calculators processing of data is done in 4-bit registers in BCD code. Analog data and numbers are usually coded as binary numbers in octal or hexadecimal form.

Data has the same structure as instructions and is also stored in the central memory. The instruction word, however, is subdivided into three groups which basically contain three different types of information (Fig. 2.2):

The operation field contains the actual instructions for the processor, to specify which operations are to be performed.
The operand field specifies where the necessary operands for the operation can be found.
The modification field contains specifications which allow calculation of the effective operand addresses. This permits the use of various types of addressing.

When one tries to determine the efficiency of a control computer, the available instruction set is an important decision criterion. This includes the instruction format, type of addressing, instruction repertoire, and the capability of microprogramming.

2.2.1 Instruction Format

The instruction word is subdivided into individual functional components. Instructions with a length of 1 to 4 bytes are commonly used. In Fig. 2.3 an example is given to show the instruction formats of a control computer with a multiregister structure. In operational instructions, for example, "No Operation" or "Stop," only the operation field is used. With a register reference instruction information contained under a memory address is manipulated. Generally the location

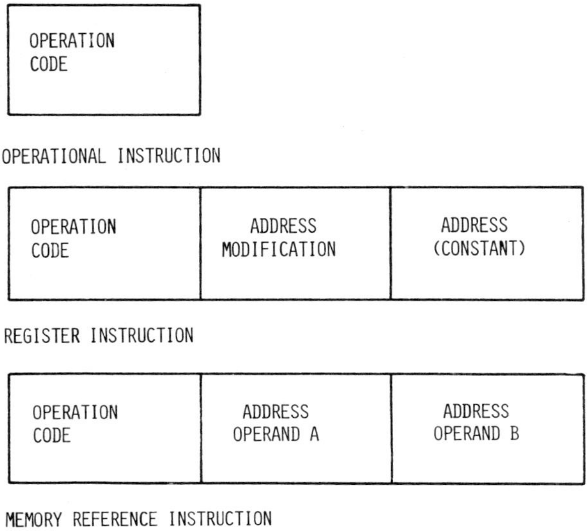

FIG. 2.3 Examples of instruction formats for a process control computer with multiregister structure.

where the second operand and the result are stored is specified. In a memory reference instruction the addresses of both operands are described and the result is stored in a specified location. As mentioned previously, the address in an instruction word can be modified. There are various modes of addressing; they are distinguished by their final or effective address (Fig. 2.4):

1. Direct Memory Addressing (absolute addressing): The content of the operand field represents the effective operand address.
2. Indirect Addressing (address substitution): The operand specifies the memory location or register in which the effective operand address will be found. It thus points to the operand indirectly with the help of at least one further address.
3. Relative Addressing: The operand specifies only part of the complete address. The base address to be added can be found in a register or in the memory. There are three different types of relative and base addressing methods.
   a. Indexed Addressing: The address in the instruction word is added to the content of the index register specified by the instruction. The result represents the effective memory address.
   b. Addressing with a Base Address (paging, mapping): The memory is divided into individual areas of equal length.

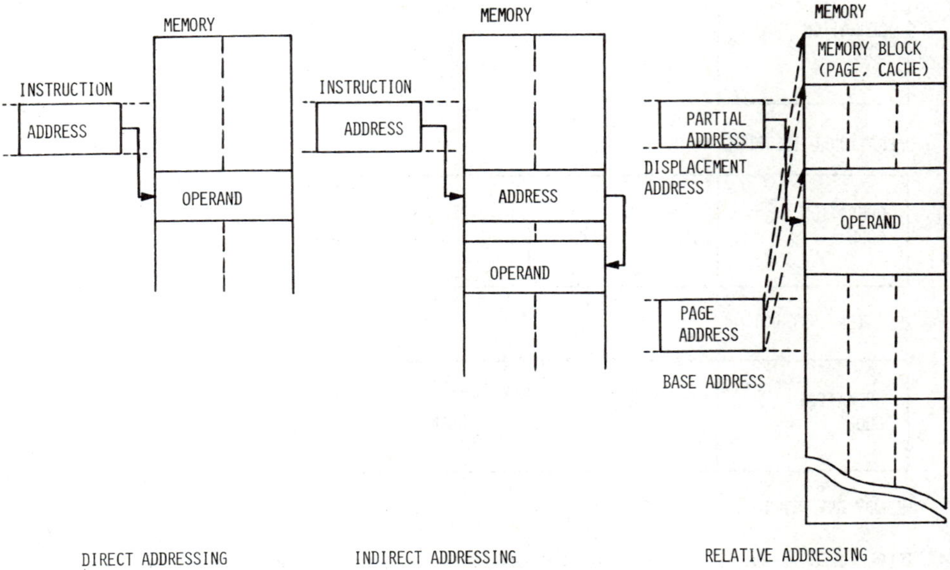

FIG. 2.4 Different addressing principles.

The base address specifies the area or page. The relative address of the instruction word specifies the addressed memory location within this area or page. The sum of both is equal to the effective address in memory.

c. Relative Addressing to the Instruction Counter: The content of the instruction counter serves as the base address. The address in the instruction word is added to or subtracted from the content of the instruction register. The result is the effective address.

2.2.2 Instruction Set

Although the instruction sets vary greatly for different control computers – ranging from 50 instructions for small computers to over 500 for large computers – they can be divided in the following groups (Fig. 2.5):

Transfer instructions which control the transfer of data.
Processing instructions include arithmetic, logic operations, and shift instructions.
Skip instructions for branching in the program. The execution of a skip can be made dependent on conditions.

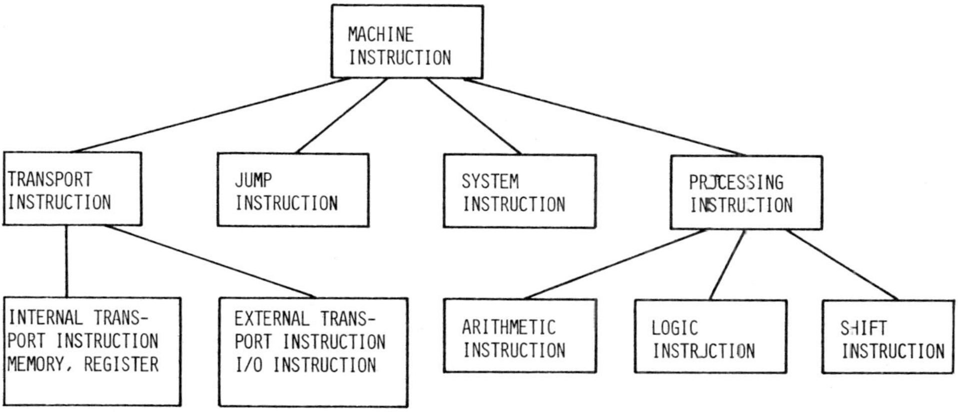

FIG. 2.5 Instruction set of a process control computer.

System instructions for control of the processor (computer). In general, these instructions are reserved for system programs.

2.2.3 Microprogramming

For a microprogramming control computer the instruction set can be made very flexible, i.e., each machine instruction is generated by a series of statements located in the microprogram memory. Any number of complex machine instructions can be combined into such a microprogram. Figure 2.6 shows the simplified diagram of a central processing unit with microprogramming capabilities. At the beginning of the execution of an instruction the operand field of the instruction is brought into the address register of the microprogram memory. The corresponding microinstruction is then transferred to the microinstruction register. With the help of a decoder, the operation code of the microinstruction is converted into the appropriate control signals for the control unit, which then executes one elementary step of the instruction. Thus for the execution of one instruction, several microinstructions are produced; they are generated and executed in sequential order.

With the help of the address field of the microinstruction the next microinstruction is selected, which in turn initiates the execution of the second elementary step. The concept of microprogram control can be thought of as a computer within a computer, whose program for control of the registers, memory, and arithmetic units are stored in a specific memory with extremely short access time.

In some control computers users can write their own instructions. Thus they can combine special instructions or entire routines in one

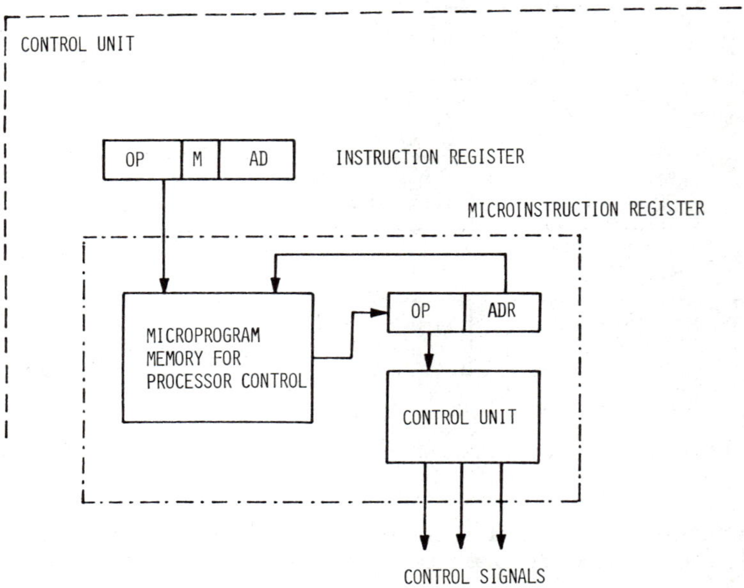

FIG. 2.6 Structure of a microprogrammable central processing unit.

instruction, i.e., floating-point arithmetic can be done by a single instruction. Elementary steps are then controlled by the microprogram. The execution speed increases through the use of a microprogram memory with very short access time. In addition, overall programming is also simplified.

2.3 REGISTER ORGANIZATION

In the introduction to this chapter the central processing unit and its registers were presented. One differentiates between registers to which the user has access through instructions (so-called accumulators) and internal registers which are used by the executing program and whose content can only be scanned by the program. Generally, the central processing unit contains several accumulators, which serve as temporary direct access buffers and as working registers. The arithmetic and logic unit usually operates on the content of an accumulator rather than accessing the memory directly. After execution of an arithmetic or logic operation, the temporary results are also stored in these registers. Only the final result is transferred to the memory (Fig. 2.7). In addition to these all-purpose accumulators other registers are provided for special functions:

1. A program counter controls the execution of the program. Its content specifies which instruction in memory is to be executed next. The content of the program counter can be changed by several instructions (for example, shift or skip instructions). Thus the sequence of execution of the program can be controlled depending on conditions.
2. The instruction code is stored in the instruction register. The CPU always interprets its content as instruction code, which is decoded by the instruction decoder.
3. The status of the CPU is recorded in the status register. For example, after the execution of an arithmetic operation, the corresponding bits, such as carry, overflow, sign, etc., are set. These bits can be interpreted and used for conditional skips; thereby branching in the program is made possible.
4. Index registers allow for indirect addressing and facilitate processing of sequentially stored data.
5. Stack registers simplify the management of subprograms. These registers, however, are not independent of each other, but are tied together to form a register set. Generally, storing (pushing) and reading (popping) of data is possible only in the top register of a stack. The address of the information on the top of the stack is indicated by a counter, the stack pointer.
   During a write operation (push) the content of the register is stored at the location of the stack indicated by the stack pointer

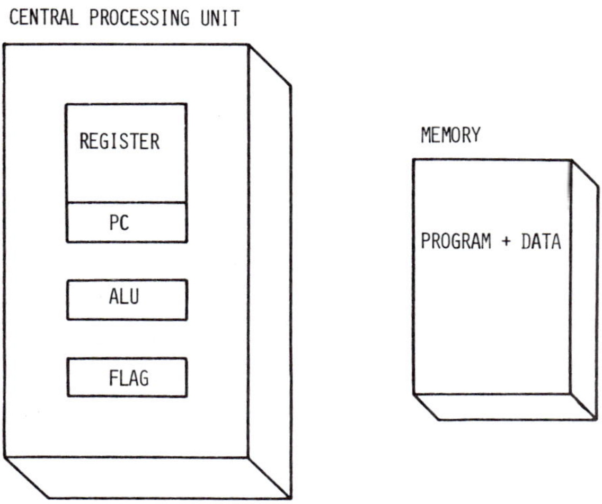

FIG. 2.7 Register-memory architecture of a conventional process control computer: PC, program counter; ALU, arithmetic logic unit; FLAG, status register.

(Fig. 2.8). The content of the stack pointer is then automatically incremented and indicate the next empty memory location. A read operation (pop) proceeds in a similar manner. The stack pointer is decremented and points to the last word written in the stack. The content of this cell is then read out. The data organization of the stack corresponds to a last-in-first-out (LIFO) strategy and can be very useful for executing service routines for peripheral equipment. The main program's return address and register contents can be stored in the stack and will be restored when a return to the main program is made.

The architecture (Fig. 2.9) of some more advanced microprocessors has a register organization different from that of traditional control computers. The central processing unit (CPU) does not contain separate registers. However, the registers are integrated in the so-called work space of the central memory. An internal register, the work-space pointer, integrated into the CPU chip refers to the corresponding register area in memory. It contains the storage address of the first register word in memory and is linked, together with the 15 following words of the register set to which the processor has access. If a different register set is needed, the work-space pointer can be loaded with the address of the new work space. The administrative effort required to change programs is thus reduced considerably. Storing or saving registers, which was necessary in the traditional register storage architecture, is no longer needed. Furthermore, any number of registers can be combined; the number is limited only by

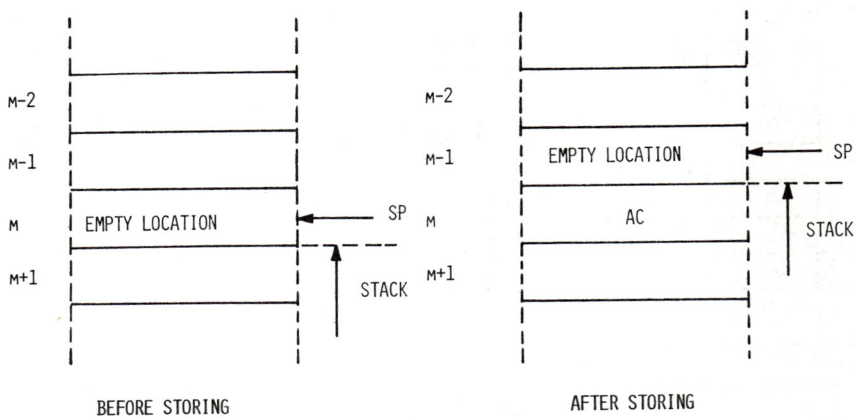

FIG. 2.8 Push operation with a stack: SP, stack pointer; AC, contents of accumulator.

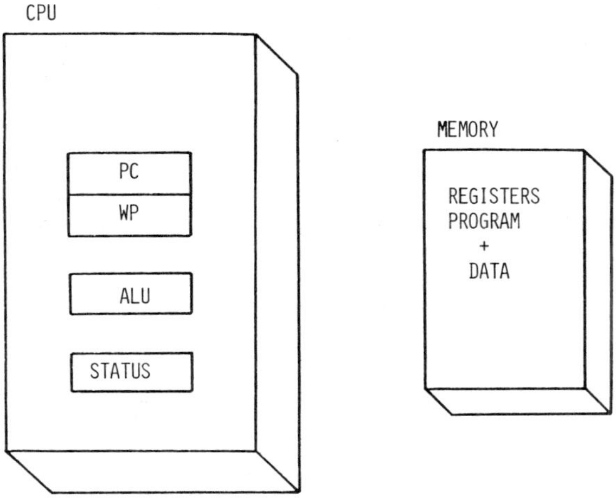

FIG. 2.9 Architecture of a modern microprocessor: CPU, central processing unit; PC, program counter; WP, workspace pointer; ALU, arithmetic logic unit.

the size of the working memory. Other advantages result from this architecture (Fig. 2.9). With two-word (double-word) instructions any number of operations can be done with two words stored in different locations. With an ADD instruction which contains two addresses in its operand field, the content of one addressed word is added to the other addressed word and the result is stored at the location of the second word. A computer with conventional architecture needs a series of instructions for the same operation: first an operand is loaded into a register, and then the content of another register is added to it. The result which remains in the register must be transferred to the corresponding memory location by another instruction. This computer architecture not only requires more memory space, but it is also more laborious to program.

2.4 INTERRUPT STRUCTURE

An essential feature of the control computer is its capability to respond to a spontaneous service request of the computer system itself or of the peripheral equipment. If a service request occurs, the computer interrupts the running program. At this point it must be decided who requested the interrupt, and which interrupt service routine is to be activated. The new service program, however, is started only if it has a higher priority than the one being interrupted. In the other case, the interrupt request is placed in a queue to be serviced at a later time. The arriving interrupt signals can be classified according to the meaning or priority of the messages which initiate them. The classification is as follows:

1. Interrupt Requests Indicating Machine Defects: In situations such as power loss, parity errors, or defect of the memory protection, continuation of the running program is impossible. The computer

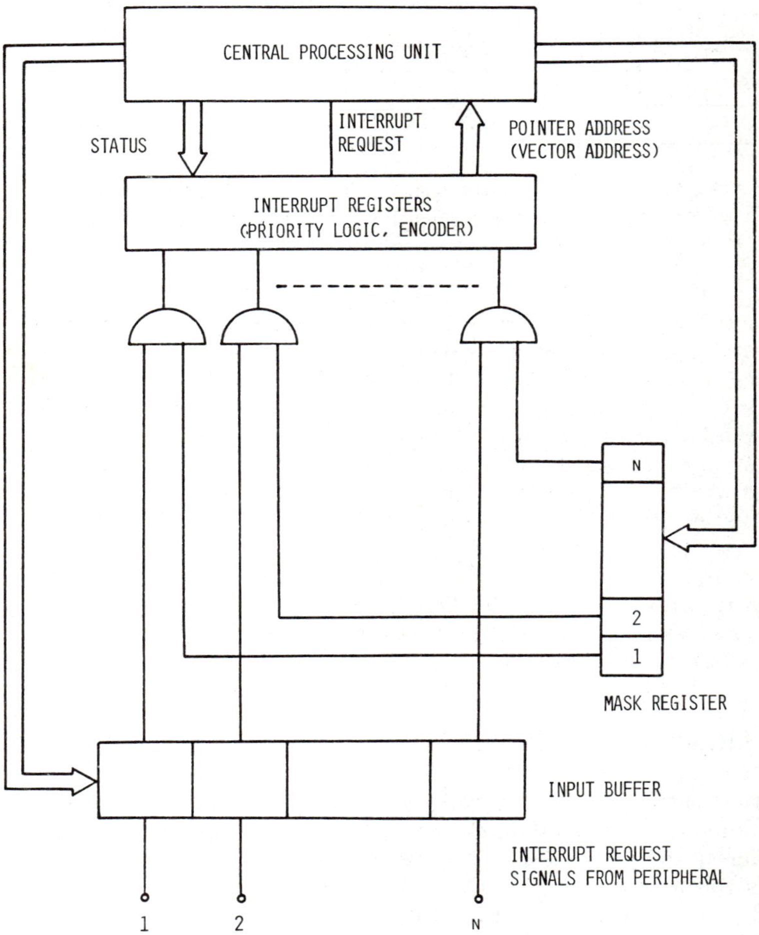

FIG. 2.10 Principal design of an interrupt unit.

will be interrupted unconditionally and a program will be activated for a restart.

2. Interrupt Requests Indicating Internal Events: An invalid instruction code or the end of a data transfer initiates this interrupt. Depending on the priority the main program is briefly interrupted.
3. Interrupt Requests Indicating External Events: Interrupt signals from the process periphery, such as the opening or closing of a process control switch, belong to this group. They are the actual alarm messages issued to the control computer and are answered by a special user program.

Without this interrupt capability the computer would have to constantly poll the peripheral to inquire about significant process changes, for example, about the drift of a valve setting (temperature, capacity, etc.). An interrupt signal notifies the computer of a change immediately, and the computer then calls the corresponding service program. The time to answer an interrupt together with the number of possible interrupt levels are other factors that determine the efficiency of a control computer. The architecture of interrupt units is discussed in more detail in Chaps. 6 and 7. Only its principal function is explained here with the help of Fig. 2.10. It shows the general design of an interrupt unit with several levels of interrupts. Such a design is usually found in large process control computers. Microprocessors often use a simpler design. They do not possess an interrupt unit but have only one interrupt level or one common input line through which several peripherals can request a service by means of an OR circuit (Fig. 2.11). The operating system analyzes the origin of the signal by testing a corresponding status register, and then it determines the priority of the request. The interrupt signals are first buffered in an

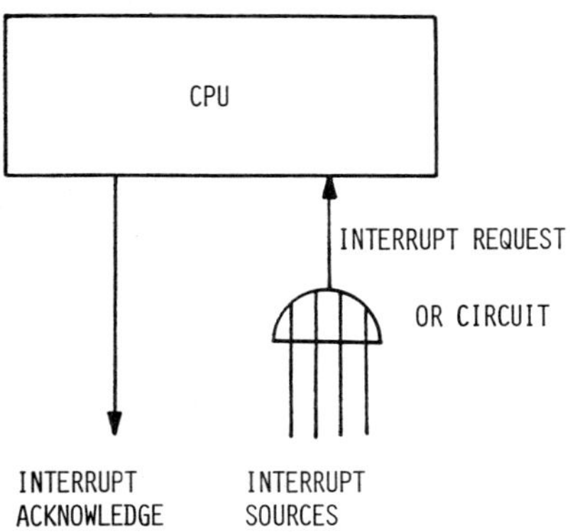

FIG. 2.11 Microprocessor with one interrupt level.

interrupt register (Fig. 2.10). A mask register, whose content may be altered by special instructions, controls further processing of the interrupts with the help of gates. This allows it to temporarily mask out certain interrupt signals. In the interrupt unit the priority of the interrupt is then determined. The signal having the highest priority initiates an interrupt request in the central processing unit and transfers to it the address of an interrupt vector. Further processing of the interrupt signal is handled differently in computers of different manufacturers. Processing of an interrupt signal by the CPU is described here in more general terms:

1. First the actual status of the interrupted program will be saved by storing the content of the instruction counter and all registers in specified memory locations.
2. The instruction counter is loaded with the address of the required service routine (interrupt address vector). This address was placed there after the occurrence of the interrupt in the interrupt unit by a decoder.
3. The specified interrupt service routine, which begins at this address, will be executed. Further interrupt signals of lower priority can be locked out by masking during this execution phase.

In most cases the interrupt-handling section of the operating system is called up at this point, which then activated the corresponding user program.

2.5 INTERNAL BUS CONFIGURATIONS

The functional units of the process control unit (Fig. 2.1) are connected to each other by multiple lines through which data, addresses, instructions, and control characters are transferred. In microcomputers and several control computers these communication paths are combined by a system bus. To this the individual components of the CPU are connected under equal access conditions.

In the architecture of process control computers basically two structures have become popular:

The universal bus
The multiple bus

The universal bus (Fig. 2.12) serves as common communication line between the CPU, the memory unit, and the input/output unit. The functional units of the computer may be connected in different configurations to this single bus. Essentially, a data transfer is possible between all attached units, as long as all interfaces possess the necessary intelligence. This means that the same bus can be used for

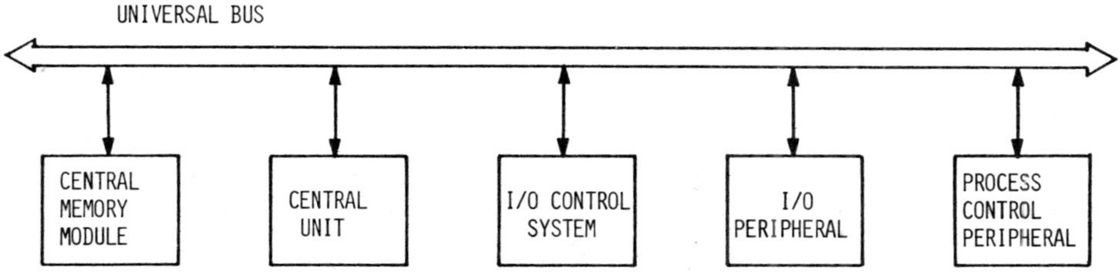

FIG. 2.12 Computer with universal bus system.

input/output operations as well as for memory access. Because of this bus architecture the CPU does not need special instructions for communication with peripheral equipment. The interfaces of the external devices obtain registers and addresses. These addresses are handled like memory locations. An instruction to store data in memory is the same as that which performs an output instruction to a peripheral. However, it uses the specific address of the peripheral. Similarly, a load instruction initiates an input.

When several functional units are attached in parallel to one line the data transfer must be controlled by a bus controller. Usually the CPU is the active element and coordinates the data communication over the bus. Also, for equipment which possesses direct memory access (DMA) capabilities, the data transfer is initiated by the CPU. Because all peripheral units are connected to the same line, it must be guaranteed that only the addressed data source and the unit requesting the data transfer occupy the bus during a data transfer.

In the two-bus system (Fig. 2.13) the central memory and in most cases the faster peripheral memory with high data transfer rates are connected to their own memory bus. All slow peripheral devices are

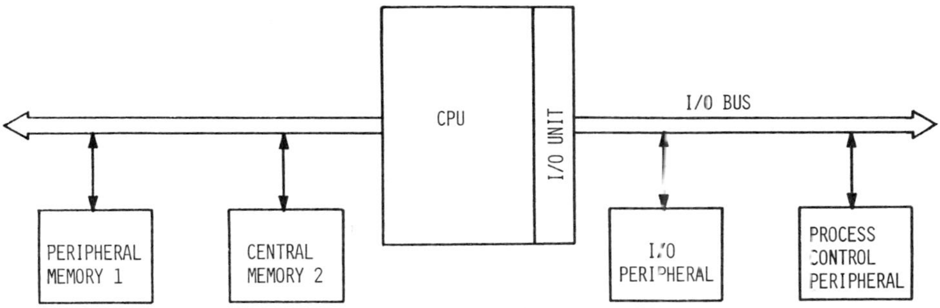

FIG. 2.13 Computer with a multiple- (two) bus system.

connected by interfaces to the input/output bus of the control computer. The memory bus has a substantially higher transfer rate than the I/O bus, therefore it is suited for fast-access peripheral such as mass storage devices. In this arrangement data exchange with the process peripherals will not be impeded by data transfer between the CPU and the memory. Thus a fast reaction time to an interrupt request is guaranteed. There are two different basic circuits applied to assure unobstructed traffic between communicating units of a computer. One uses open collector gates and the other tristate output gates (Fig. 2.14).

The open collector is the same as a wired OR circuit, which means that only the state LOW is active. If the output lines of all peripherals are inactive, the pull-up resistor creates the logic state HIGH. This resistor, however, limits the line capacity. It creates a relative large time constant which negatively influences the transfer speed of the bus.

The tristate technique possesses a third state. When turned off, the line has a high impedance and is thus separated from both potentials: the logic HIGH or LOW. Here operational amplifiers with selected hysteresis characteristics are used as receivers. Thus interference resulting from changing voltage levels can be eliminated. The parallel input gates are activated by a central control signal which usually is generated by the decoded address of the particular input/output device. The transfer rate here lies between 1×10^6 and 5×10^6 words per second. When using such high frequencies, however, the line must be terminated with a resistor having a low electrical resistance to avoid reflections on the bus.

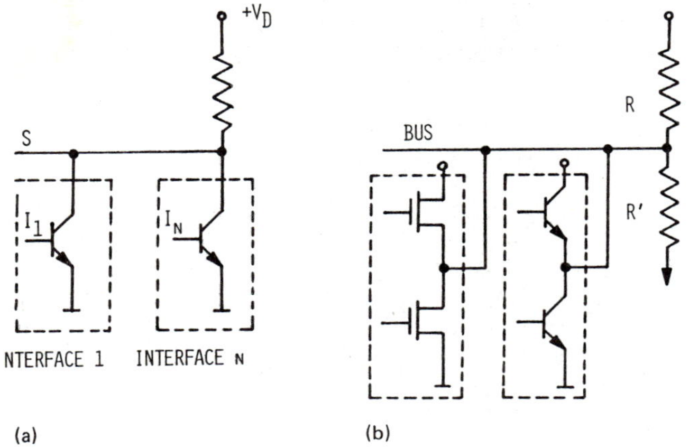

FIG. 2.14 Interfacing methods used in a bus: (a) interfacing with open collectors and (b) interfacing with a tristate circuit.

The bus structure of most microprocessors have many similarities. Nevertheless, microprocessors can, in general, not be replaced with each other since the design of the connector pins is not yet compatible. There are attempts being made to standardize a bus with 100 lines, the S100 bus, as an universal system. Thus microprocessor chips of various makes having the same data format could become interchangeable through special adapters. Essentially, the bus of microprocessors can be subdivided into four functional units:

Data bus with 8, . . . ,16 lines for byte or word transfer of data or instructions
Address bus with up to 16 lines for addressing 64K bytes of memory locations
Control and signal lines
Voltage supply and ground lines

The assignment of the bus to the various functional units is done according to the master-slave principle. This means that the CPU as an active element occupies the system bus and addresses the passive peripheral to make communication possible. If a passive element wants to exchange information,it notifies the CPU through an interrupt. The CPU then makes the connection according to the priority of the attached equipment.

2.6 INPUT AND OUTPUT ORGANIZATION

The data transfer between the CPU of a control computer and the attached peripherals is called *input/output* (I/O) *transfer*. The design of the input/output hardware, the so-called I/O control unit (Fig. 2.1), also influences to a great extent the efficiency of the computer.

The I/O unit controls data transfer between the CPU and the peripheral equipment. There is a large number of process I/O devices as well as operator I/O equipment. These devices can be very different in regard to their function, signal levels, and speed. Attempts are being made, however, to standardize the interfaces between the I/O control unit and peripheral equipment. The various kinds of input/output organization can be roughly divided into three classes:

1. Program-controlled input/output using no interrupt (Fig. 2.15): Here the central processing unit is responsible for controlling the entire data transfer, which usually is done with the help of an input/output buffer. With the aid of a program, a communication path between the CPU and a selected peripheral is created and data transfer is initiated. In this phase the computer is performing data transfer exclusively.

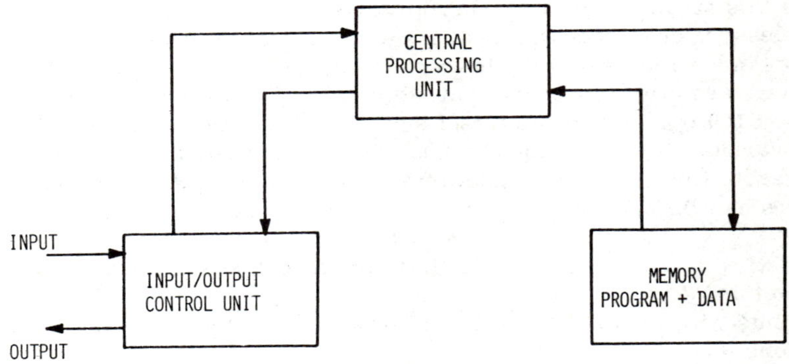

FIG. 2.15 Program-controlled input and output.

2. Direct memory access (DMA) (Fig. 2.16): Here the data transfer proceeds without control by the central processing unit. The DMA control unit establishes a direct connection between the memory and the peripheral equipment. Thus high data transfer rates are obtained. Here the computer has only the task of initializing the block-type data transfer. The autonomous interface channel (DMA controller) controls and monitors the data transfer, which can proceed – so to speak – simultaneously with the execution of a program.
3. Program-controlled input/output using interrupt control: The process control computer is capable of servicing an external data transfer request from a peripheral device by interrupting the running program and devoting time to the device.

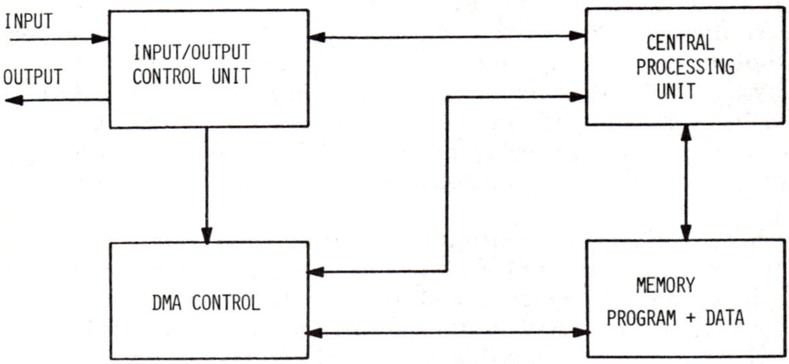

FIG. 2.16 Input and output under direct memory access control.

All peripherals are connected to a common signal line through which the computer is notified when a device is able to send or receive data. The origin of the request signal can be analyzed by a service routine or a logic circuit. If several peripheral devices interrupt almost simultaneously, the order of priority in which the devices should be served must be determined. The CPU can make its service available to a unit dependent on the present priority status of the unit, which is recorded by a status word in a status buffer. Each bit of the binary status word represents the status or a function of the unit.

The organizational structure, where data transfer is done over a central bus or channel, is found mainly in minicomputers and microcomputers. Larger control computers often have an autonomous input/output unit which operates parallel to the central processor. There are two different types of autonomous input/output units:

1. Autonomous channel units activated by the CPU transfer data in blocks. They are similar to the DMA organization in the central bus system.
2. An input/output processor interprets the user program which resides in the main memory and then controls the required data transfer. If this processor has its own memory, then an autonomous I/O computer exists. This memory can be used for storage of intermediate results, especially for buffering of data.

2.7 MEMORY ORGANIZATION

One of the important components of a control computer is its internal central memory also called *working memory*. It is part of the CPU (Fig. 2.1). It can be directly addressed by machine instructions and is capable of storing instructions as well as data. Part of this memory is usually occupied permanently by the operating system and another part may contain the entire user program or segments of it. In larger computers the control memory is usually extended by external storage devices (mainly disks and tape storage units). They are able to store large amounts of data and thereby reduce memory cost. The original storage medium of the central memory consisted of ferrite cores but is presently being replaced by semiconductor devices. As a matter of fact, most modern micro- and minicomputers are equipped with semiconductor memory. The efficiency of a memory is described by the following parameters:

Capacity
Access time
Cycle time
Power dissipation

The capacity of the memory, i.e., the number of storage locations in which words can be stored, is limited by the possible address space of the instruction. With a word length of 16 bits or 2 bytes (common with control computers) a maximum of 64K words can be addressed. Since instructions and data are stored in and retreived from memory, the working speed of the computer depends on the memory access time. It is understood to be the time needed for reading or writing a particular word into or out of memory, or more exactly formulated, the time required from the beginning of addressing a memory location until the end of the read or write operation. In some storage devices information is destroyed during the reading operation. After reading it must be rewritten or regenerated into the same memory location. For such read and write memory this regeneration time is added to the access time. The result is the cycle time, which is equal to the shortest time period between two sequential read or write operations.

In Fig. 2.17 various memory technologies are shown. Since the introduction of a new semiconductor technology in 1970, when for the first time a read/write memory (RAM) was constructed using the field effect transistor with a leading P channel (PMOS), the importance of semiconductor memory has increased steadily. They are used not only for making computer memories, but they also serve as an universal basic element for storing data in electronic equipment. The present state of development is as follows:

1. The storage capacity of read-only memories in MOS technology has already reached 64K bits per chip.
2. Complementary MOS technology (CMOS) with little power dissipation and a high noise immunity is used as RAM. When connected to a small buffer battery it has the same abilities as the core storage; the stored information can be kept for a longer time even during power loss.
3. Field-programmable logic arrays (FPLA) compete with ROM. Especially when the microprocessor or the process control computer is used to solve complicated logical functions, this basic element will become popular.
4. Charge-coupled devices (CCD) are characterized by their high integration density and low cost per bit. Access to information, however, is only sequential.
5. A similar storage organization is found in the magnetic bubble memory, which is already available with 100K bits in a dual-inline package.

In Fig. 2.17 the access time of various types of memories is shown as a function of cost. The specifications refer to the storage component only; peripheral driver or control logic are not included. Applications and performance of different types of memory are listed in Table 2.1.

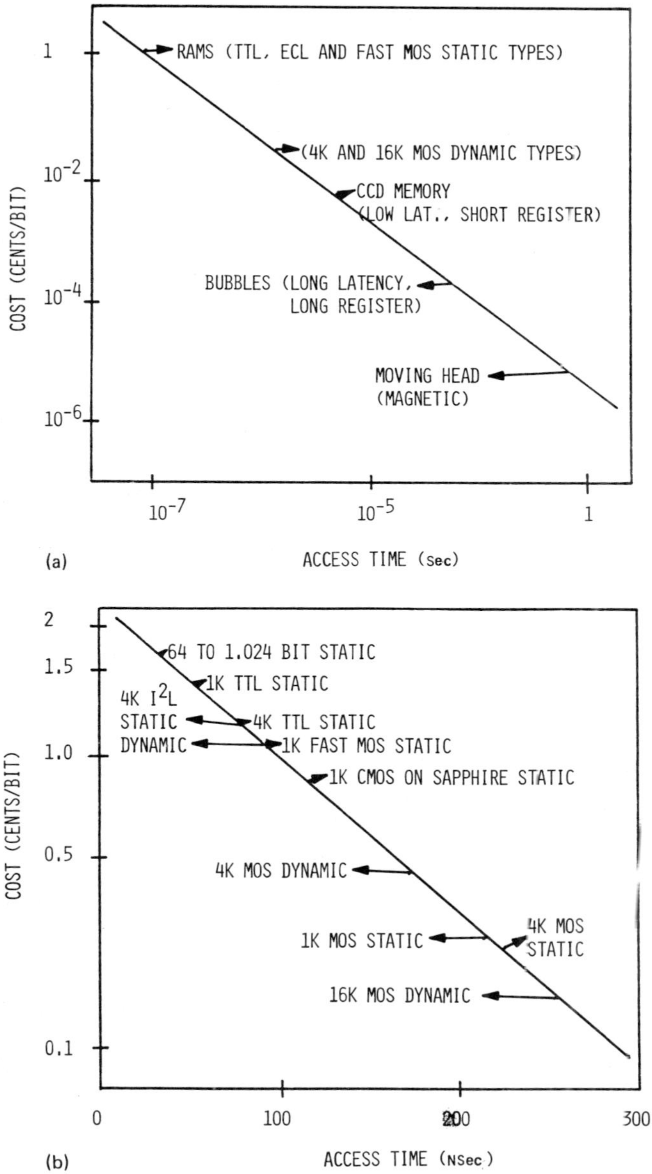

FIG. 2.17 Characteristic performance of different memory types. (a) Overall performance of memories. (b) Read and write spectrum. (Reprinted from *Electronics*, January 20, 1977.)

TABLE 2.1 Applications and Performance of Memory Types

Application	Device types	Access time (nsec)	Active power dissipation per chip (mW)	1977 cost (cents per bit)
Large mainframe	4K nMOS dynamic	150-300	450	0.15
	16K nMOS dynamic	200-350	500-700	0.08
Small mainframe, microprocessor based	4K to 16K nMOS dynamic	150-300	450-700	0.15
	4K I^2L dynamic	90	500	0.25
	4K I^2L static	70-100	500	0.25
	4K nMOS static	150-550	350-500	0.20
	1K SOS/CMOS static	150	4	1.5
Buffer, cache, etc.	256 bit ECL static	10-30	600-800	2.0
	1K TTL static	40-100	600-800	1.5
	1K nMOS fast static	70	600-800	1.0
	4K I^2L static	75-100	400	0.25
Peripheral	1K nMOS static	150-500	300	0.3
	4K nMOS static	150-500	350-500	0.2
	4K I^2L static	90	500	0.25
	4K nMOS dynamic	150-350	450	0.15
	16K nMOS dynamic	200-350	500-750	0.08
	8K EAROM	500	500	1.5
	8K EPROM	400	450	1.5
Small storage (1-5 megabits)	CCD memory	100 μsec		0.03
	Disk	100 msec		0.01
Large storage (5 megabits plus)	Bubbles	10.0 msec		
	Disk	10.0 msec		0.01

Source: Reprinted from *Electronics*, January 20, 1977.

Next we discuss the design of the more recent technologies of memories. For the design of a memory there are two basic principles being used (Fig. 2.18 and 2.19). In a bit-oriented memory the storage cells are arranged in the form of a matrix. The rows as well as the columns are addressed and decoded with the help of the x address and y address, respectively. When signals coincide in a row and column, the correct storage cell is addressed. The line- or word-

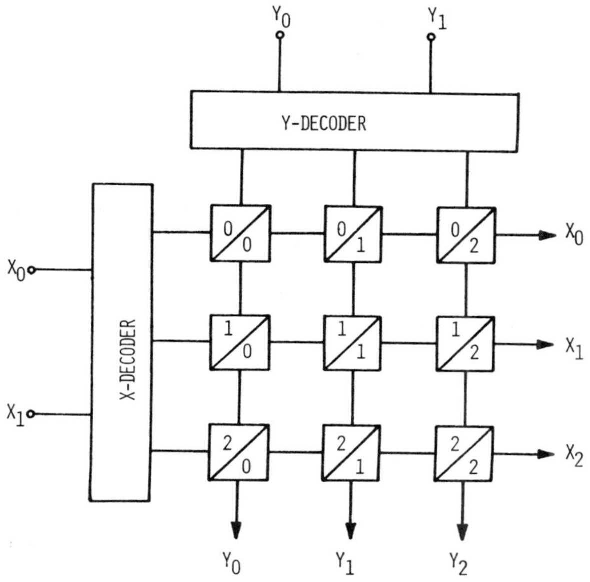

INPUT SIGNALS				SELECTED MEMORY LOCATION
LINE X_1	DIV. X_2	LINE Y_1	DIV. Y_2	
L	L	L	L	0/0
L	L	L	H	0/1
L	L	H	L	0/2
L	H	L	H	1/0
L	H	L	L	1/1
L	H	H	H	1/2
H	L	L	L	2/0
H	L	L	H	2/1
H	L	H	L	2/2

FIG. 2.18 Structure of a bit-oriented memory.

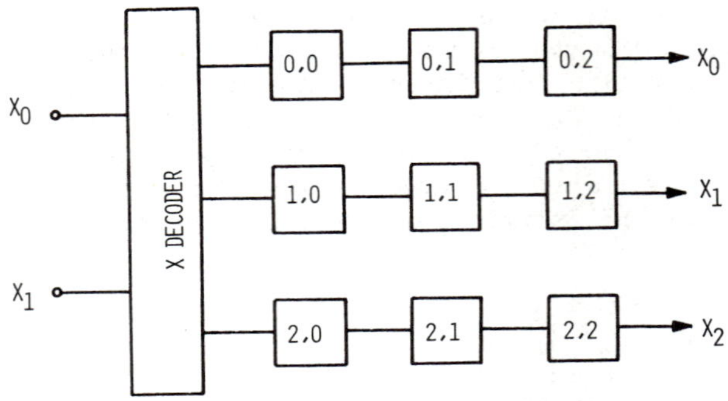

FIG. 2.19 Structure of a line- or word-oriented memory.

oriented memory is organized much more simply. With each address an entire line is selected; no direct access to an individual storage cell is possible.

Semiconductor memory is classified by the manufacturing method used to produce it (see also Chap. 5):

Bipolar memory (TTL, ECL, I^2L)
Unipolar memory (MOS)

The bipolar elements are characterized by short access time and relatively high power dissipation (Table 2.1). This limits the integration density on a chip. Therefore, microprocessors use memory built predominantly from MOS technology, which has low power dissipation. However, allowance must be given for a longer access time.

2.7.1 Read/Write Memory (RAM)

Memory to which information can be written or read from is called random access memory (RAM). The read/write memories can be subdivided according to their storage techniques:

Static memories
Dynamic memories

Static memories contain bistable flip-flops as storage elements. These flip-flops are made from field-effect transistors (FET) using NMOS-channel techniques (Fig. 2.20). In the design shown both transistors T3 and T4 are functioning as resistors. Transistors T1 and T2 act as a switch in the flip-flop. When a positive logic signal HIGH is applied to the input I2, the transistor T1 is shortened and transistor T2 is open. When the HIGH signal is removed from the input I2 the

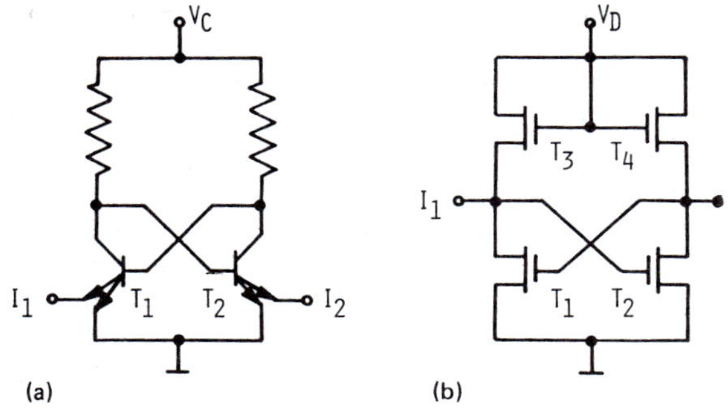

FIG. 2.20 Principal structure of bipolar and unipolar memory cells: (a) bipolar TTL-memory cell and (b) unipolar FET-memory cell.

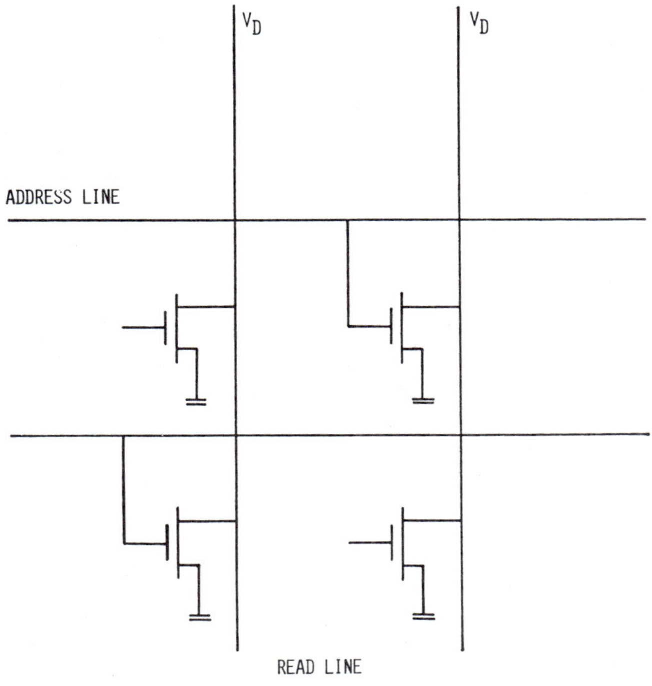

FIG. 2.21 Mask-programmed ROM.

logic information is kept. The content of the memory can be changed by applying a HIGH signal to input I1 and thus turning off T1. This state also remains stable if the input signal is removed. The information remains statically stored after completion of the write cycle and can be retrieved through input I1 or I2.

In the dynamic memory the information is stored in parasitic capacitors of the integrated MOSFET circuit. The capacitor formed by the control electrode (gate) and the n channel, which is separated by an isolation layer of metallic oxide, is used as the actual storage element. The discharge time constant of this device is approximately a few milliseconds; thus the stored information on the parasitic capacity must be cyclically refreshed to its original value by a voltage supply. In some currently available basic memory units the refresh logic is already integrated in the memory chip. It refreshes the stored charge with an additional system cycle. In Fig. 2.22 the principle of such a dynamic RAM memory is shown. Extremely high packing densities can be obtained with this simple design. Read/write memories, which are based on semiconductor technology, have the disadvantage of losing the information as soon as power loss occurs. They are therefore classified as volatile memories.

2.7.2 Read-Only Memory (ROM)

Software which may not be erased or changed, for example, firmware, is often stored in a read-only memory (ROM). These memories are programmed by the manufacturer or user and retain their information even when the power is turned off. Since the information can only be read, the individual ROM storage cell is of simple design and requires little energy. In comparison to RAM it has a higher information density.

Read-only memories are classified by the programming method used for writing the information into the memory:

1. Read-Only Memory (ROM): The memory content is permanently written into the memory during the manufacturing process.
2. Programmable ROM (PROM): The user can permanently program the PROM once by using a special programming device.
3. Erasable PROM (EPROM): The user can reprogram this memory. However, before reprogramming, the memory chip must be erased, usually with UV light.

In PROMs which can be permanently programmed by the user, the address and read lines are connected to the outside connector pins via fusible links. With a special programming device the user can separate the corresponding link and irreversibly retain the information in memory.

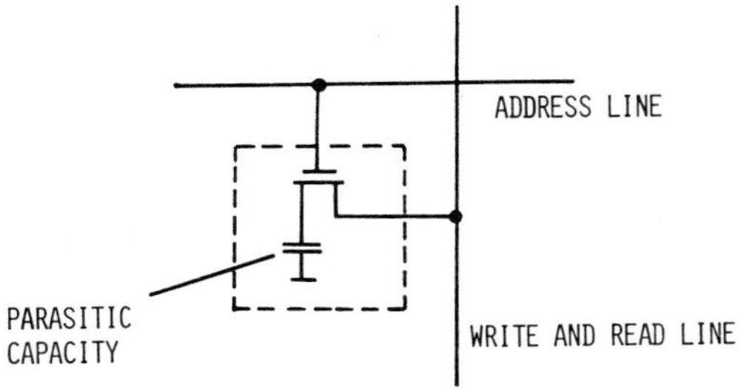

FIG. 2.22 Dynamic unipolar random access memory.

A basic element of the reprogrammable EPROM is shown in Fig. 2.22. During programming, a charge is induced by a high-intensity field on an isolated control electrode, a so-called floating gate, thereby controlling the conductivity of the DS channel. The presence or absence of the charge on this floating gate indicates presence or absence of information. With shortwave UV light the charge on the control electrode can be removed utilizing recombination processes on the semiconductor chip. Thus the programmed information can be erased, and the cell is now available for storing new information.

This type of memory is used predominantly for the development of microprocessor systems. When the application is completed and tested, the content of the EPROM is copied into the less expensive ROMs. The only reoccurring cost to the manufacturer is the expense for the programming mask.

REFERENCES

1. A. W. Burks, H. H. Goldshine, and J. von Neumann, Preliminary Discussion of the Logical Design of an Electronic Computing Instrument, U.S. Army Ordnance Department Report, 1946.

BIBLIOGRAPHY

Harrison, T. J., *Handbook of Industrial Control Computers*, Wiley-Interscience, New York, 1972.

Johnson, C. D., *Process Control Instrumentation Technology*, Wiley, New York, 1977.

Lauber, R., *Prozessautomatisierung*, Springer Verlag, Berlin, 1976.

Martin, T., *Prozessdatenverarbeitung*, Elitera Verlag, Berlin, 1976.

Martin, W., *Mikrocomputer in der Prozessdatenverarbeitung*, Hanser Verlag, Munich, 1977.

Mielentz, P., *Der Prozessrechner mit seinen Koppelelementen*, VDI-Verlag, Dusseldorf, 1974.

Osborne, A., *An Introduction to Microcomputers, Vol. I.*, Adam Osborne and Associates, Berkeley, Calif., 1976.

Terminals and Communication Handbook, Digital Equipment Corp., Maynard, Mass., 1980.

3

MEASUREMENT SIGNAL GENERATION AND CONVERSION

3.1 INTRODUCTION

When controlling technical processes there is a multitude of data originating from measurements. They have to be transferred to the computer by an economical and suitable method. Depending on the process, the measured parameters can be of different nature. Because of the inherent advantages which electrical measurements offer (accuracy, high sensitivity, and easy transfer of signals), it is preferable to convert the measurements taken by transducers into electrical units. This chapter discusses the properties of such transducers. Within the framework of this book only some important methods of transforming physical properties into electrical signals will be discussed. The reader interested in becoming more familiar with these methods should study the more detailed literature.

3.1.1 Properties of Transducers

A transducer is a physical device whose output signal represents the static dynamic behavior of its input parameter. In the simplest case the output is an electrical signal (current, voltage), whereas the input parameters, for example, light, pressure, temperature, are not. During the measuring process the transducer should be operated in such a manner that its output signal is an exact image of the measured input parameter. The influence of additional factors such as noise must be eliminated (Sec. 8.3). This problem will be discussed later. When converting measured parameters into electrical signals, one distinguishes between active and passive transducers. The active transducer transforms energy received from the measured object during the measuring process. In extreme cases this may affect the state of the measured object sufficiently to falsify the reading. Examples of active transducers are thermocouples operating on the Seebeck effect principle or displacement or velocity transducers operating on the induction principle.

The passive transducer obtains energy from an external current source. Therefore, it does not adversely affect the measuring object. All methods which measure the change of an electrical resistance employ passive transducers.

One further differentiates electrical transducers by their ability to process analog or digital signals. There are few transducers which directly generate digital signals, for example, incremental displacement pickups. Here only a level adaptation of the input signal to the computer is necessary. Most signals, however, are in analog form. These must be transformed with the help of an A/D converter into digital signals. Further details on A/D conversion methods will be discussed in Sec. 5.3.

3.2 CONVERSION OF SIGNALS FROM MECHANICAL TRANSDUCERS TO ELECTRICAL SIGNALS

3.2.1 Displacement Transducers

In many processes absolute or relative changes of a distance have to be measured and the obtained values have to be converted into suitable process signals. Typically physical measurement units, such as velocity, force, pressure, etc., are obtained from displacement transducers. For the measurement these parameters specific displacement pickups have been developed.

Limit Switches

The simplest method for sensing a traveled distance is to use a limit switch. It indicates that a certain distance has been covered by the object to be observed. Since a limit switch does not generate an absolute measurement value but only a digital signal, it can be directly connected, with the help of a voltage-level transformer, to the input unit of the computer. This signal may, for example, interrupt the program to actuate or interrogate the internal time clock of the computer. More details on the control of interrupts can be found in Chap. 6. Problems of interfacing and level changes will be discussed in Chaps. 5 and 8, respectively.

Proximity Switches

Another type of limit switch is the proximity switch. It is used when a direct contact between the moving part and the limit switch is not practical or when the use of an optical method is not advisable because of dirt or other environmental influences. The proximity switch operates on the principle of resistance change by induction. To the

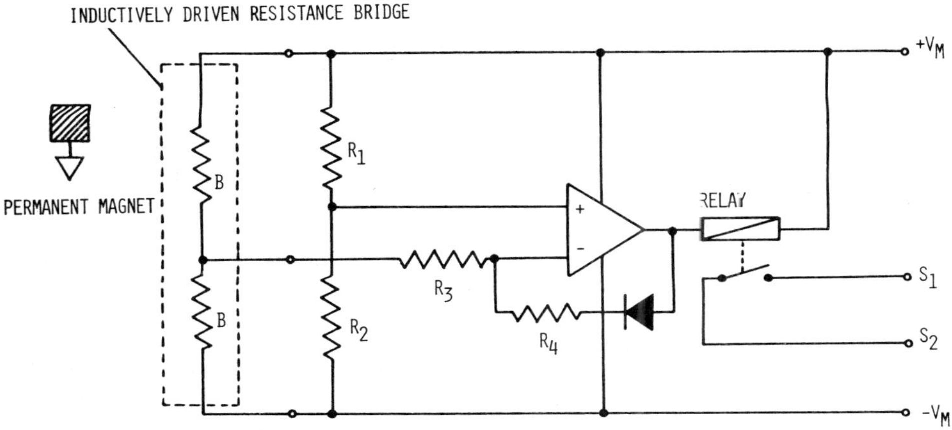

FIG. 3.1 Proximity switch containing an inductively driven resistance bridge: B = magnetic induction; V_M = bridge supply voltage.

moving part of the machine to be observed a permanent magnet is attached. The switching signal is generated when a change in magnetic induction is registered by the sensor. The basic circuit of such a proximity switch is shown in Fig. 3.1. The measuring section consists of two magnetically alterable resistors which are connected to a differential amplifier (Sec. 5.2). In the feedback line of the circuit a diode is located with which the desired switching hysteresis can be obtained. This differential amplifier controls a relay which switches the terminals S1 and S2. When the relay is closed, a current flows across its contact, and thus it is possible to actuate the interrupt line of the computer.

Displacement Measurement with Analog Transducers

If one is interested in measuring not only the endpoints of a distance to be traveled but also the exact value of a distance, one can use an analog displacement pickup. Here only the group of displacement pickups which use the principle of ohmic resistance change will be discussed. The ohmic resistance of a metallic conductor can be expressed by the following formula:

$$R = \frac{\rho \cdot \ell}{A} \tag{1}$$

where

ℓ = length of conductor
A = cross section of conductor
ρ = specific resistivity of conductor

It can readily be seen that the ohmic resistance R is directly proportional to the length of the conductor. The moving part of the machine to be measured can be connected directly to the sliding contact of a rheostat, the latter being connected to a current source. The measured distance is proportional to the voltage measured across the resistor X (Fig. 3.2). The measured value can be transferred via an A/D converter to the computer. When selecting the level of the supply voltage V_M, one should observe that the resistor may not be overheated since its resistance is effected by its operating temperature. Overheating would result in erroneous data.

If only small displacements are to be measured, strain gauges may be used. A strain gauge consist of a wire grid bonded to a paper substrate. The strain gauge is fastened to the work piece with an adhesive. When the work piece changes its dimension, the diameter of the strain gauge wire changes, resulting in a resistance change. The change of the electrical resistance of the strain gauge is sensed by the measuring instrument. It can be described by formula (1). Since only a very small dimensional change occurs in this type of measurement, the resistance changes is also very small. Thus a bridge circuit must be employed to measure and transfer the signal via a A/D converter to the computer. More details about the bridge circuit are discussed in Sec. 3.5.2. Other analog transducers operate on capacitive, inductive, or piezoelectric principles. Closer details will not be covered here.

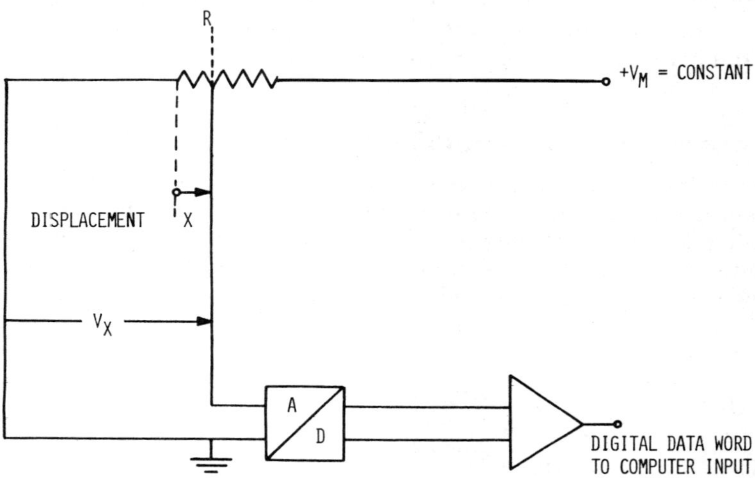

FIG. 3.2 Ohmic displacement pickup.

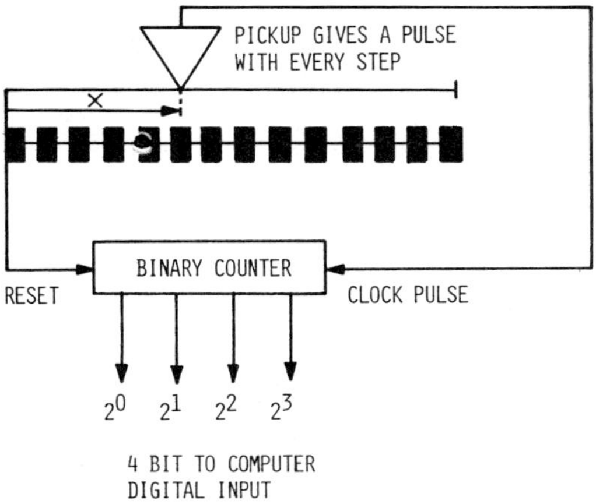

FIG. 3.3 Digital displacement pickup.

Displacement Measurement with Digital Transducers

One incremental transducer will be discussed as a typical example of a digital displacement transducer. The measuring scale of the transducer is divided into individual elements of constant length. The principle of operation is based on the fact that the elements are counted by the moving pickup during the measuring procedure. The accuracy of measurement depends on the length and the number of the elements. Incremental sensors can operate on mechanical, inductive, capacitive, or optical principles. Figure 3.3 shows such an example. When the sensor passes a measuring mark, it generates a pulse which will be counted by a binary counter (Sec. 5.6). When the sensor is located at the beginning of the scale at the zero point, the counter is reset. The output of the counter can be transferred directly to the computer by means of the digital input.

3.3 OPTICAL SENSORS

Optical sensors are used both for processing of digital signals and for light intensity measurements. They transform optical signals into electrical information.

3.3.1 Photoelectric Switches

An important tool to transmit digital information is the photoelectric switch using a photocell. In a circuit employing such a switch the interruption of a light beam is registered. This device actually is an-

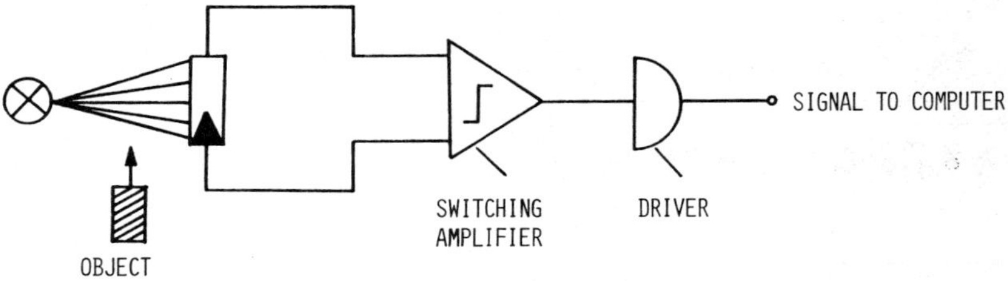

FIG. 3.4 Principle of a light-activated switch.

other type of contactless switch. Figure 3.4 shows the basic circuit. When light illuminates the transducer, the amplifier generates no signal. When the light is interrupted, the amplifier goes to saturation and the driver and level transformer render an active signal to the computer (Sec. 5.2). The amplification characteristics of the amplifier is not linear but has a hysteresis. Below a certain threshold of input voltage level, it generates a very small output signal. Above the threshold, it generates a very high signal level. The driver has the task of forming the signal into a suitable shape so that the input level and rise time of the signal is acceptable for further processing (Sec. 5.4.3).

Light-Actuated Resistors (LDR)

Very commonly used optical sensors are photoresistors. Contrary to ohmic resistors they consist of semiconductor material. When light falls on the semiconductor sensor, the photons with an energy level greater than the energy gap of the semiconductor create new electrons and holes and the conductivity of the material increases. In a photo-

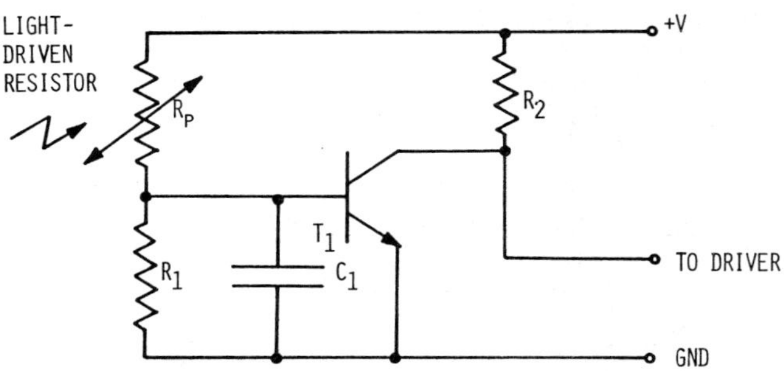

FIG. 3.5 Light-activated switch with light-driven resistor. (From Siemens Halbleiter Schaltbeispiele, courtesy of Siemens AG, Munich.)

resistor the resistance thus changes from several megaohms when dark to a few hundred ohms when intensively illuminated. Figure 3.5 shows a simple light-activated switch employing a photoresistor. When light falls on the photoresistor, R_p becomes $<R_1$ and the NPN transistor is turned on. Thus the output signal of the light-activated switch is very low. When light is removed from R_p, the basis of the transistor T_1 becomes more negative and is turned off. Thereby the output voltage rises to a very high positive value. The capacitor C has the task of reducing transient side effects which occur during partial illumination.

Active Photosemiconductors

Active optical transducers are the photodiodes, phototransistors, and photothyristors. These elements have a higher sensitivity and switch higher frequency signals than the photoresistors. Another advantage is their very small dimension. Presently they are being used more and more frequently in modern interface circuits. Active photosemiconductors are turned on and off by light that falls into the barrier layer. When light falls on the photosemiconductor, an increase in reverse current is sensed. A more detailed explanation will not be within the scope of this book. The interested reader should survey the literature listed at the end of this chapter.

Phototransistors can be used as fundamental building elements of photoelectric switches and are often incorporated directly into the logic circuits. Figures 3.6 to 3.9 show examples for some logic families.

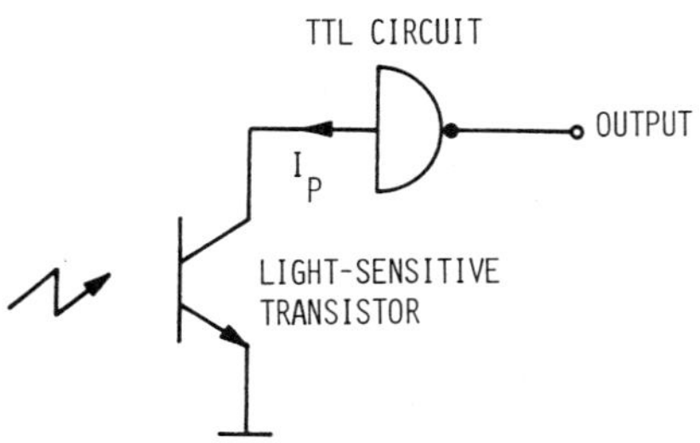

PHOTOCURRENT I_P	OUTPUT LEVEL
>0.9 mA	H
<0.2 mA	L

(EXAMPLE FOR A LIGHT-SENSITIVE TRANSISTOR SIEMENS BPX 80)

FIG. 3.6 TTL interface. (From Siemens Halbleiter Schaltbeispiele, courtesy of Siemens AG, Munich.)

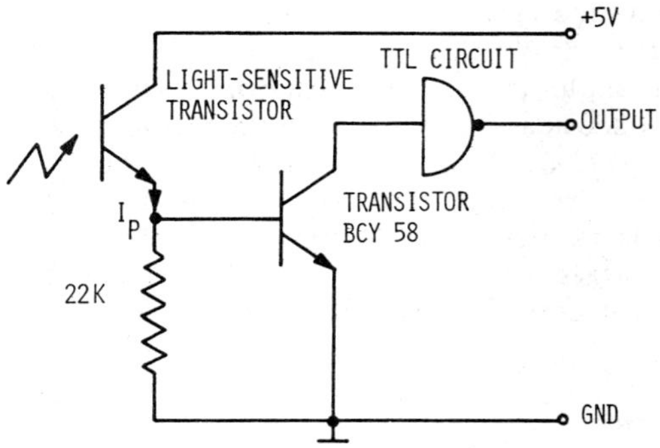

PHOTOCURRENT I_P	OUTPUT LEVEL
>40 μA	H
<20 μA	L

(EXAMPLE FOR BPX 80)

FIG. 3.7 TTL interface using an amplifier stage. (From Siemens Halbleiter Schaltbeispiele, courtesy of Siemens AG, Munich.)

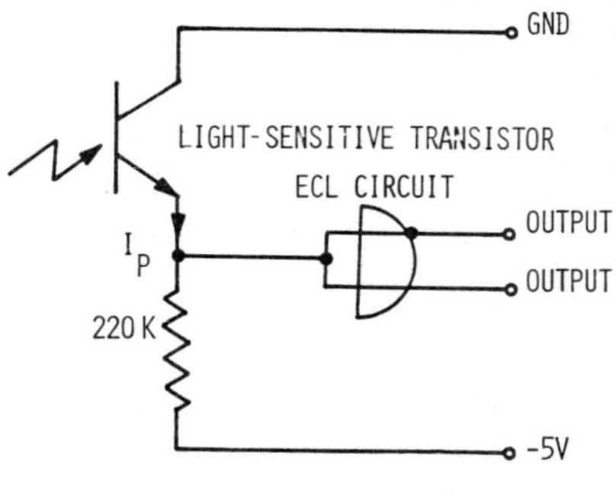

PHOTOCURRENT I_P	OUTPUT LEVEL
>18 μA	L
<16 μA	H

(EXAMPLE FOR BPX 80)

FIG. 3.8 ECL interface. (From Siemens Halbleiter Schaltbeispiele, courtesy of Siemens AG, Munich.)

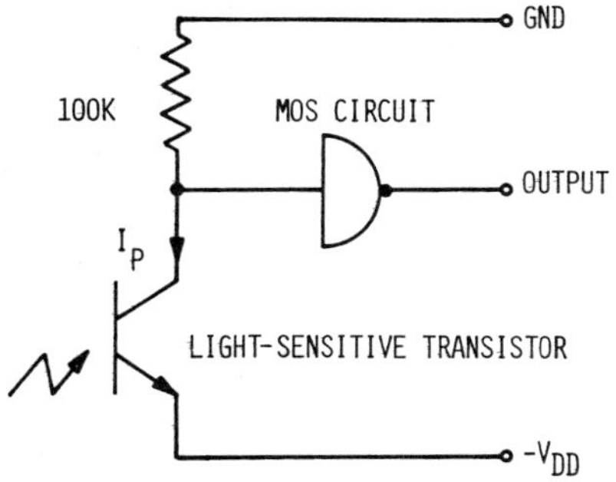

PHOTOCURRENT I_P	OUTPUT LEVEL
>100 μA	L
< 15 μA	H

(EXAMPLE FOR BPX 80)

FIG. 3.9 MOS interface. (From Siemens Halbleiter Schaltbeispiele, courtesy of Siemens AG, Munich.)

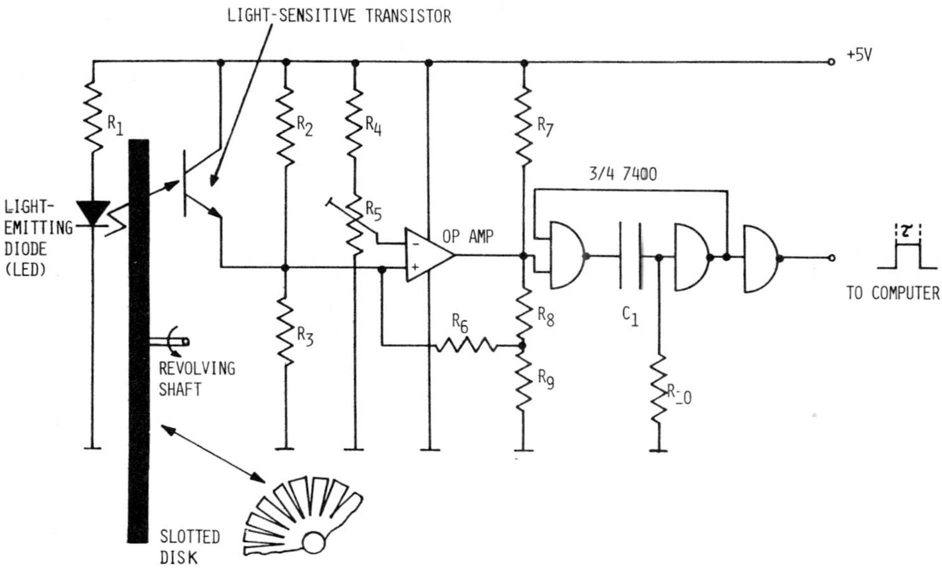

FIG. 3.10 Measuring revolutions with a light-sensitive transistor. (From Siemens Halbleiter Schaltbeispiele, courtesy of Siemens AG, Munich.)

To measure the rotational speed (revolutions per minute) of small motors and other rotating parts, photosensors are commonly used. For this purpose a rotating disk having a light surface with black measuring marks at its circumference is mounted on the rotating object to be measured. A light beam is generated by a light-emitting diode, falls on the disk, and is reflected to the phototransistor. The black marks interrupt the light beam periodically. Figure 3.10 shows the circuit of the measuring system. During illumination the noninverting input of the operational amplifier (Sec. 5.2) will be more positive in regard to the adjustable threshold of the inverting input, and a high output signal is produced. The amplifier activates the three NAND gates of the monostable flip-flop and thereby produces well-defined output pulses proportional to the revolutions per minute (rpm). The delay of the monoflop is obtained by the RC components and can be selected within certain ranges. The values shown here would give a pulse of 10-μsec duration. The pulses can activate a binary counter whose content is read at regular time intervals by the computer and represent the rpm of the rotating shaft (Sec. 5.6).

3.3.2 Light Intensity Measurement

If it is necessary to transmit light intensity information to the computer, photodiodes operated in reverse direction can be used as a transducer. These elements generate a photocurrent when illuminated. This current can be amplified and sent to the computer with the aid of an A/D converter (Sec. 5.3). When very small dc currents are encountered, stability problems may occur. This will be discussed in Sec. 3.5.2.

3.4 TEMPERATURE TRANSDUCERS

3.4.1 Thermocouples

The most important temperature transducer is the thermocouple. It operates on the Seebeck principle. With this principle an electromotive force is generated at the interface of two different metal wires across which a temperature difference exists between their two opposite ends. The level of electromotive force depends on the temperature difference and the location of the two metals in the electromotive table. Figure 3.11 shows the principle of a thermocouple transducer. The voltage difference between both ends of the wires is determined by the equation

$$V_{th} = K(T_2 - T_1) \tag{2}$$

The values for K can be taken from tables. T_2 is the temperature to be measured, and T_1 is a known reference temperature, for example, the temperature of ice water.

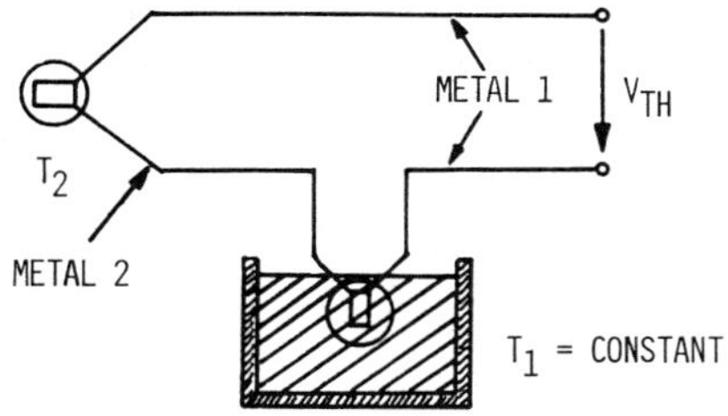

FIG. 3.11 Principle of temperature measurement by thermocouple: T_1, constant reference temperature; T_2, temperature at measuring point.

For industrial applications where many thermocouples must be connected to the computer, only one reference transducer may be used. Figure 3.12 shows such a measuring circuit. All connector terminals for the transducer wires of the measuring circuit are connected to one or more groups of base plates made from brass or on other good heat conductor. The brass plates assure that the temperature on the board is the same at all thermocouple terminals. All plates are interconnected with a reference thermocouple, one leg of which is usually held at the reference temperature. The reference voltage obtained in this manner is also transferred to the computer

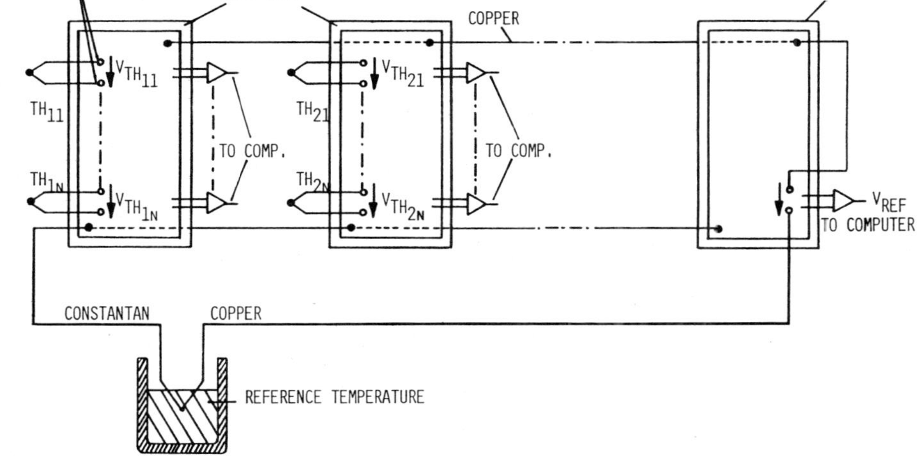

FIG. 3.12 Multitemperature measurement using one reference source.

and is used to relate all other thermocouples readings to the reference temperature. The thermocouple terminals are electrically insulated from the brass plates.

The output voltage of thermocouple is not proportional to the temperature change. Thus the measurement has to be linearized with the help of software by using the following binomial curve fit:

$$T = a_0 + a_1V + a_2V^2 + a_3V^3 + \cdots \tag{3}$$

The constants a_0, ···, a_n can be obtained from the supplier of thermocouples. They are very specific for different materials.

3.4.2 Resistance Thermometers

Since metallic resistors are temperature dependent, a measurement of the temperature can be based on a resistance change. For a certain temperature range the resistance change is proportional to the temperature change. Equation (4) shows the relationship.

$$R_T = R_{T_0} [1 + \alpha_0(T - T_0)] \tag{4}$$

where

R_T = resistance at temperature T
R_{T_0} = resistance at reference temperature T_0

α_0 = temperature constant, valid in the range of T_0

The temperature constant α_0 for metals is positive. The resistance increases in value with increasing temperature. Thus, metals behave like PTC resistors. Resistors made from semiconductor material show negative characteristics. They behave like NTC resistors. It is possible to build different types of resistance thermometers for different temperature ranges. The range depending on the material being used. The values for the temperature constant α_0 can be taken from tables. For the measurements of the temperature change, a bridge circuit is used (see Sec. 3.5.2).

In addition to the methods for temperature measurement discussed above, the method which senses the radiation dissipated from a warm object should be mentioned. This method determines the temperature of an object from the composition of the emitted color spectrum. It can be applied within a temperature range of almost 0 to 2000°C. In addition, improved infrared measuring techniques offer the possibility of measuring temperatures between the range of 20 to 500°C.

3.5 PREPROCESSING OF SIGNALS EXTERIOR TO THE COMPUTER

The electrical signals transmitted by most transducers are not suitable for direct input into the peripheral data channel of the computer. This is particularly true for direct current in the range of microvolts to millivolts as well as for measurements where a resistance change is recorded. Such signals should be directly amplified at the location of the measurement.

3.5.1 Signal Amplifiers

In order to assure transfer of a strong signal and to minimize the influence of noise, an amplifier is often required. Here a distinction is made between directly coupled dc amplifiers and so-called modulation amplifiers, of which there are various types.

Direct-Coupled DC Amplifiers (Sec. 3.5.2)

In most measuring systems operational or instrument amplifiers are used. When direct-coupled dc amplifiers are installed, the problem of zero drift may occur. This means that when the input of the amplifier is short-circuited the output voltage is different from zero. This is due to external influences such as changes in temperature or changes in power supply voltage. In particular, when very small input voltages are measured, the drift can be greater than the useful signal. In order to reduce drift as much as possible, most modern dc amplifiers are manufactured as large-scale integrated circuits (Sec. 3.5.2). Through appropriate selection of the integrated basic components, as well as the compactness of the circuit on the substrate, cancellation of drift can be achieved. The advantage of direct-coupled signal amplifiers lies in the broad frequency range, for which they can be used. They can be applied for dc signals as well as for signals in the megahertz range.

Modulation Amplifiers

In older measuring equipment a chopper-amplifier is often used; it operates on the modulation principle (Fig. 3.13). With this amplifier the problem of drift is eliminated first by multiplying the signal voltage with a square wave (Fig. 3.14). The resulting ac voltage is then amplified with the help of a wide-frequency-band ac amplifier, and afterward, it is demodulated. Thus the ac amplifier can be designed for the much higher amplitudes of the multiplied voltage. This makes it possible to easily separate the useful signal from the noise. A low-pass filter is used to remove the remaining noise components from

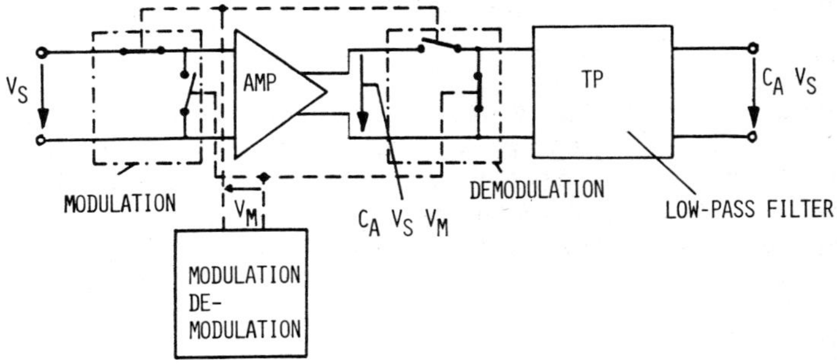

FIG. 3.13 Principle of chopper-amplifier.

the modulated voltage. At the output of the filter the amplified signal appears. Figure 3.14 shows the principle. The switches used for the modulator and demodulator generally work on the field-effect principle. They are usually integrated analog switches (Sec. 3.5.5).

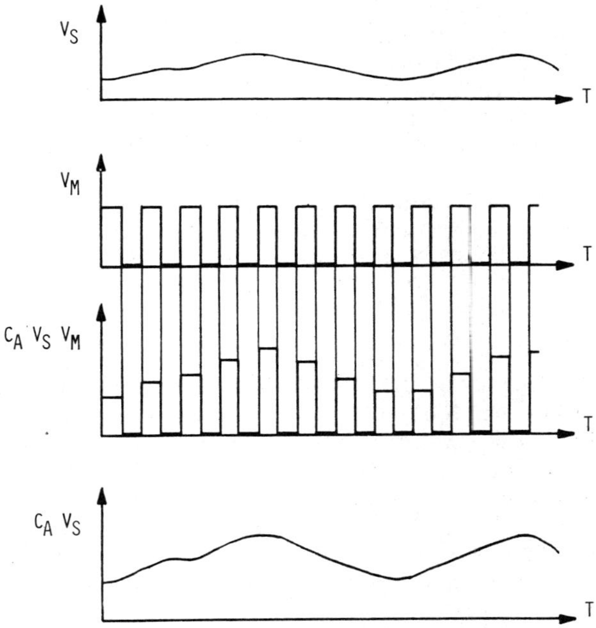

FIG. 3.14 Voltage diagram of a chopper amplifier.

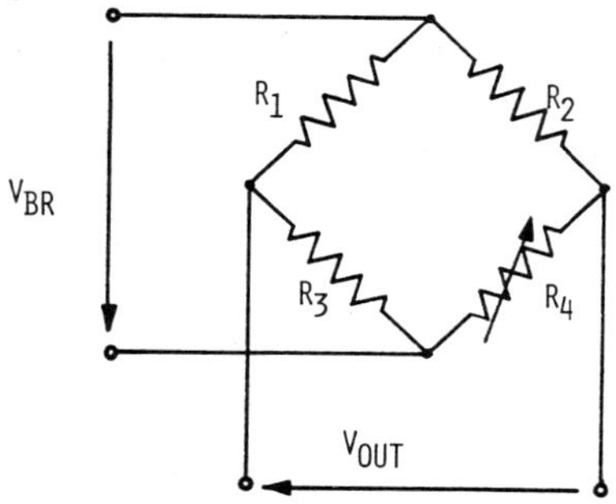

FIG. 3.15 Wheatstone bridge: V_{BR}, bridge supply voltage; V_{OUT}, output voltage.

3.5.2 Bridge Circuits

Bridge circuits are used where the change or the actual value of a resistance or impedance must be measured. In this brief introduction, only a bridge circuit to measure the actual resistance will be discussed. The most commonly used instrument employs a Wheatstone bridge. The output voltage of the bridge is approximately proportional to the resistance change ΔR. Figure 3.15 shows the bridge circuit and Fig. 3.16 its equivalent circuit. It is easy to prove that the highest sensitivity is obtained when all the resistors in the bridge are equal ($R_1 = R_2 = R_3 = R_4 = R$). The change of R_4 in relation to the measured value R_3 is ΔR. For the output voltage one can then write approximately

$$V_o = V_B \frac{\Delta R}{4R + 2\,\Delta R} \tag{5}$$

For $\Delta R << R$, the following approximation is obtained:

$$V_o = V_B \frac{\Delta R}{4R} \tag{6}$$

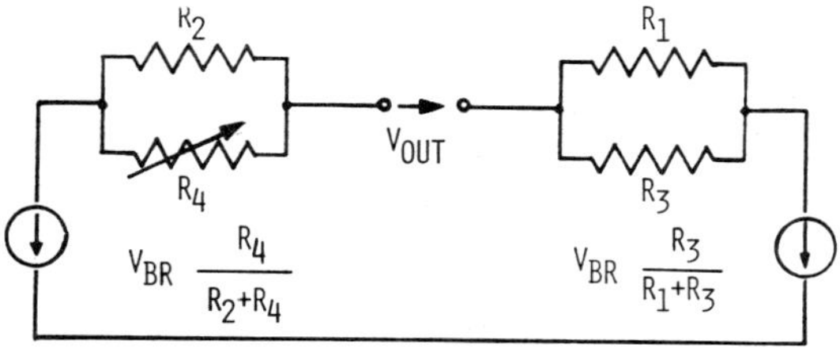

FIG. 3.16 Equivalent circuit diagram of Wheatstone bridge.

It can readily be seen that it is necessary to keep the supply voltage V constant. Otherwise, the results will be inaccurate. The bridge can be operated by the direct measurement principle or by the zero-adjustment principle. In the first case, the bridge is initially adjusted to zero by short-circuiting the input signals. Here the output signal from the measurement is proportional to the resistance change ΔR and can be brought to the computer via an A/D converter. When using the zero-adjustment method, one branch of the bridge is connected to a variable precision resistor. During measurement the output voltage of the bridge is adjusted with this resistor to zero. When properly dimensioned, the change of the resistance is an indication of the measured value. The zero-adjustment method is well suited for offline measurements. It is very accurate. For online measurements the direct method is better suited because the output signal is easier to use for further processing.

3.5.3 Analog Multiplexers and Analog Switches

When developing an interface circuit, the problem of switching analog signals often arises. In addition, galvanic isolation of the measuring equipment from the computer is necessary. Formerly relays were used for this task (Sec. 8.3). Here mechanical wear and low operating frequency may result in considerable difficulties. Modern analog switches consist of integrated MOS transistors, which have an extremely low resistance when turned on. The value lies between 50 and 250 Ω. In the off-state mode the resistance has a value of approximately 1000 MΩ. In addition to the switching circuit, a TTL driver may be integrated. Galvanic decoupling of the transducer from the computer may be obtained by a photoisolator (Fig. 3.17).

Where many analog transducers are used, an analog multiplexer should be employed to keep installation cost low. This unit sequentially switches one analog input signal after another to the common input line of the computer. In principle, there exist two multiplexing methods.

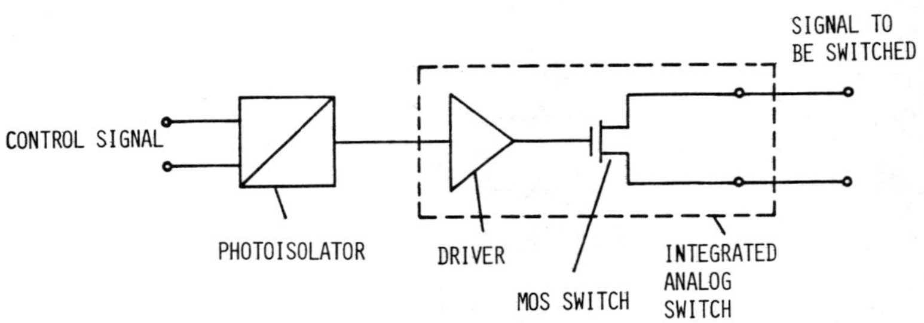

FIG. 3.17 Principle of an integrated analog switch.

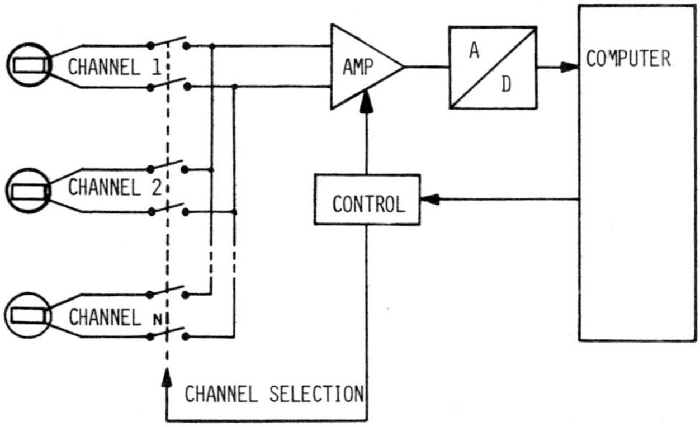

FIG. 3.18 Low-level multiplexer. (Reprinted from Rembold et al., by courtesy of Marcel Dekker, Inc.)

Low-Level Multiplexer

A multiplexer can be directly connected to the measuring channels (Fig. 3.18). This solution is rather inexpensive since cable and amplifier costs can be saved. Only one common line leads from the multiplexer to the input channel of the computer. This channel contains the amplifier and an A/D converter in its peripheral. This circuit called a *low-level multiplexer*, can only be realized when the transducer has a low impedance and a high signal level. Only in this case, the signal can be switched without undue distortion and interference and brought to the computer. Unfortunately both prerequisites are not applicable to many transducers.

High-Level Multiplexer

If there is a very weak signal with a small amplitude, it must be amplified before it is switched and sent via the data channel to the computer (Fig. 3.19). The multiplexer and the A/D converter are connected to these amplifiers. Both of these units are usually built into one component; thus an A/D converter with a built-in multiplexer is obtained. This unit is called a *high-level multiplexer* and is used in applications where exact measurements must be taken.

Analog measuring systems have become a less costly factor in large distributed process control systems within the last years. Cost savings were obtained by using multiplexers and reducing the cost for signal lines and peripherals. However, this less flexible circuit only allows processing of data for less time-critical applications. Time-critical alarm signals still must be brought to the computer through separate input channels.

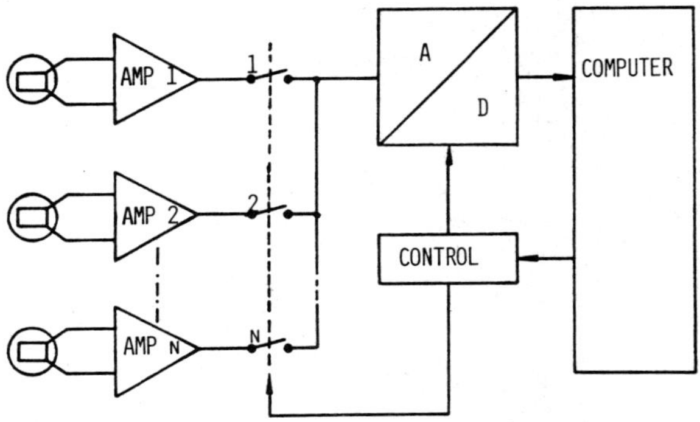

FIG. 3.19 High-level multiplexer. (Reprinted from Rembold et al., by courtesy of Marcel Dekker, Inc.)

3.5.4 Intelligent Bus Systems (Chap. 9)

With the introduction of less expensive intelligent processors and the general trend in cost reduction of hardware, analog multiplexing is more frequently employed in intelligent distributed bus systems for measuring applications. Utilizing their flexibility they can be adapted

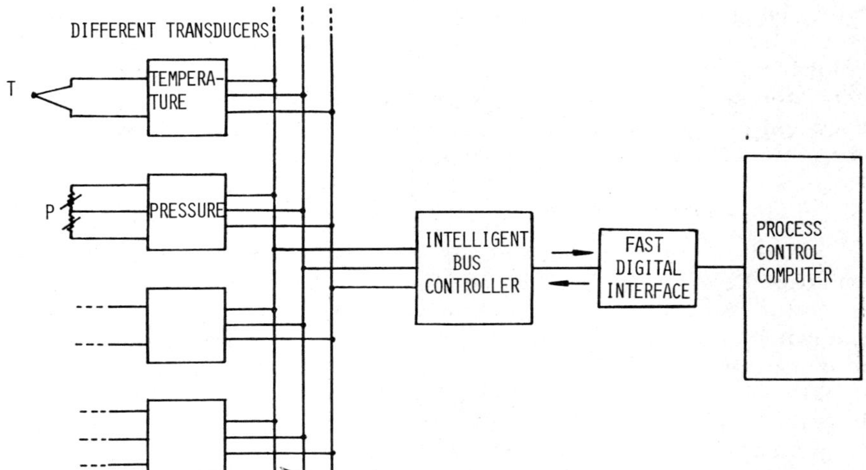

FIG. 3.20 Principle of data acquisition with the help of an intelligent bus system.

easily to a variety of different tasks. Thus it is also possible to connect time-critical processes to the bus system. A very high data transfer rate over the transmission lines to the computer can be obtained by digitizing the data and proper control of the transmission procedures (Fig. 3.20). Details are given in Chap. 9.

BIBLIOGRAPHY

Benedict, R. P., *Fundamental of Temperature, Pressure and Flow Measurements.* Wiley, New York, 1977.

Fribance, Auston, *Industrial Instrumentation Fundamentals*, McGraw-Hill, New York, 1962.

Huelsmann, L. P., *Operational Amplifiers: Design and Applications*, McGraw-Hill, New York, 1971.

Kronmüller, H. and Barakat, F., *Prozessmesstechnik I*, Springer Verlag, Berlin, 1974.

Rembold, U., M. K. Seth, and J. S. Weinstein, *Computers in Manufacturing*, Marcel Dekker, New York, 1977.

Samal, E., *Elektrische Messung von Messgrössen*, AEG-Handbuch, Vol. 17, Elitera Verlag, Berlin, 1174.

Sheingold, D. H., *Transducer Interfacing Handbook*, Analog Devices, Norwood, Mass., 1980.

Siemens Halbleiter Schaltbeispiele, Hrsg. Siemens AG, Bereich Halbleiter, Munich.

Unger, H. G. and Schultz, W., *Elektronische Bauelemente und Netzwerke 1*, Verlag F. Vieweg & Sohn, Braunschweig, 1968.

Wentz, Eric S., Digital Revolution in the Test Lab, *Machine Design*, *3* (1976), 144-148.

4

DATA COMMUNICATION

4.1 FUNDAMENTALS OF DATA TRANSMISSION

4.1.1 Transmission Rate

In data communication the rate at which data can be transferred between computers and between the computer and the process is of utmost importance. The data transfer rate depends on the equipment used and the response time needed to perform process control. Fundamentally there are two different units of measurement used to describe the transfer rate. These are the bit rate and baud rate.* At this point we will not further discuss the fundamentals of the communication theory which describes these definitions. Instead, a very simple description of these units will be given.

A character to be transferred is represented by a string of a unique signal pattern. In digital communication the bits transmitted over a line are represented by electrical pulses of equally long information units. If we define T_1 as the shortest possible signal element by which a character or part of a character can be transmitted over a line, then the baud rate is equivalent to (see Fig. 4.1)

$$\text{Baud rate} = \frac{1}{T_1} \text{ baud}$$

Typical baud rates are

Telex: 50 baud
Datex: 200 baud
Public communication lines: 400 baud
600 baud

---

*A bit is a unit of information and a baud a unit of signaling speed.

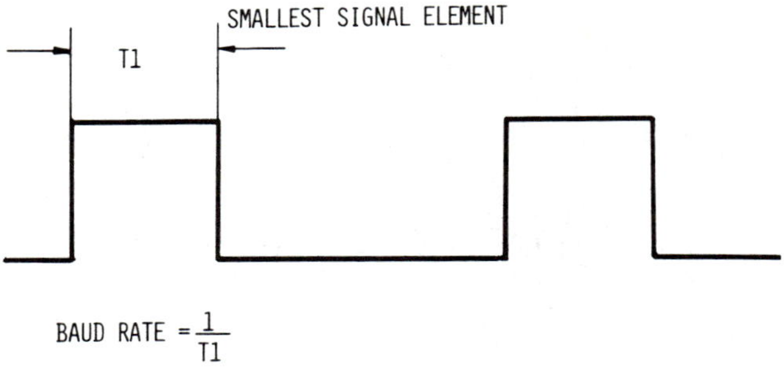

FIG. 4.1 Definition of baud rate.

Private communication lines: 1200 baud
2400 baud
4800 baud
9600 baud

Bits per second define the total number of information units transferred in one second; thereby all redundant bits needed for the transmission of a character are included.

To explain this concept more closely, we will take a look at the procedure for transferring one character. One baud is the reciprocal value of the smallest time element in the character. The bit rate, however, is determined from the sum of all bits necessary to initiate and complete a transfer of a character from one communication device to another. This includes the bits for the character itself, control bits, and parity bits. The baud rate is equal to the bit rate only when all time intervals are of equal length and all signal pulses are information pulses.

In data communications both definitions are used. The baud rate is important in telephone communication; it specifies the type of communication line to be employed. The computer user is more interested in the bit rate since it informs him or her of the magnitude of information to be transferred.

4.1.2 Characteristics of Transmission Lines

The capacity of the communication line depends on the possible transfer speed and the physical properties of the line. In this section the properties of communication lines will be discussed. They are especially important for interface technology. Within the framework of this book the theory of data communication is not discussed. The reader interested in learning more about this field should make use of other literature on the subject.

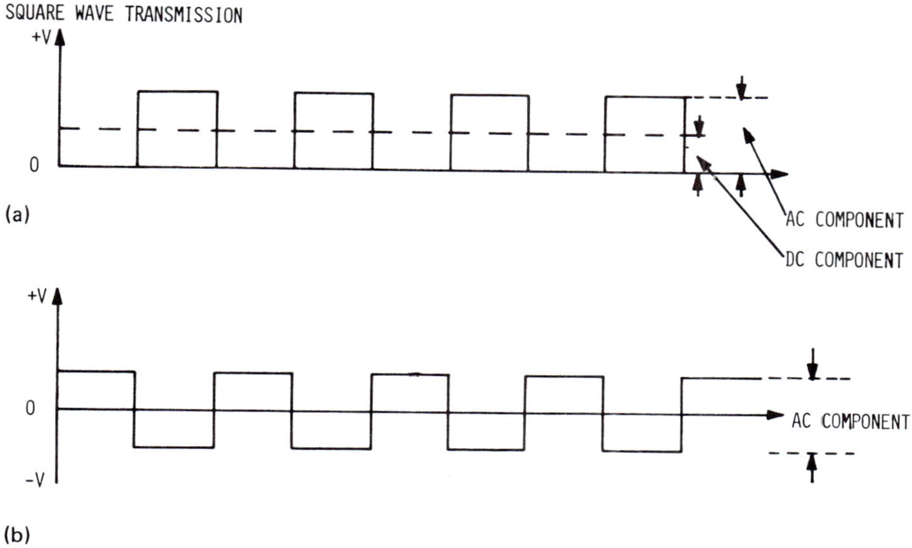

FIG. 4.2 Different methods to transmit square waves (a) with dc offset and (b) without dc offset.

Fundamentally it is possible to transfer data with the help of square or sine waves with or without a dc offset (Fig. 4.2). A square wave typically consists of a sine wave having the fundamental frequency as this wave and a multitude of superimposed harmonics. With the help of the Fourier theorem, it is possible to represent a square wave with sine and cosine functions (Fig. 4.3). Thus both communication techniques can be described with the same theoretical background.

Line Impedance

A segment of a communication line can be described with the help of an equivalent circuit element consisting of a series resistor R', a shunt conductance G', a series inductance L', and a shunt capacitance C' (Fig. 4.4). The line over its length can be visualized as a series of such similar circuit elements. The characteristic impedance of the line is

$$\text{Characteristic impedance } Z = \sqrt{\frac{R' + j\Omega L'}{G' + j\Omega C'}} \qquad (1)$$

The imaginary parts $j\Omega L'$ and $j\Omega C'$ of this equation describe the ac losses during data transmission. They depend on the transmission fre-

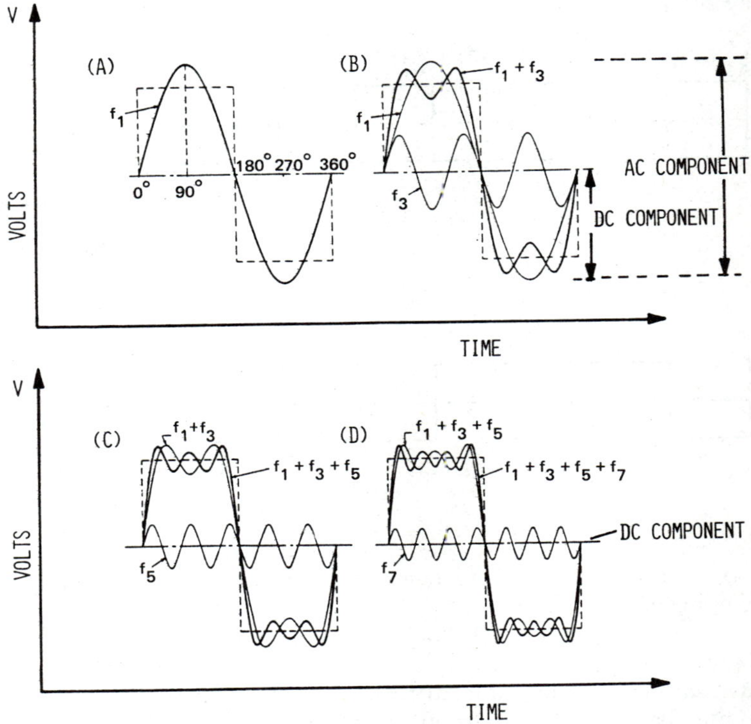

FIG. 4.3 AC and dc components of a square wave.

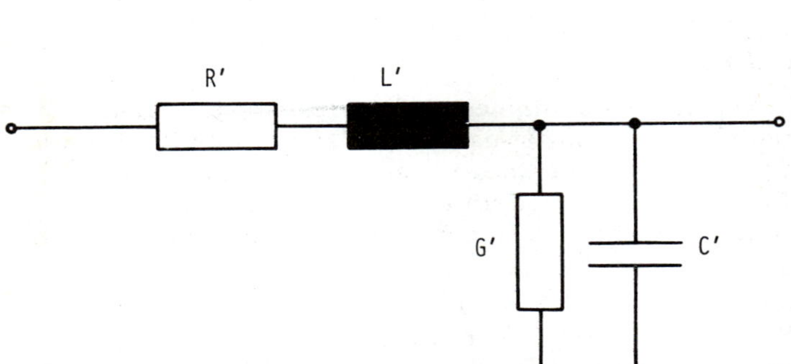

FIG. 4.4 Equivalent circuit of a data transmission system.

quency. The real components R' and G' describe the dc losses in the cable. A wave to be transmitted will always be affected by the characteristic line impedance. This phenomenon can be described by the current and voltage which are effective in the travel direction of the wave.

$$\vec{Z} = \frac{\vec{U}_x}{\vec{I}_x} \tag{2}$$

Signal Attenuation

A signal transferred over a communication line will be attenuated. The magnitude of attenuation depends on the physical properties of the line and the transmission frequency. *Attenuation* is defined as

$$D = 10 \ln \frac{\text{input power}}{\text{output power}} = 20 \log \frac{\text{input voltage}}{\text{output voltage}} \text{ dB} \tag{3}$$

Figure 4.5 shows the attenuation of a communication signal for various communication lines. Signals are attenuated more in twisted pair cables than in coaxial cables. The attenuation increases with the transmission frequency.

The distortion of the square wave is the result of different attenuation of the higher harmonics (Fig. 4.3 and 4.5). The fundamental ac component, which has the same frequency as the square wave, is attenuated the least. Attenuation of this wave is easily calculated. The result renders a rough estimate of the overall attenuation of the signal. With higher frequency the higher harmonics are increasingly at-

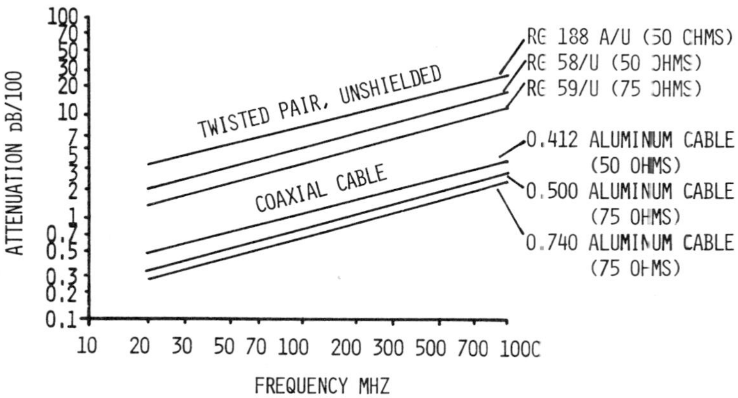

FIG. 4.5 Attenuation characteristics of various types of cables.

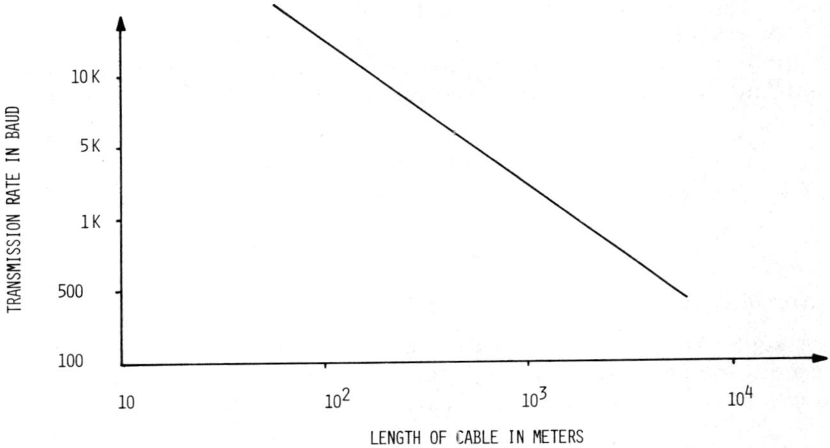

FIG. 4.6 Maximal possible transmission rate over twisted pair cables.

tenuated until only a sine wave appears at the end of the communication line. This sine wave is superimposed over the dc component. Since the dc component is only affected by the ohmic resistance of the line and the ac component is attenuated in relation to the length of the line, it is possible that the line voltage reaching the receiver is periodically above or below the threshold voltage of the receiver. To

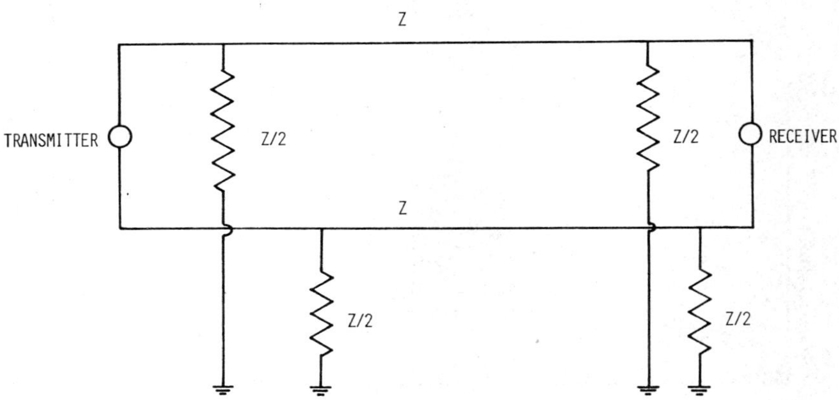

FIG. 4.7 Elimination of reflections with the help of grounding resistors.

avoid such difficulties, the operation of the line is limited by the maximum transfer frequency.

The influence of the length of the line on the transmission rate of a square wave can be seen in Fig. 4.6.

Line Reflection

To avoid reflection in the line the communication system must be terminated (grounded) at both ends by a resistor whose value is about equal to the characteristic line impedance (Fig. 4.7).

When lines of different impedances are employed, reflection at the interface between both lines can also occur. The same holds true for branched lines. Often in communication systems a reflection echos back and forth, causing considerable noise. During layout of the system this must be taken into consideration.

4.2 DATA CHANNELS

4.2.1 Unmodulated Data Transmission

It is not a simple matter to select a data communications system. Particularly when data must be transferred over long distances, factors such as transmission velocity, type of data, type of cable, routing of the cable, transmission distance, noise, and costs must be considered. When data are transferred over long distances it must be made certain that they remain free of communication errors. A simple linear parity check, where only one parity bit is used, is not sufficient in most cases. All too easily two bits can be altered simultaneously. In this case a cross-check or another method must be employed.

Parallel Data Transmission

The computer architecture suggests word parallel transfer of information (Fig. 4.8). However, this method of data transfer mainly is employed between components of the internal computer system. Here ribbon cables are popular as transmission media. Because of the short distances, in general, the communication is done with square waves. A certain amount of noise immunity in the ribbon cable can be obtained by grounding every second conductor (Fig. 4.9). Also a twisted pair of cables is often used for parallel data transfer. In general, a ground conductor is run parallel to the twisted pair of cables. In addition, they are shielded by a metal foil. The possible data transfer rate of square waves decreases considerably in both types of cables with distance (Fig. 4.5). For example, the maximum possible transfer rate of twisted pair cables is approximately 20,000 bits sec^{-1}. If higher transfer rates are desired, coaxial cables must be

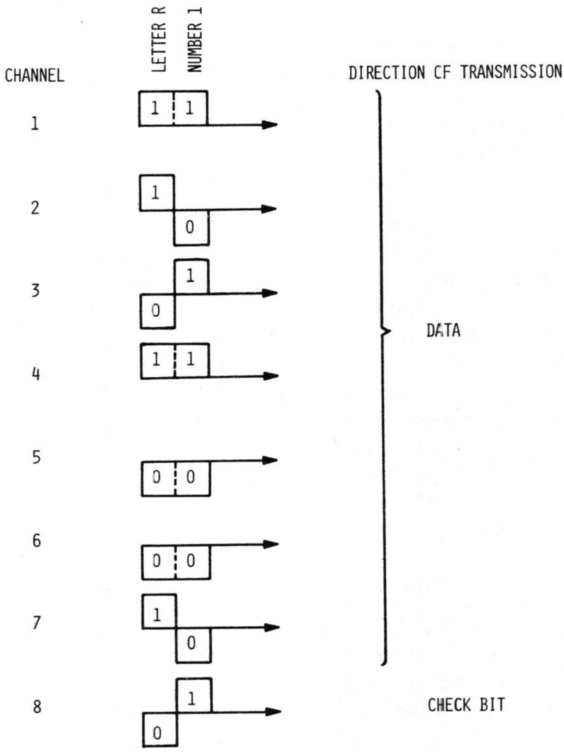

FIG. 4.8 Parallel data transmission.

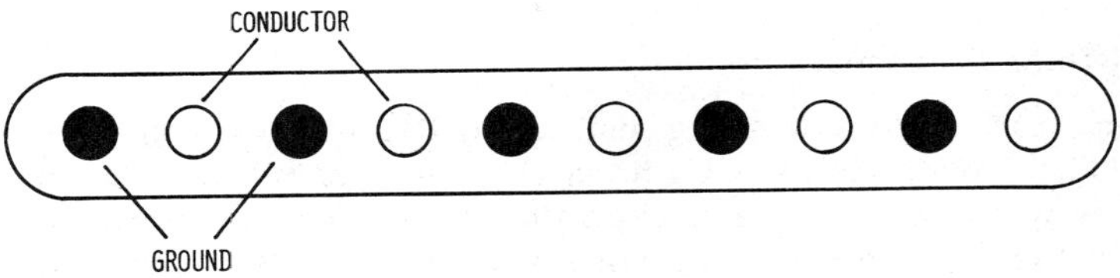

FIG. 4.9 Ribbon cable.

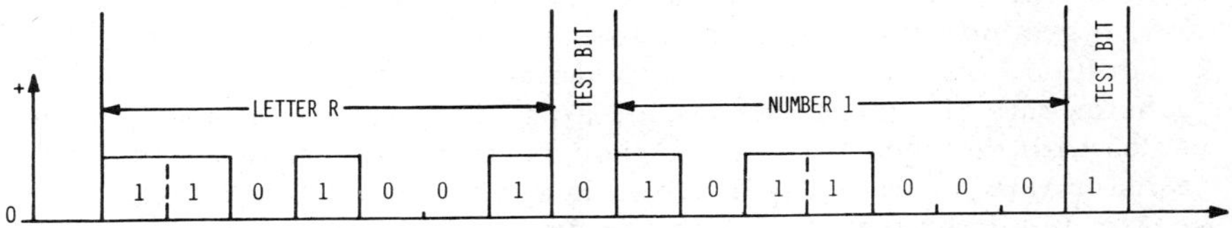

FIG. 4.10 Return to zero serial data transmission.

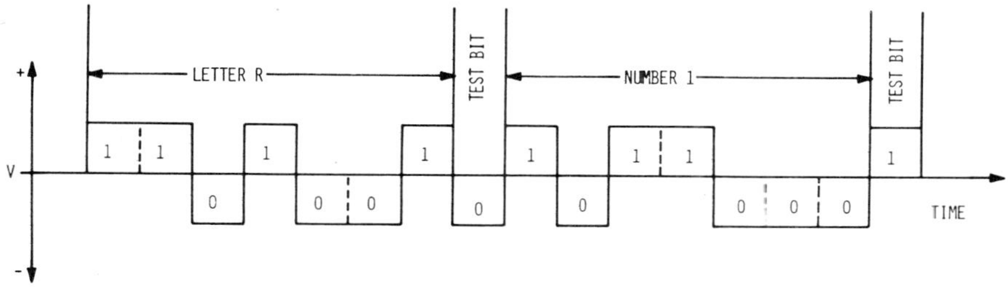

FIG. 4.11 Bipolar serial data transmission.

employed. They allow the transfer of square waves at a rate of 500,000 bits sec^{-1} over longer distances.

Serial Data Transmission

At distances over 100 m the line costs for parallel transfer become so high that serial transfer is more economical. Single-ended (return-to-zero) current mode as well as double-ended (bipolar) current mode is used (Fig. 4.10 and 4.11). Here it is necessary to utilize a multiplexer which transforms parallel information coming from the computer into serial information. When a single byte is transferred by the serial method, the possible transfer rate is reduced by a factor of 1/8. In addition, the switching time of the multiplexer must be added. At the receiver input, the data is transformed from serial to parallel information.

Decoding of serial information at the receiver in general requires more complex circuits than are needed for decoding of parallel information. The principle of serial data communication is sketched in Fig. 4.12. Seven keys and a rotary switch operating as parallel/serial converter are located at the transmitter. The receiver contains a serial-to-parallel converter made up of a rotary switch and seven signal lamps. For example, if the letter R is transmitted, on the transmitter side a motor is started, which drives the rotary switch. Simultaneously a start signal is sent to the receiver, which starts its motor for the rotary switch. Both rotary switches now operate synchronously. The individual positions of the parallel keys of the transmitter, which contain the code of the letter R are transferred in serial mode to light up the lamps of the receiver. Only those lamps which obtain the logic signal HIGH are turned on. The synchronous operation of the coding and decoding circuits in general is done with the aid of electronic circuits. When this transmission mode is used, distortion of transmitted signals must remain within certain tolerances. Otherwise, the receiver will not get the correct decoding information.

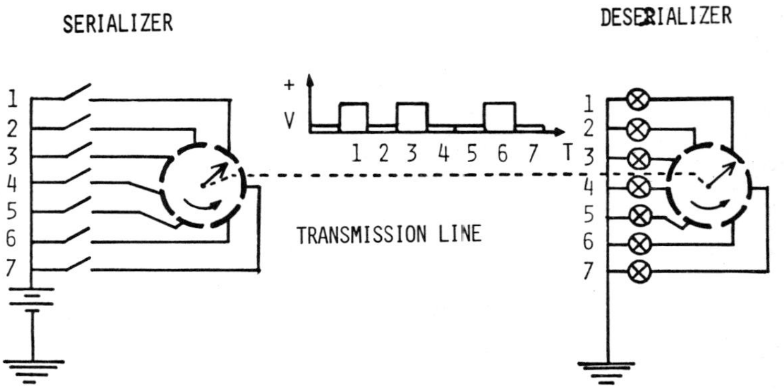

FIG. 4.12 Principle of serial data transmission. (Reprinted from Ref. 2 by courtesy of Marcel Dekker, Inc.)

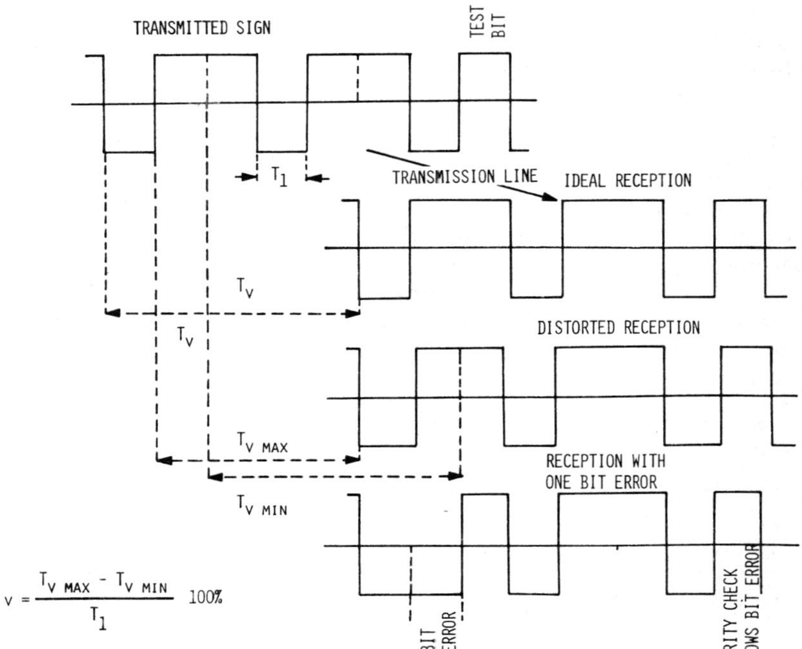

FIG. 4.13 Transmission error resulting from signal distortion. (Adapted from Ref. 1.)

The delay of the transferred signal at the receiver end is T_V. If synchronization at the receiver end is delayed by less or equal to this time T_V, the signal can be correctly identified. Often, however, there are distortions (jitter) of the bit pattern, which cause possible misinterpretation of a character (Fig. 4.13) [1]. The jitter is defined as follows:

$$V = \frac{T_{v,max} - T_{v,min}}{T_1} \; 100\% \tag{4}$$

In practical applications jitter should not be more than 20 to 30 percent.

Synchronous and Asynchronous Operation

Information transfer can be carried out by two different methods: one is the synchronous and the other the asynchronous method. With the synchronous method, before initiation of the data transfer, the receiver and transmitter must be synchronized, and then the transfer of the data strings occurs (Fig. 4.14). This method is the most effective one of the two since the use of control bits is reduced to a minimum.

In asynchronous communication control bits must be utilized for the transfer of each character (Fig. 4.15). In the idle state the communication line is at a HIGH level. The transfer of a character can be announced by a negative control signal. With this signal it is possible to synchronize transmitter and receiver. After transmission of the character and a stop bit the transmitter and receiver go offline again. A transfer of a new character can take place any time. This transfer method, for example, is used by the very slow teletype, which often serves as a data input/output terminal for a computer. This type of data communication, however, is extremely inefficient since, depending on the mode of operation, only 60 to 70 percent of the bit string is used for actual data transfer.

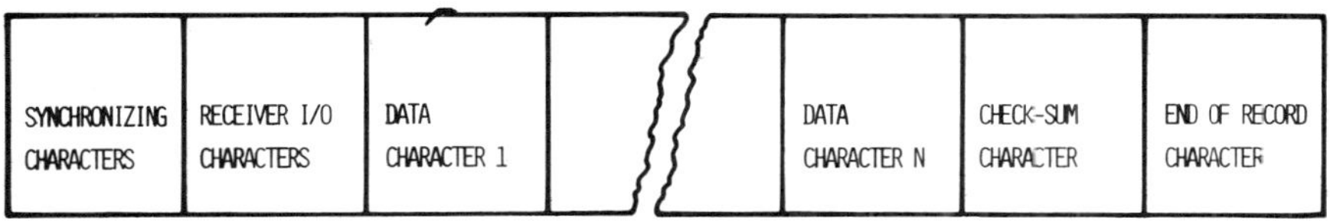

FIG. 4.14 Synchronous data format.

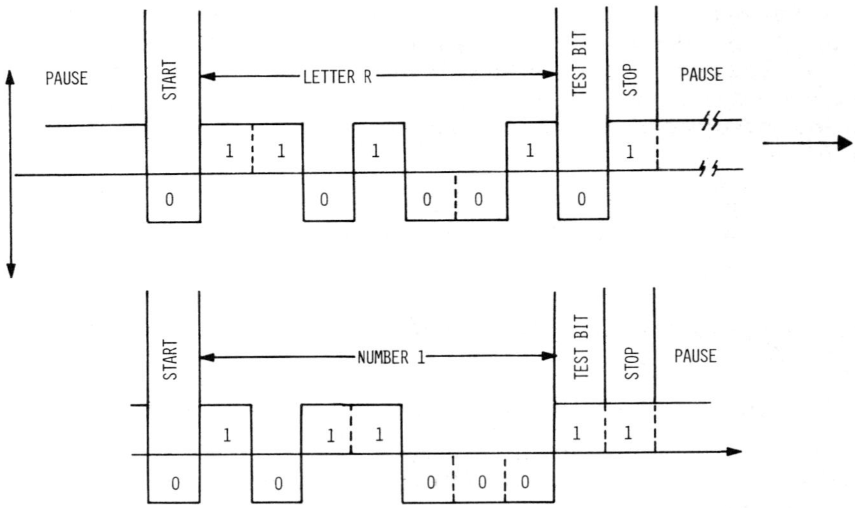

FIG. 4.15 Asynchronous data transmission.

Transmission Lines

There are various possible ways of using data channels. With the simplest method data is only transferred in one direction. It is called the *simplex* method (Fig. 4.16). The control logic for data communication is very simple in this case. If better utilization of the line is desired, the half-duplex method can be employed. In this case infor-

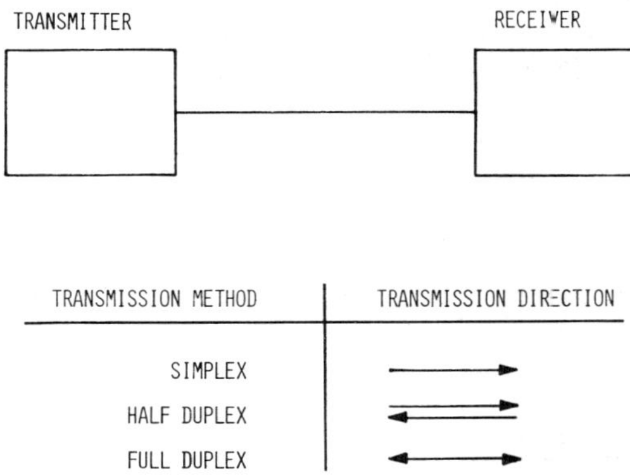

FIG. 4.16 Types of communication circuits.

mation is transferred in both directions on one line, although at different times (Fig. 4.16). The control logic is somewhat more complex than that needed for the simplex method. In duplex operation information can be transferred simultaneously in both directions (Fig. 4.16). This is only possible, however, with very complex and expensive control logic.

4.2.2 Modulated Data Transmission

Modem

For long distances square waves are unsuitable for data communication. For this reason square waves are transformed into sine waves with the help of a modulation technique at the transmission side (Fig. 4.17). The information then is transferred over a telephone cable to the receiver. By using a demodulator the receiver transforms the information back into the original square waves. The equipment which performs the modulation and demodulation is called a *modem*. Most modems operate with two lines using the half-duplex method (Fig. 4.16). The design of the modem determines the time needed for switching directions. This time lies between 0.5 and 100 msec and averages about 20 msec. The time loss can be reduced by a simplex operation with four lines.

Telephone cables have been designed to transmit the human voice. Good voice communication is possible in the frequency range of 300 to 3400 Hz. Thus this range is available for data communication by the ac method (Fig. 4.18) [1]. There are different modulation methods in use. They will be described in the following paragraphs.

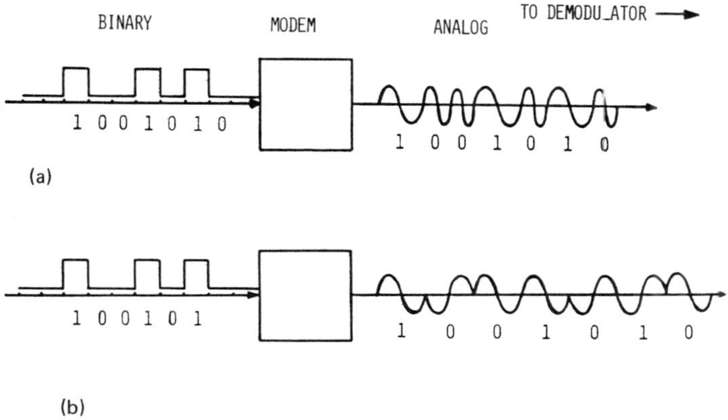

FIG. 4.17 Modulation methods for data transmission: (a) frequency modulation and (b) phase-shift modulation.

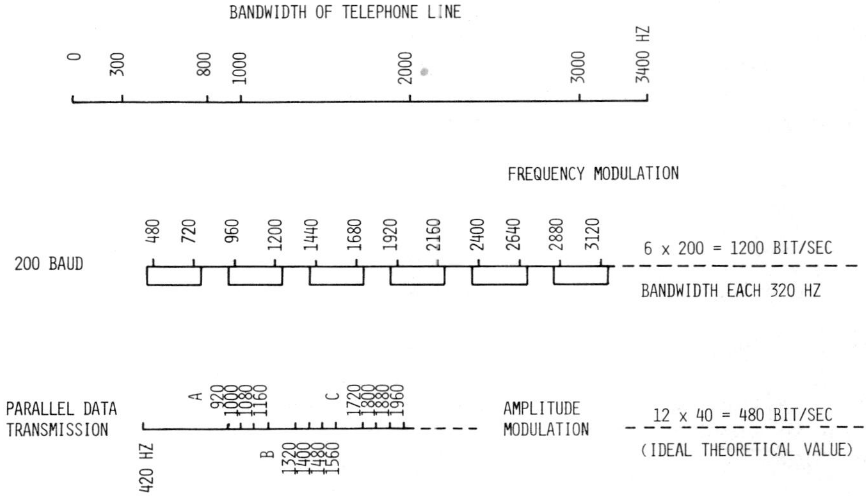

FIG. 4.18 Transmission bands of public telephone lines.

Frequency Modulation

Frequency modulation is mainly used for transmitting data to or from slower peripherals (Fig. 4.17 and 4.19) [1]. A transfer rate of up to 1800 bits sec^{-1} is possible. Figure 4.17a displays the principle of frequency modulation. For example, to a low-frequency band a logic level of HIGH is assigned, and to a high frequency band a logic level of LOW. This transfer method is fairly immune to noise, since with amplitude limitation distortion of the amplitude does not play a major role. This is the method used most often.

Phase-Shift Modulation

The second most common modulation technique is the phase-shift method (Fig. 4.17 and 4.19). When shifting from one logic level to another, a phase shift of, for example, 180° is carried out. Modems designed for this method can operate on the principle of four or eight phase shifts. This method is also quite immune to noise and therefore is used more and more frequently.

Communication Multiplexer

If several peripherals are to be connected to a computer, it is often expensive to provide an individual communication line for each unit. For this situation it is useful to employ one common transmission line provided with a multiplexer. There are basically two multiplexing methods available today.

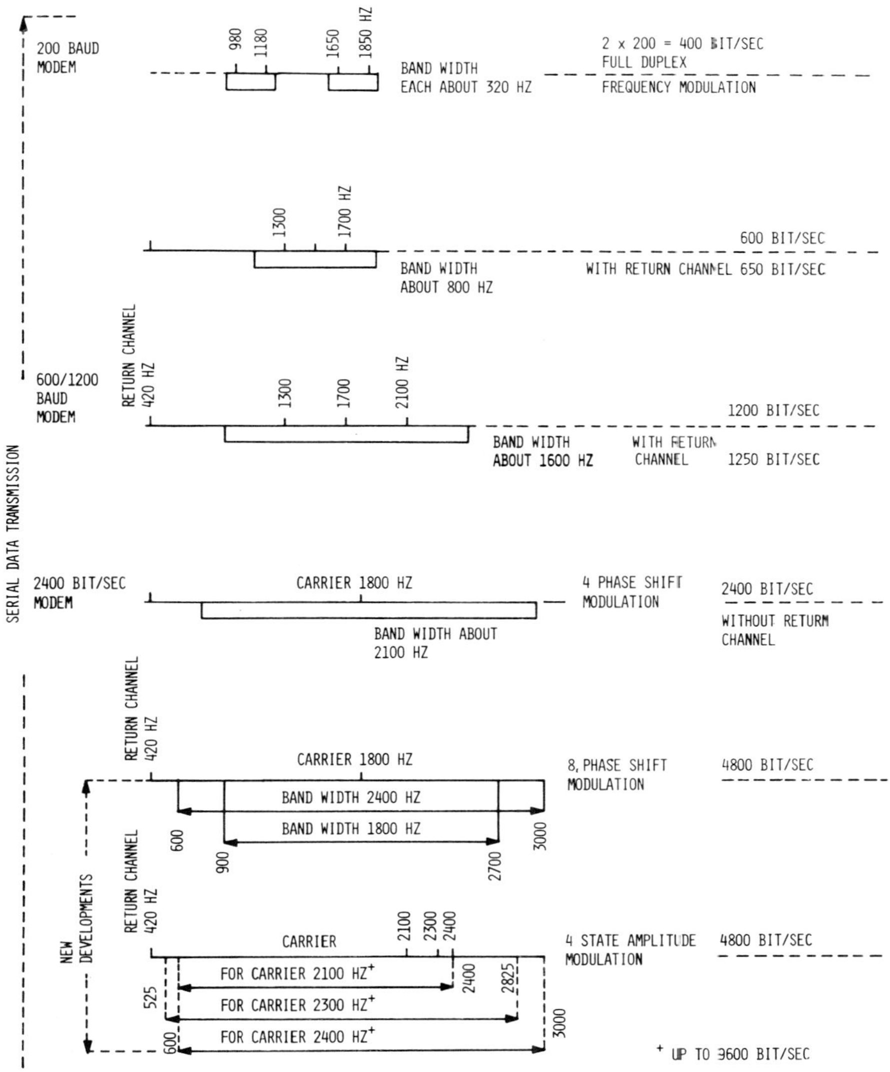

FIG. 4.19 Assignment of frequency bands to different transmission modes.

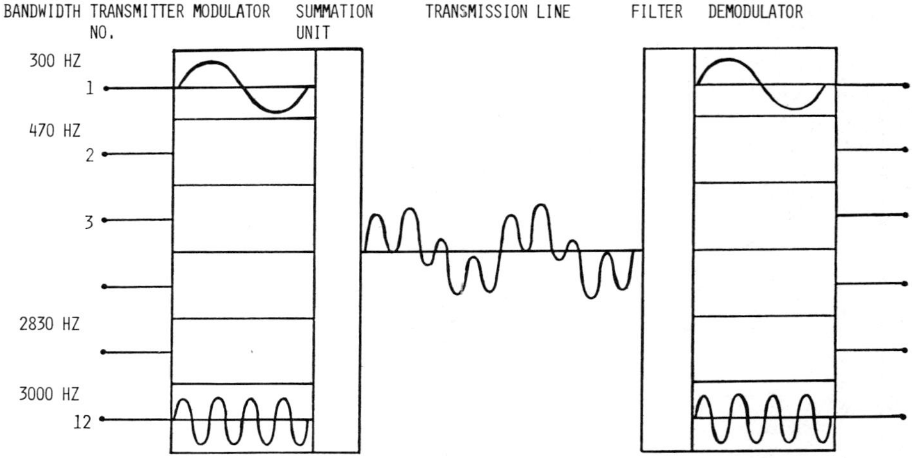

FIG. 4.20 Principle of frequency-division multiplexing. (Reprinted from Ref. 2 by courtesy of Marcel Dekker, Inc.)

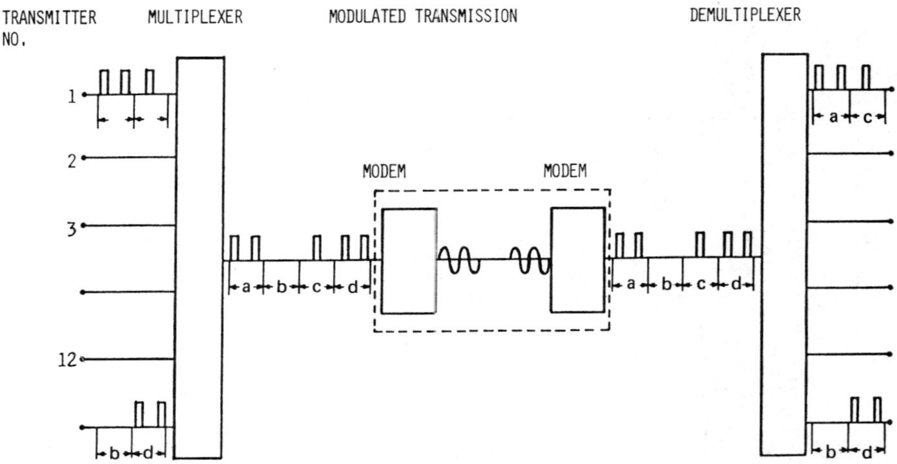

FIG. 4.21 Principle of time-division multiplexing. (Reprinted from Ref. 2 by courtesy of Marcel Dekker, Inc.)

Frequency-Division Multiplexing. In frequency-division multiplexing each peripheral has assigned to it a specific bandwidth of the entire available telephone communication band (Fig. 4.18). Digital information originating from a given peripheral is transmitted within its designated band. The modulated analog signals of the various peripherals are added and the information is sent over a common line to the receiver. Individual frequencies of each peripheral are filtered out and sent to the demodulator which reconstructs the original digital information (Fig. 4.20) [2].

Time-Division Multiplexing. In time-division multiplexing the digital information of the connected peripherals is sent sequentially in a nested mode over the common communication line (Fig. 4.21) [2]. At the receiver end the information is unnested and reassigned to individual channels. This method is more complex than frequency-division multiplexing. It can be used, however, for modems designed for data transfer rates of up to 9600 bits sec^{-1}. This method also permits the connection of more data channels than the previous method allows.

4.2.3 Line Drivers and Receivers

When data has to be transferred over a line, a driver must be employed on one end and a receiver on the opposite end of the line. The driver converts, for example, the TTL level of the transmitter to the voltage level needed for operation of the communication system. The receiver reconverts this level back to the voltage level used by the receiving equipment. The threshold of the receiver commonly is provided with an adjustable hysteresis in order to suppress noise. The main problems with communication lines in interface technology are ground potential, noise, reflection, signal distortion, attenuation, and cross talk. When selecting driver, signal lines, and receivers, all these problems must be taken into consideration. In the following sections several commonly employed communication circuits for unbalanced (asymmetrical) and balanced (symmetrical) lines will be discussed.

Unbalanced Lines

Drivers. These lines consist of either a single wire, where the ground is used as a return lead, or a twisted pair of wires, with one line grounded. With a single communication line a very inexpensive communications system can be constructed (Fig. 4.22a). Its disadvantage is the possibility of picking up inductive and capacitive noise. It also may have difficulties with different ground potentials. The maximum transfer distance is approximately 30 m. By using twisted pair cables the transfer distance and velocity can be increased

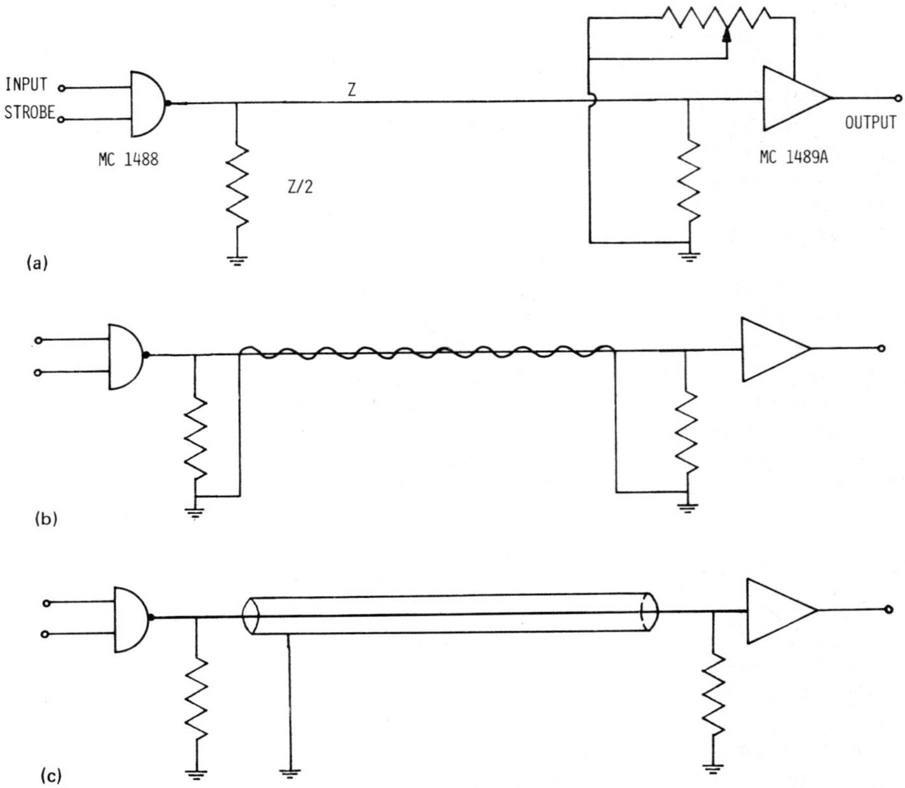

FIG. 4.22 Unbalanced signal transmission lines: (a) single wave, (b) twisted pair cables, and (c) coaxial cable.

(Fig. 4.22). Possible noise interference is reduced by the shield. A further improvement may be obtained by using a coaxial cable (Fig. 4.22c). This cable should have a line impedance of 50 to 200 Ω. It has low losses and is relatively insensitive to noise.

Unbalanced communication lines use voltage drivers (Fig. 4.23). Here a signal is transmitted when a supply voltage is applied to the output leg of the driver transistor. The voltage source is grounded by activating the transistor. There exist three different types of voltage drivers:

The active pull-up driver
The active pull-down driver
The totem pole method

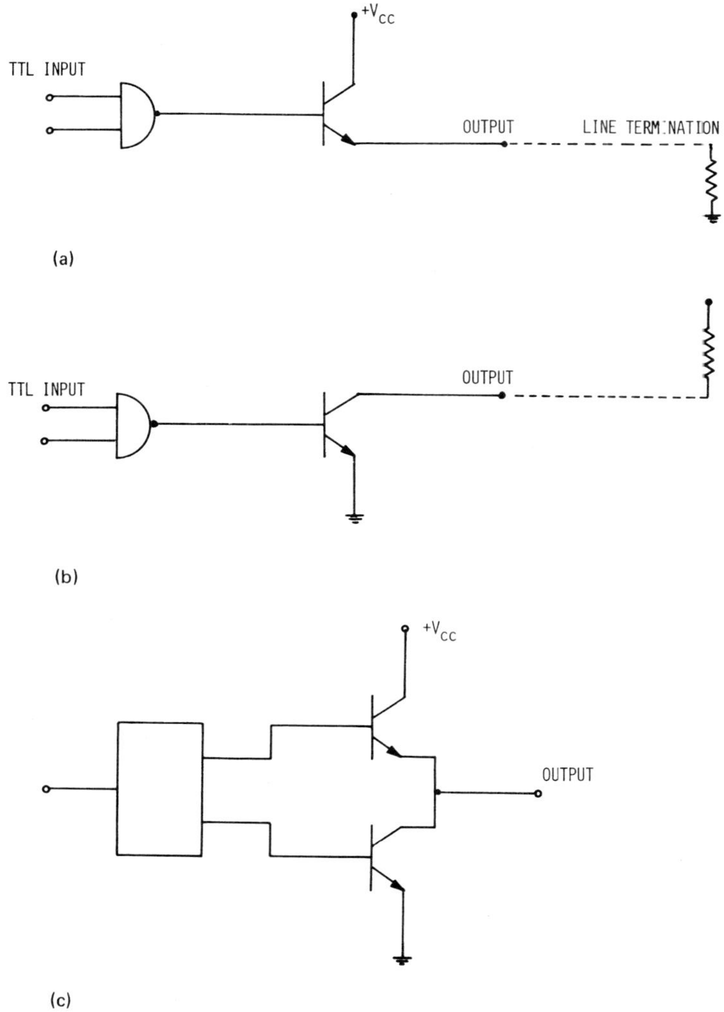

FIG. 4.23 Voltage line drivers: (a) active pull down, (b) active pull up, and (c) totem pole.

Generally, these drivers use TTL signal levels. It is necessary that they remain relatively insensitive to noise. The totem pole driver is a combination of the active pull-up and active pull-down driver. It has a low ohmic output in both switching modes and permits very high switching frequencies. This driver, for example, is used for lines

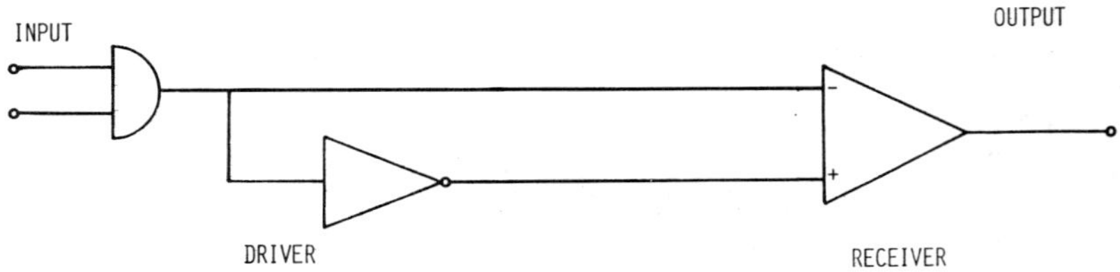

FIG. 4.24 Dual-line voltage driver for balanced lines.

operated to EIA specification RS-232 C (the voltage level for logic LOW lies between 5 to 15 V and for logic HIGH between -5 to -15 V). Typical circuits to operate under EIA specifications are SN 75150 N for the transmitter and SN 75154 N for the receiver.

When using voltage drivers, it must be made certain that the supply voltage is below the maximal allowed level of the integrated-circuit family used. The driver current is determined by the capacity of the circuit selected. It may be as high as 100 mA. Many communication systems require that the transmitter and receiver can selectively be turned on and off. This can be accomplished by employing an AND gate at the input of the component to be switched (Fig. 4.24).

The advantages of unbalanced lines are low cost, simple logic circuits, simple to turn on and off, and the need for only one power supply (except when in use for EIA specifications). The disadvantage is high susceptibility to noise. For this reason they can only be used for relatively short distances.

Receivers. Basically, it is possible to use TTL gates on the receiver side. However, they must be carefully adapted to the communication system to eliminate the possibility of transferring noise. An improvement in noise susceptibility can be achieved by the use of a Schmitt trigger, which has a much higher hysteresis. Figure 4.22a shows a circuit in which the noise immunity can be set by a potentiometer.

Balanced Lines

The most commonly employed data transmission line in interface technology is the twisted pair cable. Depending on its diameter and insulation material, as well as the twisting method, the impedance ranges from 50 to 200 Ω. When the line is correctly terminated (grounded), the low impedance improves in susceptibility to noise. A further advantage is a uniform impedance over the entire length of the line. The noise immunity of the twisted pair cable can be improved by using a metal shield, but thereby its capacitance goes up and it loses its uniform impedance properties. With twisted pair cables it is very easy to obtain common-mode noise rejection. Noise usually is induced symmetrically on both lines and can be eliminated at the receiver end by using a differential amplifier (Sec. 5.2).

The drivers for balanced lines work in either the current or voltage modes described in the following sections.

Voltage Drivers. For operating a balanced line by the differential method an inverter must be connected to the output of the driver (Fig. 4.24). Thus the actual logic level of signal to be transmitted is on one line and the inverted signal on the other line. When the logic level changes, the differential amplifier at the receiver senses a change of polarity. The common-mode rejection of the receiver reduces all noise which appears simultaneously on both lines. There are many different voltage drivers used to adapt TTL circuits to electrical components, such as relays, lamps, and transmission lines. They usually can be operated in a range of 300 to 500 mA.

Current Drivers. Differential current drivers are commonly employed for distances from 30 to 2000 m (Fig. 4.25); they have a very high common-mode noise rejection and therefore are insensitive to noise [3]. The output impedance of the driver as well as the input impedance of the receiver are high. For this reason the line must be terminated with appropriate resistors at both ends. Typical transmission lines can be built with the help of the TTL circuit SN 75109 as a driver and SN 75107 A as a receiver. These circuits were developed for a twisted pair of cables. The receiver can distinguish a small difference in voltage level of the signal in the presence of high noise, changing temperature, and changing supply voltage. The driver has two outputs operated at constant output current. Either of the two outputs are turned on depending on the logic input signal. The receiver still can differentiate voltages of a magnitude as low as 25 mV.

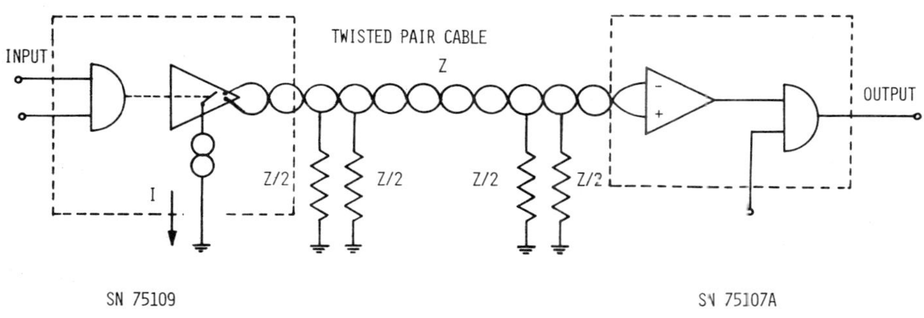

FIG. 4.25 Dual-line current driver and receiver.

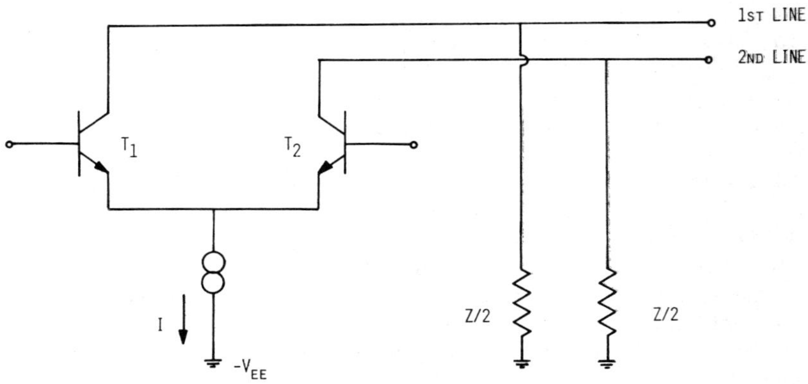

FIG. 4.26 Current driver with one current source and unidirectional current flow.

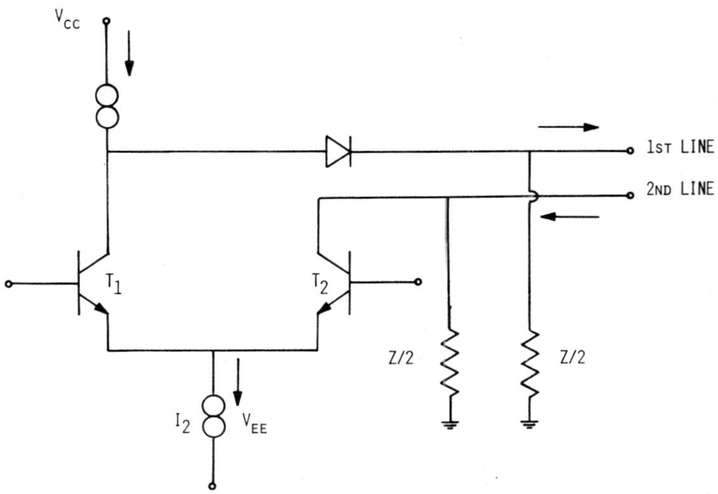

FIG. 4.27 Current driver with current source and sink and bidirectional data flow.

It measures the polarity in order to recognize the state of the logic signal. These drivers are available in various circuit technologies. Figure 4.26 displays a driver with one current source [3].

The circuit seen in Fig. 4.26 can be improved by a second current source (Fig. 4.27). If T_2 leads and T_1 is open, current flows into one lead and out of the other. If T_2 is open and T_1 conducts, then the ground potential is connected to both lines [3].

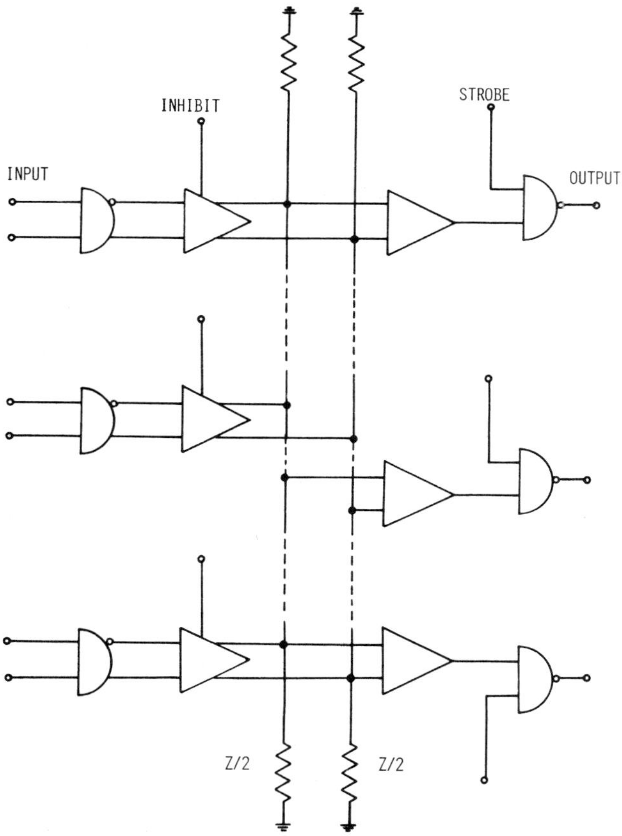

FIG. 4.28 A possible circuit to connect and disconnect peripherals in a party line.

The advantages and disadvantages of current drivers for symmetric lines can be summarized as follows:

Advantages

Very high transmission velocity
TTL compatible
Commonly used power supply
Simple low-impedance line operation
High common-mode noise rejection
Very insensitive to noise
Low power consumption of driver
Long line operation (up to 2000 m)
Good signal reception properties
External on/off control
Suitable for party-line operation
High receiver output impedance
High temperature stability

Disadvantages

Two power supplies required
Symmetric lines necessary

Example of a Driver/Receiver System. In multiuser systems connected to a common bus it is often necessary to selectively turn peripherals on or off. This can be done with the party-line circuit shown in Fig. 4.28. Each gate can be turned on or off individually so that it is possible to connect each driver with each receiver [3].

4.3 COMPUTER BUS ARCHITECTURES

Information exchange between the peripherals and the computer generally occurs over a common bus. The central processing unit, mass storage (memory) device, and peripheral equipment are connected to this bus. In the universal bus system the central memory also communicates over the bus. In simpler systems, particularly those of older design, the entire data communication takes place over the bus, which is usually controlled by the central processing unit. No direct data communication is possible between the various peripherals. In newer computers, individual peripherals, when authorized, can request the bus and exchange information with other peripherals, the memory, or the central processing unit (see Chap. 2). The lines of the bus basically have three different functions (Fig. 4.29) [2]. These may be classified as follows:

1. Data transfer
2. Address transfer
3. Control signal transfer

Some computers transfer the control signals over the data line. However, in this case signals must be designated by the central processing unit as control signals. Information transfer over the bus is usually done with the help of parallel lines, either byte or word parallel. In some computers these lines can be used unidirectionally and in others bidirectionally. Bidirectional bus systems are fast but costly. The design of the bus influences the operation of the computer. For example, for a critical process with many interrupts, a bus structure must be selected which is fast enough to report difficulties with the control system and to initiate an immediate correction. The bus structure, however, influences the complexity of the interfaces. In a system where the computer directly addresses the interface of the peripheral over a dedicated channel, a special address decoder for the interface is not required. If there are several peripherals attached to one line, however, an address decoder is necessary. The most common bus systems for process control computers are discussed here.

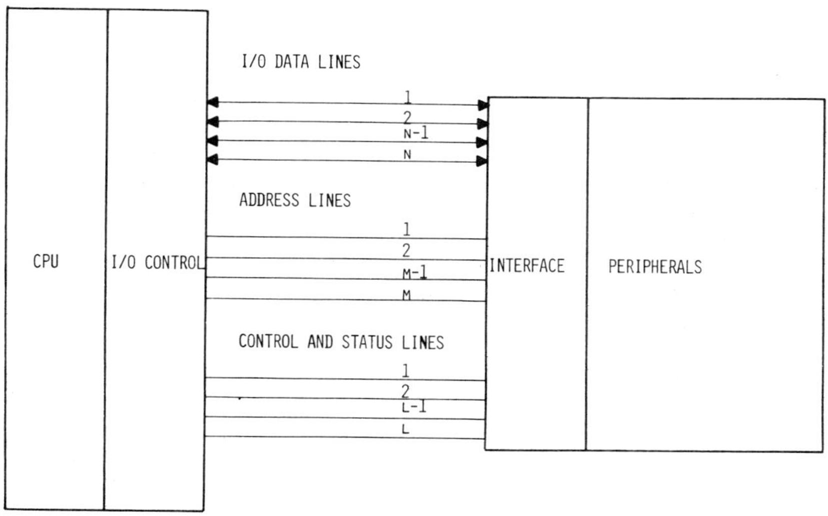

FIG. 4.29 Bus structure of a process control computer: N = number of bits in computer words; M = number of bits in address section of I/O instruction; L = number of control lines.

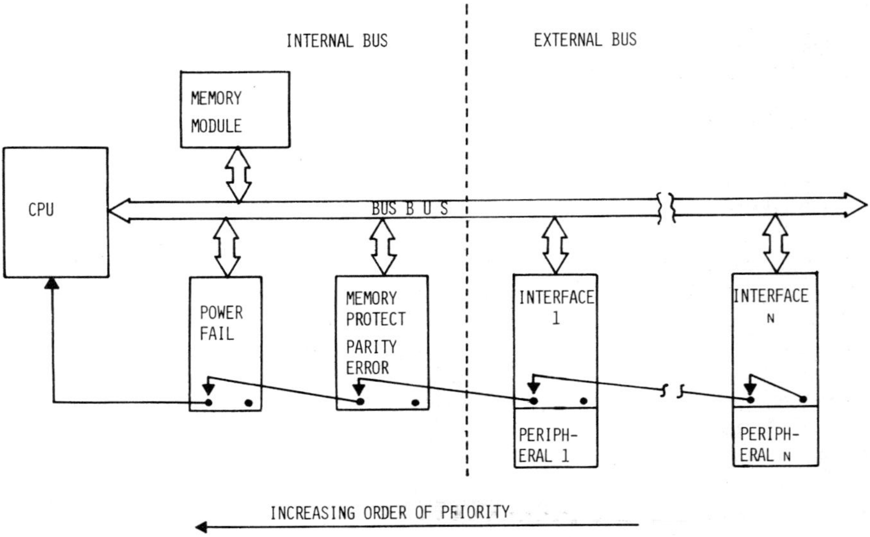

FIG. 4.30 Universal bus with I/O priority structure.

4.3.1 The Universal Bus

In the universal bus system the peripherals are connected to the extended internal bus by input/output interfaces in sequence of their priority. Active peripherals (master) are able to exchange information with the central processing unit, central memory, and other peripherals. Reliable data transfer over the bus is only possible over a distance of up to 10 m. If longer distances are to be covered, bus amplifiers must be installed. The internal bus was described in Sec. 2.5. To each interface an address is assigned. The computer communicates with the peripheral by addressing the interface. The peripheral with the highest priority obtains the lowest address, and so on. The input/output instructions are not specific to any peripheral (Fig. 4.30).

It is possible to connect one interface to several peripherals. They then obtain all the same hardware priority. When an interrupt service request is placed, however, there must be the possibility of distinguishing between the various peripherals connected to that particular interface. The priority of the interrupt is then determined by software routines. Addressing external peripherals is done in exactly the same way as internal memory addressing. External peripherals may be active (master) or passive (slave). Active peripherals are able to initiate data transfer and exchange information with other peripherals. After completion of data transfer, the receiver places an end-of-transmission signal on the bus. The characteristics of the universal bus are

There is fast data communication between peripherals as a result of short transfer distances and the use of only one instruction.
During high transfer rates the bus can be easily blocked.
Complex interfaces and a address decoder are required.
It responds quickly to interrupts.
The same interfaces apply for all peripherals.

4.3.2 The Two-Bus System

In the two-bus system the working memory communicates with the CPU over a separate bus. Its basic structure was explained in Sec. 2.5. Here also the peripherals are connected to the extension of the internal bus. In this system, direct data exchange between peripherals is impossible. Data originating from one peripheral must first be transferred to the central processing unit and then to the other peripheral. Therefore, there are two instructions necessary. This appears to be a considerable disadvantage compared to the one-bus system. However, in many two-bus systems the instruction cycle is shorter than in the one-bus system. The particular characteristics of the two-bus system are

Relatively slow data exchange between peripherals, because data transfer takes place via the CPU.
The possibility of blocking the input/output bus by data transfer between the CPU and the working storage has been reduced.
A less complicated interface can be used for the low-speed peripherals.
Quick response to interrupts.
No standard interface.

4.3.3 Daisy-Chain System

In this system all peripherals are connected to one bus. The peripheral with the highest priority is located the closest to the central processing unit. Thereafter, the peripheral with the second-highest priority is connected, and so on (Fig. 4.31).

If a peripheral is to be addressed, the central processing unit places the address and a request signal on the bus. The first peripheral determines whether it is its own address; if yes, it blocks the line to all other peripherals and information exchange takes place between the CPU and peripheral. If the address belongs to another

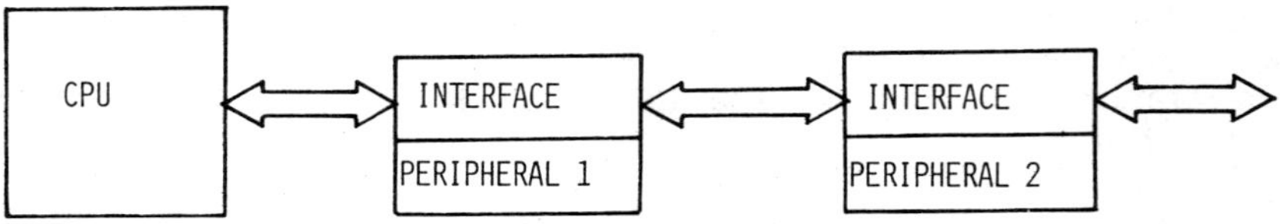

FIG. 4.31 Daisy-chain bus.

peripheral, the request signal is transferred to the next peripheral. This process is repeated until the right peripheral is found. The essential characteristics of this system are

Slow data transfer
Slow response to interrupt
Inexpensive bus
Inexpensive interface logic
Possible blocking of low-priority peripherals
No standard interface

4.3.4 Radial Bus (Channel)

In this design there are many channels directly connected to the central processing unit (Fig. 4.32). Each channel has an individual number and priority. When data transfer is initiated, no address must be issued in order to address the peripheral, because each channel can correspond directly with the CPU through the interface logic. This design offers considerable advantages, although it is not used too often in control computers because it is expensive.

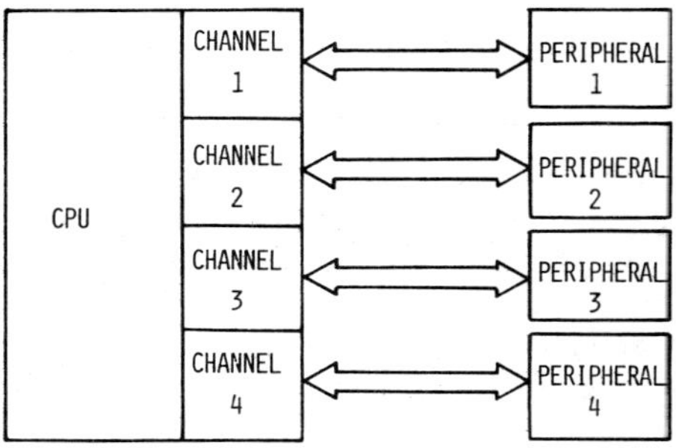

FIG. 4.32 Radial channel system.

The essential characteristics of this design are

Many hardware interrupt levels
Very fast response to interrupts
Expensive control logic in the computer
No address decoder in the interface
No standard interface

4.4 ARCHITECTURE OF COMMUNICATION NETWORKS

The increasing complexity of industrial processes requires that new computer system architectures be used for the automation of plants. The formerly used systems with central intelligence are very inflexible and expensive.

To overcome these difficulties a decentralized computer system is of advantage. Individual computers will be used at the process locations where needed and then interconnected to each other. The most often used design of distributed computer systems can be divided into four fundamental configurations:

1. Star structure
2. Ring structure
3. Bus structure
4. Network

4.4.1 Hierarchical Network

Computer systems employed in industry quite often are of a hierarchical design similar to the star structure. This structure is analogous to the organizational pyramid of a firm. Each computer at each level has its specific assignment corresponding to the task of the management at that level. In Fig. 4.33 [2] the ideal hierarchical structure is displayed. This structure was created by a systematic modular division of the control system of an organization. It can be presented as a star system having satellite systems attached to it (Fig. 4.34). In general, star systems cannot be realized as ideally as shown. They mostly grow together from existing computer installation and assume a very irregular structure. This may lead to great interfacing and programming difficulties. The problem is shown in Fig. 4.35. If the hardware and software interfaces of all computers on the particular level are equal, the total system can be put together easily. Only one type of interface and one communication package are needed. If five different computers are to be employed and an attempt is made to connect these to one system, 10 different communication packages are needed (Fig. 4.35) [2]. In practice, however, one would not inter-

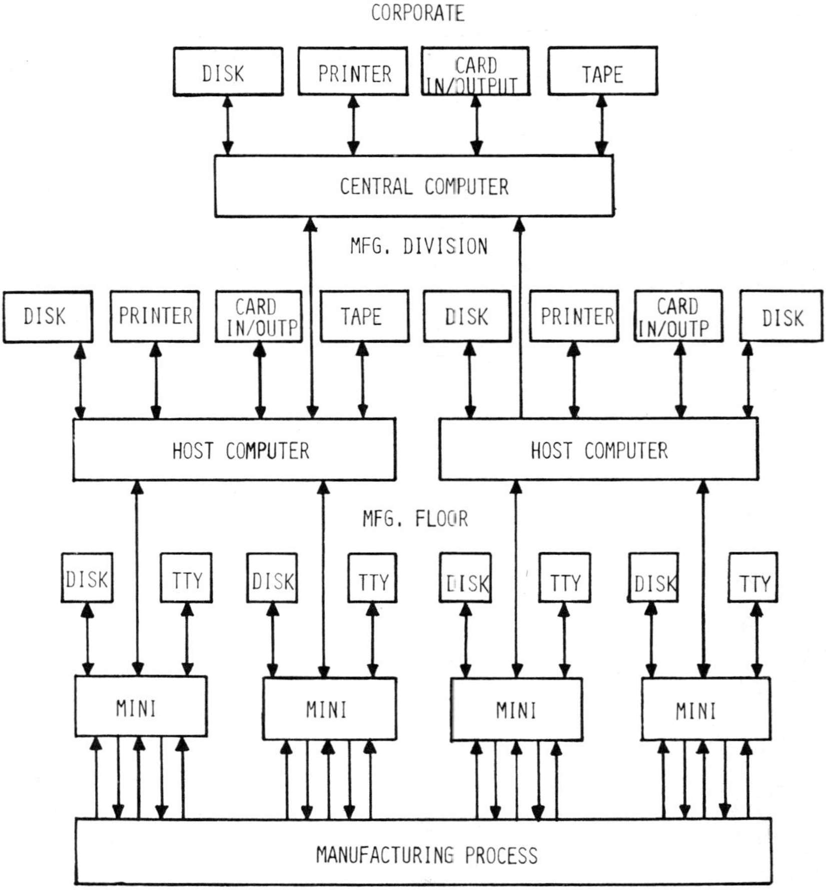

FIG. 4.33 Typical form of a hierarchical computer system in a plant. (Reprinted from Ref. 2 by courtesy of Marcel Dekker, Inc.)

FIG. 4.35 Increase of number of required communication packages with increased number of different computer models in a distributed computer network. Maximum theoretical number of communication packages is $m = n(n-1)/2$. (Reprinted from Ref. 2 by courtesy of Marcel Dekker, Inc.)

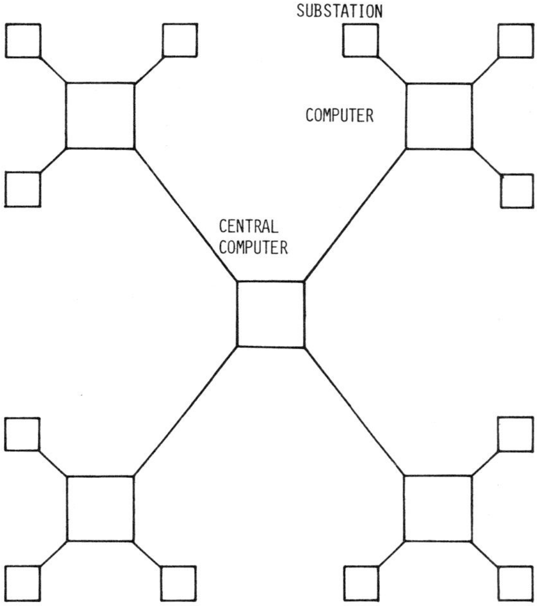

FIG. 4.34 Hierarchical computer structure with multiple substations.

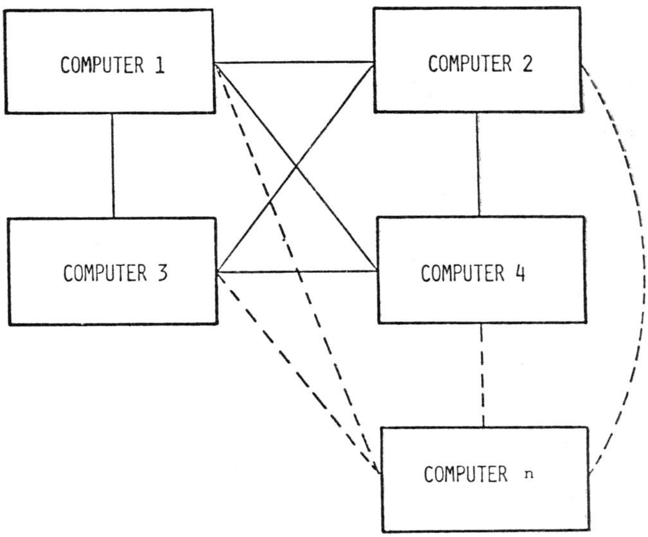

FIG. 4.35

connect all computers with each other. Therefore, several of the connecting lines shown in Fig. 4.35 could be omitted. In general an attempt is made to standardize computer equipment in a hierarchical system to avoid this problem. This is not always possible though since computer network systems often contain many subsystems purchased as complete hardware/software packages.

At the lower end of the hierarchy, communication only takes place with the next higher levels. Therefore, the design of communication hardware and software is fairly simple. The host computer at the next level must exchange information with two levels. First, it operates as a multiplexer in order to communicate with the computers at the lower level. Second, it exchanges data with a central computer further up the hierarchy. Thus, at this level the computer to computer communication and the design of hardware and software is complex. The data transfer speed in a hierarchical structure can be very high. Transfer is usually done blockwise. Congestion may occur only when several computers at lower levels intend to exchange information simultaneously with the host computer. In a well-developed hierarchical network this seldom happens.

4.4.2 Ring Structure

The ring structure offers another possibility of designing a decentralized system (Fig. 4.36). In the ring an intelligent interface is required for each computer. The information flow in the ring can be done by different methods. In one method two computers wanting to exchange data must make the ring available for themselves. There must be a master scheduler to prevent other computers from accessing the ring while two computers exchange data. A message is sent from one computer and arrives at the first intelligent interface. This interface investigates the message to determine its destination address. If this address does not match the address of its computer, it is sent to the next computer. If it matches that computer's address, the computer copies the message and attaches an acknowledge signal to it, indicating that it was accepted. This message is returned to the ring. Finally, the message and the acknowledge signal are arriving at the originating computer. If the acknowledge signal is missing, the message is placed again on the ring. This happens as often as directed. If a problem occurs, the operator is notified. The ring structure can be very slow because the communicating computers occupy the ring and thereby lock out the lines for other computers. This problem can partially be solved by a multiplexing procedure (lazy susan). With this method data is divided into equally long blocks and placed on the ring at constant intervals. Between each data blocks there are blanks. Data communication between other computers can be performed by placing the message in suitable block sizes into the blanks at prescribed intervals. In this case additional bits must be attached

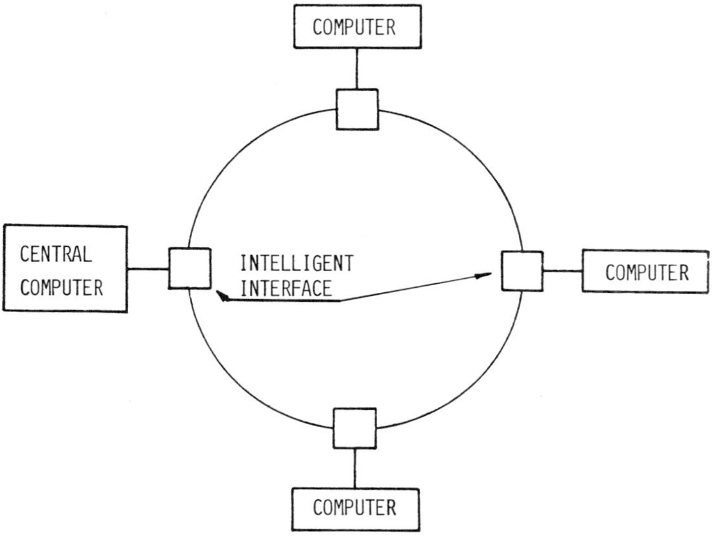

FIG. 4.36 Ring system.

to the message in order to indicate whether or not the blank spaces are occupied. A master scheduler is not needed; however, there must be a controller responsible for synchronizing the information as soon as the flow of data travels once around the bus. If many computers are connected to the bus, the data flow can be very slow and message exchange may be delayed. In the ring structure a provision must be made for a bypass switch at each interface for automatic disconnection of its computer from the bus if a defect occurs. Thus the communication system will remain intact.

4.4.3 Bus Structure

Another data communication channel is the bus (Fig. 4.37). It can be considered as an open ring. Here again, a master scheduler is needed to control the data traffic. If information is to be transferred, the requesting computer notifies the scheduler, which puts the request into a queue with other requests. It assigns priorities and notifies the receiver as soon as the bus is available. After completion of the data transfer, the bus is made free for data exchange with other computers. This method can be very slow when many computers are connected to the bus. By assigning different frequency bands to each set of computers wanting to communicate, the traffic problem on the bus can be improved. This, however, requires a very complex

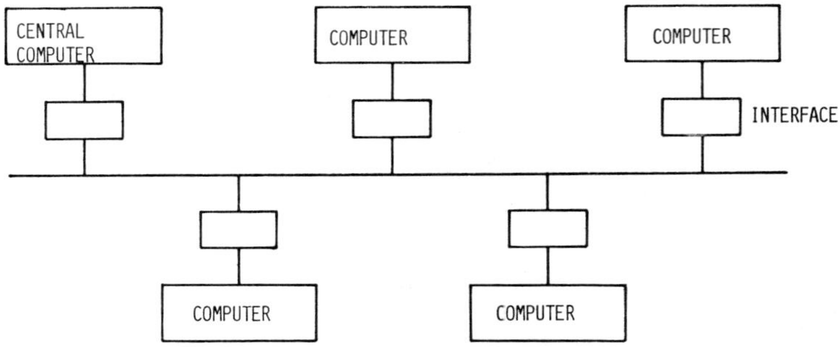

FIG. 4.37 Bus system.

master scheduler who assigns frequencies and priorities. Such data transfer systems are realized with coaxial TV cables by utilizing the complete bandwidth of the TV communication channel. They are fairly insensitive to interference and can be expanded easily. Presently there are optical communication lines being developed. Their use should greatly improve communication tasks in ring and bus systems, because noise has little effect on them.

4.5 FUTURE TRENDS IN INTERFACE TECHNOLOGY

Process peripherals and interfaces add greatly to the cost of a control computer installation. In many cases the peripheral equipment is by far more expensive than the computer itself. Originally the connection and the maintenance of process peripherals was very cumbersome. Figure 4.38 shows a typical conventional data acquisition systm still used today [4]. The numerous functions of the system are described in this figure. Most components of this system consist of discrete elements which are responsible for the high cost. Often it is not possible to accurately tailor the components purchased from a computer manufacturer to a particular application. Therefore, the tendency of overdimensioning is widespread. With the availability of linear integrated circuits, based, for example, on CMOS technology, it was possible to considerably decrease the price of many components. Today there are already integrated-circuit chips available containing a complete multiplexer, sample-and-hold unit, instrument amplifier, and A/D converter. The circuit of Fig. 4.38 can be simplified by introducing a digital multiplexer and individual low-cost D/A converters, which have a resolution of 10 to 12 bits. The improved data acquisition system with these improvements is shown in Fig. 4.39. Here a large portion of the circuit for the output of analog signals has been

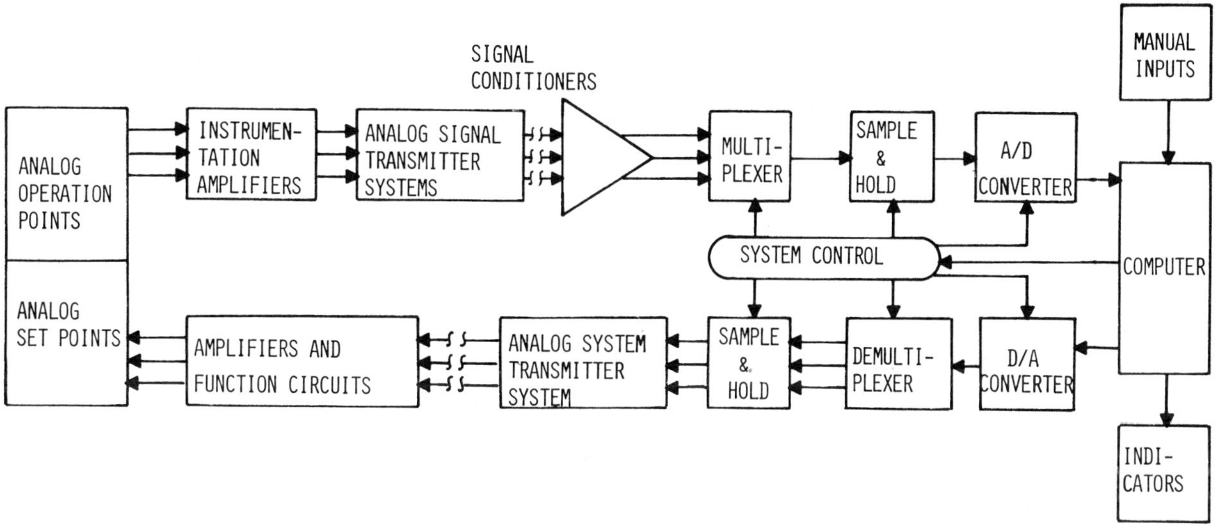

FIG. 4.38 Conventional computer control system with discrete components. (From Ref. 4, reprinted from *Electronics*, April 13, 1975, Copyright © McGraw-Hill, Inc., 1975. All rights reserved.)

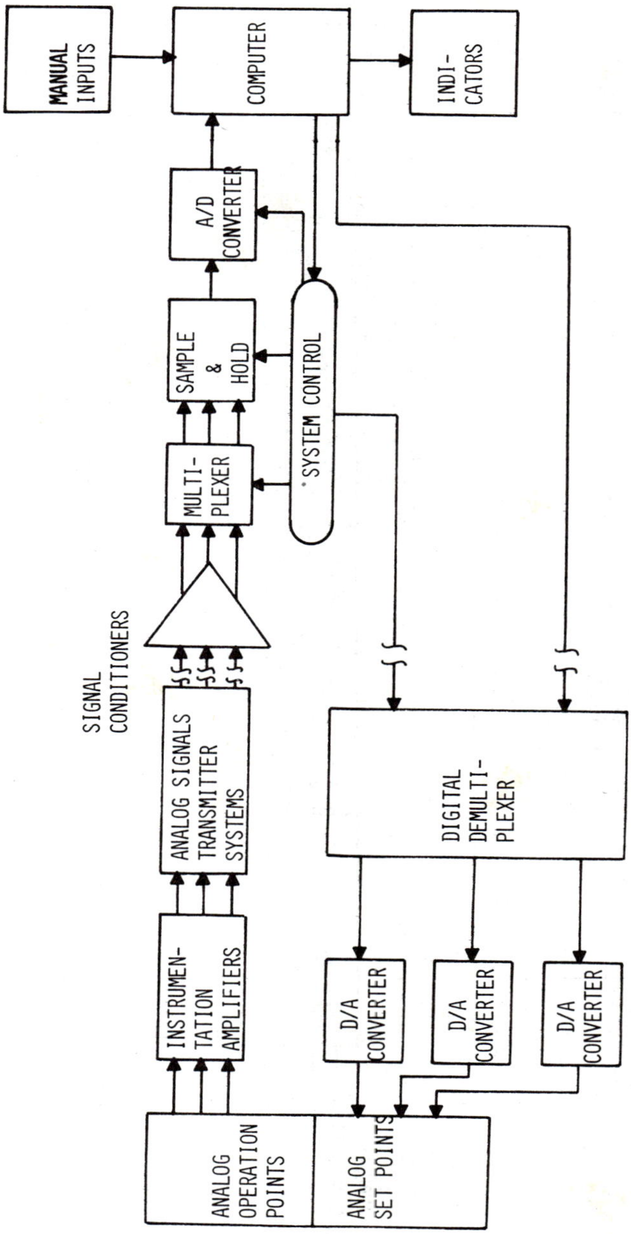

FIG. 4.39 Simplified control system with improved analog output. (From Ref. 4. Reprinted from *Electronics*, April 13, 1975, Copyright © McGraw-Hill, Inc., 1975. All rights reserved.)

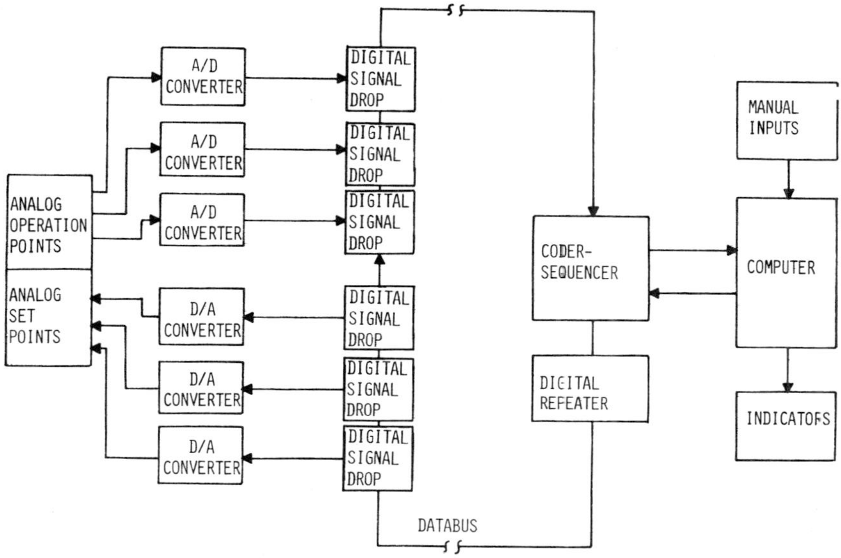

FIG. 4.40 Further improvement of the control system by use of integrated-circuit components. (From Ref. 4. Reprinted from *Electronics*, April 13, 1975, Copyright © McGraw-Hill, Inc. 1975. All rights reserved.)

eliminated. This system can further be improved when individual fast A/D converters are used for input of analog data. Thereby the sample and hold circuit and the analog multiplexer are eliminated. The A/D converter is controlled digitally and data exchange is initiated periodically by the computer (Fig. 4.40). Since time-division multiplexing of digital signals can be done easily nowadays, the task of data transfer can be reduced to handling a sequence of digital data flow, address, and synchronization pulses. Each A/D converter digitizes its analog data, adds its address to it, and sends it to the bus via a signal switch. This information travels on the bus in a circle. The bus controller identifies the data and transfers it to the computer. The measurement data is replaced by a control sign and both address and control sign again are sent back to the bus. As soon as a signal switch identifies its own address and the control sign, it knows that data has been transferred correctly. Thus the channel can initiate a new measurement. Compare this operation to that of the ring bus described in this chapter.

A further improvement of this system can be achieved by introducing individual microprocessors (Fig. 4.41). In this case small

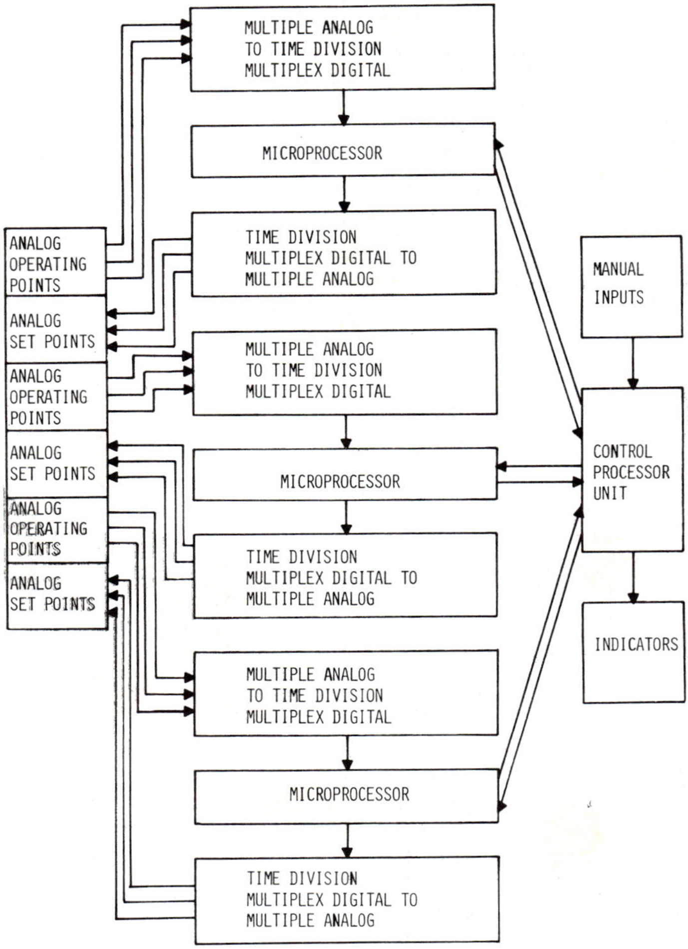

FIG. 4.41 Control system with master computer and satellite substations. (From Ref. 4. Reprinted from *Electronics*, April 13, 1975, Copyright © McGraw-Hill, Inc., 1975. All rights reserved.)

microcomputers control local measuring or control systems of the process. They contain algorithms to evaluate the measured information and to advise the control element. Process data is sent in digital form to the main computer, which utilizes this data, e.g., in an optimizing model, or stores it if required. This computer again sends instructions to the microcomputer in order to control the process optimally.

REFERENCES

1. K. Oehl, *Daten-Ubertragung and Fernverarbeitung*, Walter de Gruyter, Berlin, 1974.
2. U. Rembold, M. K. Seth, and J. S. Weinstein, *Computers in Manufacturing*, Marcel Dekker, New York, 1977.
3. S. Davis, Linear ICS for Digital Interface Applications, *Computer Design*, August 1973.
4. S. Harris, Prospects Brighten for Data Acquisition, *Electronics*, April 3, 1975.

5

FUNDAMENTALS OF INTERFACE TECHNOLOGY

5.1 INTRODUCTION

To become acquainted with the different functions of the interface that connects the control computer and the process, it is necessary to understand some fundamentals of digital circuit technology.

Until a few years ago, the necessary intelligence to operate measuring systems was concentrated in the control computer. In this early concept the measuring devices and instruments were connected to the computer via a line multiplexer. The control of the system and the actual performance of the data-acquisition task were handled in principle by software located in the central processor. The architecture of the interface therefore was relatively simple. In modern computer systems there is a trend toward decentralization of intelligence. The control computer is relieved from the basic data-acquisition task and can devote more attention to performing complicated measuring and control strategies. Control and interrogation of process sensors are done by an intelligent bus system. In general, this system contains a microprocessor whose operation is initiated by the control computer and then performs its function independently. The user of such a system should understand the design and configuration of the intelligent subsystem to be able to utilize it to its full capacity.

Within the framework of this chapter the fundamentals necessary for understanding modern interfaces will be discussed. A small selection of the wealth of available material will be given. Detailed information can be found in the listed literature.

5.2 THE OPERATIONAL AMPLIFIER

5.2.1 Modern Operational Amplifiers

When using process control computers for electronic measurement and control tasks, digital parameters, such as the position of a switch, as well as analog data such as a temperature measurement have to be acquired and processed. These parameters, which are normally made available by a transducer as electrical signals, are, in general, first amplified and then transmitted to the control computer. In modern electronic instruments operational amplifiers are widely employed to perform the necessary signal amplification. For the transmission of analog signals with a small amplitude, amplification considerably reduces the influence of noise. The operational amplifier can also be used to do arithmetic operations. There are many special-purpose circuits available to perform a desired mathematical function. All the user has to do is wire the operational amplifier so that the input signals will produce at its output the required results.

Ordinarily, the whole circuit of the operational amplifier is integrated on a chip. The main element of the chip is a differential amplifier. The voltage difference sensed at the input lines appears as an amplified signal at the output. Although the characteristics of the traditional amplifier are determined by its internal design, an integrated operational amplifier is constructed in such a manner that its performance can be programmed by external components. For this reason in this chapter the operational amplifier is presented as an integrated building block ("black box"). A detailed description of the internal design and the functions of the amplification stages in this black box exceeds the scope of this book.

By using the model of an ideal operational amplifier, we will explain its basic characteristics. Thereafter some typical applications will be discussed. Emphasis is placed on the discussion of circuits which perform basic mathematic operations.

5.2.2 The Ideal Operational Amplifier

Figure 5.1 shows a scheme of an ideal operational amplifier. In this design it is used to amplify the voltage difference sensed at the input:

$$V_D = V_P - V_N \tag{1}$$

By assuming a gain of A > 0, the voltage becomes

$$V_O = AV_D = A(V_P - V_N) \tag{2}$$

For $V_N = 0$, Eq. (2) can be rewritten as

$$V_O = AV_P \tag{3}$$

This means that the output voltage is in phase with the input voltage.

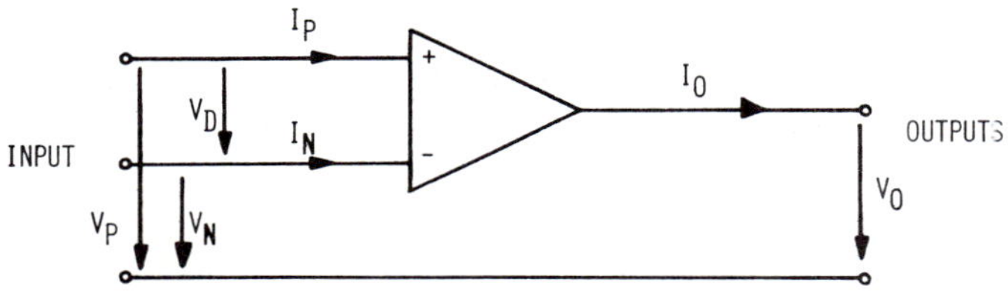

FIG. 5.1 Circuit diagram of an operational amplifier.

The corresponding input is denoted as *noninverting* or p input. In the graphic symbol of Fig. 5.1 this condition is marked with a plus sign. In a similar manner for $V_P = 0$, one obtains

$$V_O = -A\ V_N \tag{4}$$

In this case the output voltage is out of phase by 180° with the input voltage and assumes a negative value. The corresponding input is called *inverting* or n input. In the graphic symbol this is denoted with a minus sign (Fig. 5.1).

The following properties discussed are assumed for an ideal operational amplifier:

1. The gain is infinite and independent of frequency. The output voltage of Eq. (2) then becomes

$$V_O = \infty \text{ for } V_D \neq 0 \tag{5}$$

2. The input impedance is infinite, which means that there are no input currents flowing:

$$I_P = I_N = 0$$

3. The output impedance is zero, which means the output of the operational amplifier may be regarded as an ideal current source.
4. The output voltage is free of drift or noise.

In the real world physical properties prevent meeting these requirements. In the following section the operational amplifier will, however, be assumed of ideal design. Thus its performance can be studied more easily. If one wants to know the difference of the real operational amplifier from that of the ideal amplifier, the knowledge of various characteristic values is required.

5.2.3 Characteristics of Real Operational Amplifiers

In this section only the most important characteristic properties of the real operational amplifier which differ from the ideal operational amplifiers will be discussed.

The open loop gain A_{OL} of an operational amplifier without load (without feedback connection) has the following finite value:

$$A_{OL} = \frac{V_O}{V_D} = \frac{V_O}{V_P - V_N} \tag{6}$$

This property is strongly influenced by the stability of the voltage supply. Which means a very stable power supply has to be used in order to reduce the influence of possible voltage fluctuations. Furthermore, the open-loop gain is frequency dependent; it decreases with increasing frequency. Typical values for A_{OL} lie between 10^3 and 10^6 Hz. The output voltage V_O is only proportional to the input voltage V_D within the range of $V_{Omin} < V_O < V_{Omax}$. This range is called the *linear region*. If this range is exceeded the output voltage remains constant and independent of the voltage difference V_D. In other words, the amplifier is overdriven (Fig. 5.2).

Real operational amplifiers have a finite input impedance. Here only the common mode input impedance will be discussed. It is defined as

$$r_D = \begin{cases} \dfrac{\Delta V_P}{\Delta I_P} & \text{for } V_N = 0 \\[2ex] \dfrac{\Delta V_N}{\Delta I_N} & \text{for } V_P = 0 \end{cases} \tag{7}$$

Typical values for r_D are in a range of 1 to 10 MΩ. Special purpose operational amplifiers using field-effect transistors in the input stage have an input impedance of 10^3 to 10^4 MΩ.

The output impedance r_A of the real operational amplifier differs from the internal resistance of an ideal voltage supply in which r_A is assumed to be zero. Typical values for r_A are 50 to 100 Ω. In order to protect the operational amplifier from thermal overload in some cases, it is necessary to limit the output current to

$$I_O \approx 20 \text{ mA}$$

The real operational amplifier has an output voltage of $V_O \neq 0$, even if $V_N = V_P = 0$, and therefore $V_D = 0$. Thus one defines an offset voltage V_{OS} (input offset voltage) as the voltage difference which has to prevail at both inputs to make V_O equal to zero:

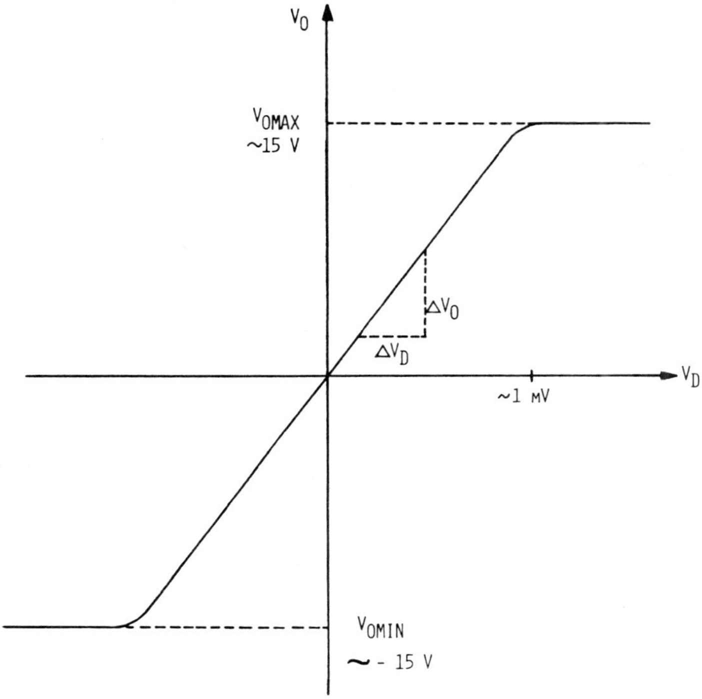

FIG. 5.2 Transfer function of an operational amplifier.

$$V_{OS} = V_P - V_N \quad \text{for } V_O = 0 \tag{8}$$

Compensation of the output voltage can be obtained with low ohmic potentiometers connected externally to the operational amplifier. It may also be necessary to compensate for the frequency dependency by means of external connectors of the amplifier. However, some operational amplifiers have an internal circuit for frequency compensation.

5.2.4 Applications of Operational Amplifiers in Linear Circuits

In the following section some of the very typical applications of operational amplifiers will be presented. The basic principles of operation have been adopted from analog computer technology. Simple mathematical operations can be performed with the help of operational amplifiers. The equations of the ideal operational amplifier will be used to illustrate these properties. For analog computing circuits the operational amplifier in most cases will be wired to have negative feedback. The function of this circuit will be explained with the help of Fig. 5.3.

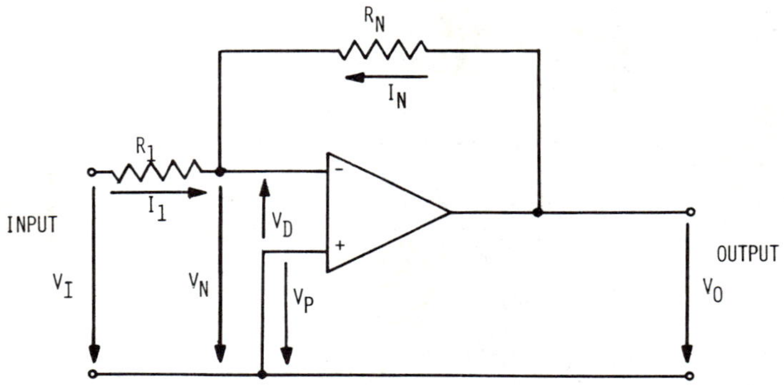

FIG. 5.3 Inverting amplifier.

Inverting Amplifier

Here the noninverting input is connected to the ground. Thus we obtain, with $V_P = 0$,

$$V_D = -V_N \tag{9}$$

If a positive voltage V_I is applied to the input, V_N immediately assumes the value

$$V_N = \frac{R_N}{R_N + R_1} V_I \tag{10}$$

This happens because V_O initially equals to zero.

The voltage $-V_N$ is amplified, and thus the output voltage V_O quickly assumes a negative value. The voltage V_O has a phase shift of 180° in regard to the voltage V_N applied to the inverted input. This voltage counteracts the voltage V_N across the feedback resistor R_N. The output voltage V_O thereby changes until the input voltage V_N for all practical purposes is zero. The fact that the change of the output voltage counteracts the input voltage is typical for the feedback behavior of the operational amplifier. By this measure the stability of the circuit is increased. The transient response is also improved and the output impedance decreases. Thus within the linear operating range of the inverting operational amplifier an output voltage prevails which holds V_N at zero. The inverting input is therefore defined as the virtual ground or the summation node.

This circuit (Fig. 5.3) can also be considered as a voltage-to-current converter. Since the resistor R_1 is connected to the virtual ground, the input current becomes

$$I_1 = \frac{V_I}{R_1} \tag{11}$$

The same current I_1 flows through the resistor R_N of the feedback loop, which in this case can be an arbitrary load resistor since it has no influence on the value of the current. It must be observed, however, that the signal source of the operational amplifier is able to deliver the desired current.

When calculating an output voltage V_O (whereby V_N is assumed to be zero), it must be taken into consideration that in the ideal operational amplifier there is no input current flowing. Thus we obtain

$$\frac{V_I}{R_1} + \frac{V_O}{R_N} = 0$$

$$V_O = -\frac{R_N}{R_1} V_I \tag{12}$$

The influence of the feedback loop on the response of this circuit can also be best shown with the parameters of the closed-loop gain:

$$A = -\frac{V_O}{V_I} = \frac{R_N}{R_1} \tag{13}$$

The parameter A is determined by the resistance ratio of R_N/R_1, it is smaller than the open-loop gain A_{OL}. For the ratio $R_N/R_1 = 1$, the operational amplifier operates as an inverter.

Noninverting Amplifier

Similar to the inverting amplifier (Fig. 5.3) there is the noninverting amplifier (Fig. 5.4). Here again with the help of the ideal operational amplifier we obtain

$$I_1 = I_N = \frac{V_O - V_I}{R_N} = \frac{V_I}{R_1}$$

$$V_O = V_I \left(1 + \frac{R_N}{R_1}\right) \tag{14}$$

and from this it follows that

$$A = \frac{V_O}{V_I} = 1 + \frac{R_N}{R_1} \tag{15}$$

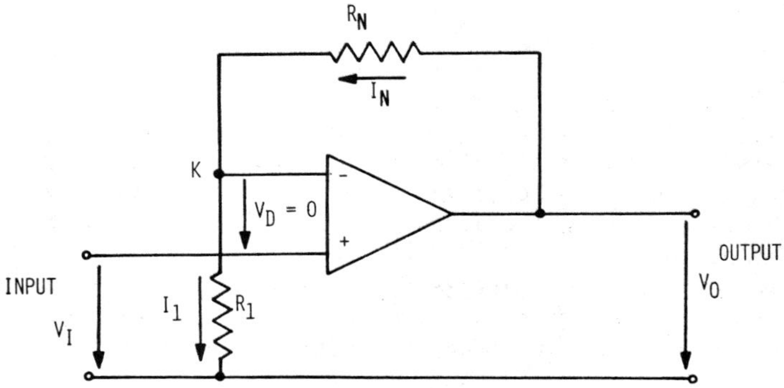

FIG. 5.4 Noninverting amplifier.

The gain of the noninverting amplifier depends on the resistance ratio R_N/R_1 as it was with the inverting amplifier. If the resistor R_N in Fig. 5.4 is short-circuited, we obtain a voltage follower. This special case is signified by a high input impedance and low output impedance. This amplifier therefore is suitable as an impedance transformer.

Summing Amplifier

The circuit of Fig. 5.5 allows the addition of several voltages with different or the same values. With the help of Kirchhoff's law applied to the currents flowing into and out of the junction points one obtains

$$\frac{V_1}{R_1} + \frac{V_2}{R_2} + \frac{V_3}{R_3} + \cdots + \frac{V_I}{R_I} + \frac{V_O}{R_N} = 0 \tag{16}$$

The output voltage V_O thus assumes the value

$$-V_O = \frac{R_N}{R_1} V_1 + \frac{R_N}{R_2} V_2 + \cdots + \frac{R_N}{R_I} V_I \tag{17}$$

Difference Amplifier

The circuit shown in Fig. 5.6 renders at its output the difference between the input voltage V_1 and V_2 weighted by a constant α. This can be shown with the help of the following derivation:

$$V_P = \frac{R_0}{R_0 + R_2} V_2 \tag{18}$$

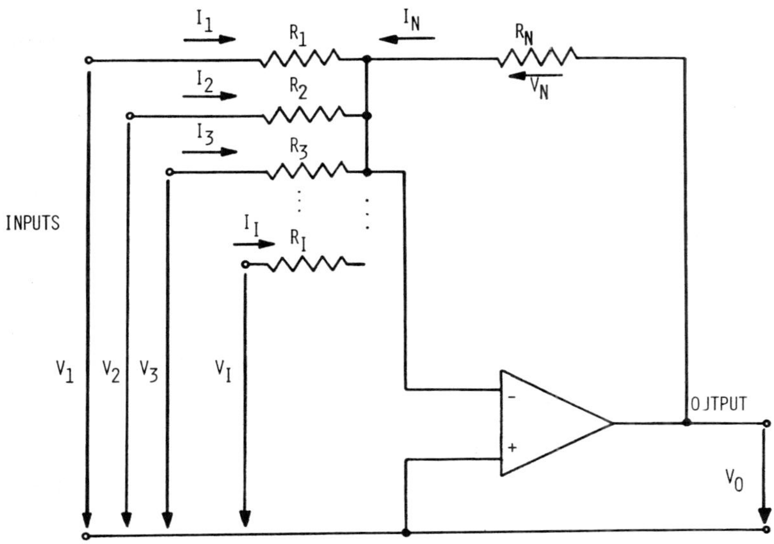

FIG. 5.5 Summing amplifier.

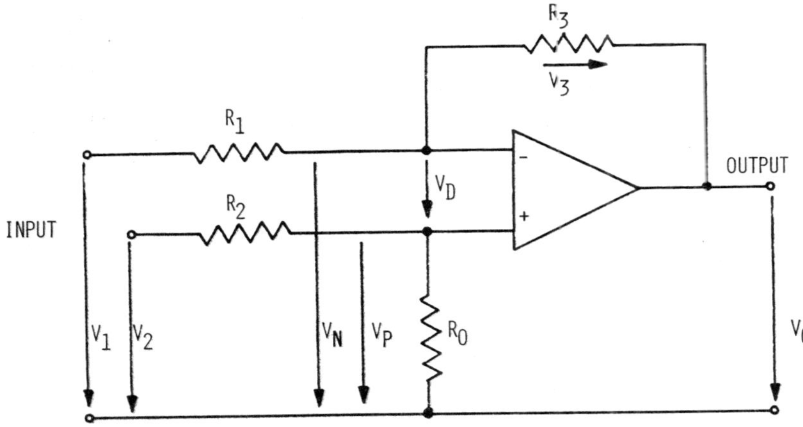

FIG. 5.6 Difference amplifier.

$$V_3 = \frac{R_3}{R_1 + R_3} (V_1 - V_O) \tag{19}$$

With $V_D = (V_3 + V_O) - V_P = 0$, it follows from Eq. (18)

$$V_P = \frac{R_3}{R_1 + R_3} (V_1 - V_O) + V_O \tag{20}$$

By setting Eq. (17) and (19) equal we obtain

$$V_O = \frac{R_0 (R_1 + R_3)}{R_1 (R_0 + R_2)} V_2 - \frac{R_3}{R_1} V_1 \tag{21}$$

and now reducing this equation with

$$R_1 = \frac{R_3}{\alpha} \qquad R_2 = \frac{R_0}{\alpha}$$

we obtain

$$V_O = \alpha (V_2 - V_1) \tag{22}$$

Integrator

In analog computing technology, the integrator and differentiator are the most important circuits. In addition, they are used for signal processing and as a function generator. The mode of operation and the derivation of the equation for the integrator are explained with the help of Fig. 5.7. Contrary to the circuits discussed up to this point in which the electrical parameters behave quasistatically, here current and voltages are functions of time. These time-dependent parameters are denoted by the parameter t.

The current flow into the capacitor can be described by the following relations:

$$\begin{aligned} I_C (t) &= C \frac{\Delta V_C(t)}{\Delta t} \\ I_1 (t) &= \frac{V_I (t)}{R_1} \end{aligned} \tag{23}$$

With $I_C (t) = I_1 (t)$ and $V_C (t) = -V_O (t)$, we obtain

$$\frac{V_I (t)}{R_1} = -C \frac{\Delta V_O(t)}{\Delta t}$$

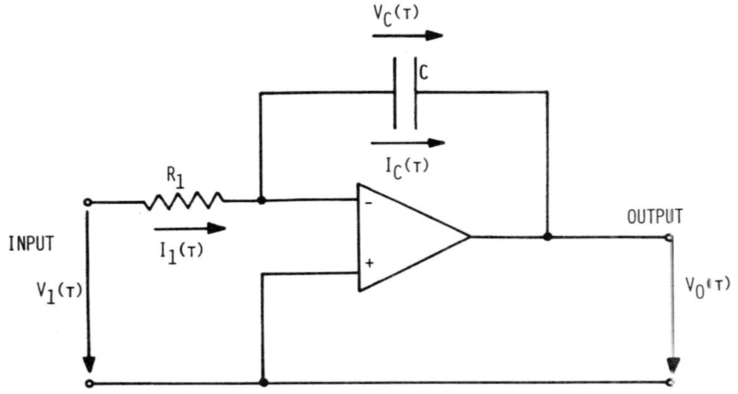

FIG. 5.7 Integrator.

and finally solved for V_O (t):

$$V_O\ (t) = -\frac{1}{R_1 C}\int V_I(t)\ dt \tag{24}$$

The characteristic properties of the integrator are determined by the time constant

$$T = R_1\ C$$

Differentiator

By interchanging R and C in Fig. 5.7, we obtain the differentiator, whose output voltage can be derived in a manner similar to that of the integrator

$$V_O(t) = -\ RC\ \frac{\Delta V_I(t)}{\Delta t} = -\ T\ \frac{\Delta V_I(t)}{\Delta t} \tag{25}$$

Comparator

The operational amplifier may also be wired as a comparator (Fig. 5.1). Here the unknown voltage V_m at the inverting input is compared with the reference voltage V_{ref} at the noninverting input. Again taking a look at the ideal open-loop amplification A_{OL}, one may notice that for a differential voltage $V_D \neq 0$ the output voltage instantaneously goes to saturation V_{sat}. Here the transfer function is not linear. We obtain binary information indicating whether or not a measured voltage is greater or smaller than the predefined reference voltage. One obtains the following relations:

$$V_m < V_{ref} \rightarrow V_O = +V_{sat} \tag{26}$$

$$V_m > V_{ref} \rightarrow V_O = -V_{sat} \tag{27}$$

5.3 DATA CONVERSION AND ACQUISITION

5.3.1 General Considerations

Data converters allow communication between the real world of analog processes and the digital world of the computer. They are used in numerous applications in industry. In automatic process control converters transform analog parameters into digital parameters and vice versa. Also, in communication technology often analog signals are first converted into digital form, sent over the communication line, and then restored at the receiver end into the original analog form (Sec. 4.2.2).

Figure 5.8 shows a complex technical process controlled by a typical computer system. This process is located in close proximity to the computer. In order to transfer a signal to the computer, it must first be converted by the A/D converter into digital form. Correspondingly, control instructions or signals issued by the computer must be converted via a D/A converter into analog form.

Physical parameters to be measured such as temperature, pressure, and flow are first acquired by suitable sensors and then changed into electrical analog values at the signal source (Fig. 5.8). In general, voltages are suitable to transfer the signal when the signal-to-noise ratio can be improved with the help of a modulation technique. Via a multiplexer the transducer transmitting the analog signals will be selected, and the signal will be sent to a sample and hold circuit, which in turn holds the sampled voltage for a short period of time. The holding time is selected in such a manner that the following A/D converter is able to convert the information completely within a desired time frame into its digital value. In this instrument configuration the analog multiplexer makes it possible that only one A/D converter has to be used (Sec. 3.5.3). It converts analog signals by the time-division-multiplexing method (Sec. 4.2.2). The sample rate for the individual channels is relatively fast in comparison to the slow-changing quasistatic process signals. The computer compares the acquired actual measured value of the parameters with a desired value stored beforehand in its memory. Resulting from this comparison, it calculates corrections for the process variables if necessary. This information is given back to the process via a D/A converter in order to induce a change in the process to comply with its desired operating state.

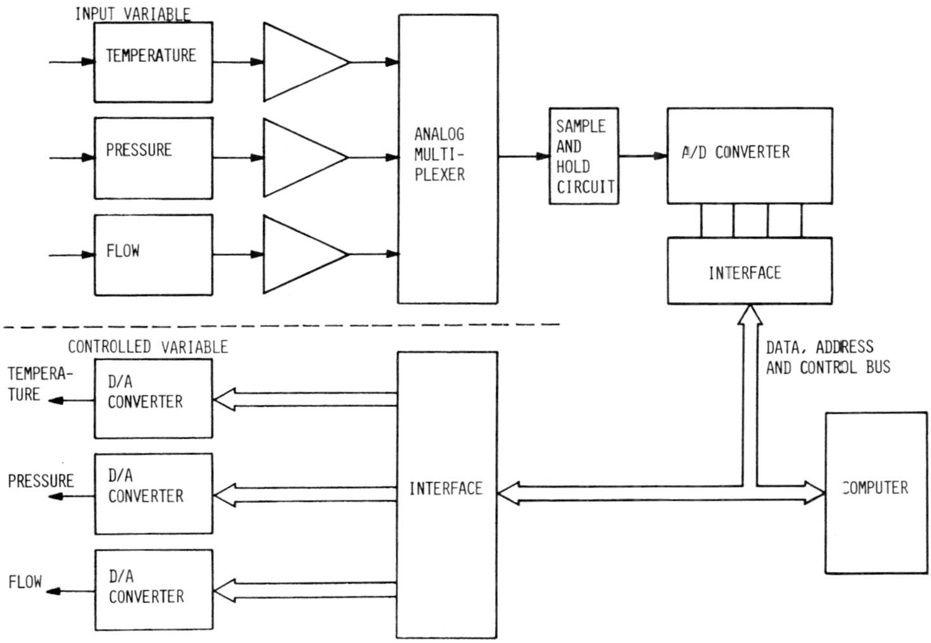

FIG. 5.8 Basic components of process control computer installations.

The user has a wide range of A/D converters to choose from. The difficulty is to select from the many types available the one that is most suitable for the particular application. The converters differ not only in their mode of operation but also in their technological design. In addition to discrete systems, there are also very highly integrated monolithic circuits available, often only tailored for a specific use. The most important parameters of the so-called standard D/A and A/D converters will be explained in the following section. A short description of the mode of operation as well as the different technology will help the user to evaluate commonly available converters and select the most appropriate specific kind.

5.3.2 Parameter Definition for Standard Converters

When selecting A/D and D/A converters, the performance characteristics of the various component must be evaluated. To simplify the understanding of data sheets a brief description of the parameters will be given.

Resolution

Resolution is defined as the relative value of the smallest possible increment of the full measuring scale, in other words, the least significant bit (LSB), which can be distinguished by the converter. If the digital value is coded in binary form, the resolution is determined as follows:

$$\frac{\text{Full scale}}{2^n} \tag{28}$$

whereby n represents the number of bits for which the converter has been designed. For example, a 12-bit D/A converter with the full scale of 10 V has the following resolution:

$$\text{LSB} = \frac{\text{full scale}}{2^n} = \frac{10\text{ V}}{4096} = 2.44\text{ mV} \tag{29}$$

Thus the maximum output voltage is

$$V_{O,max} = 10\text{ V} - 2.44\text{ mV} = 9.997\text{ V} \tag{30}$$

Relative Accuracy

The relative accuracy of the converter is determined by its linearity. In this parameter the errors resulting from the scale error or offset error are not considered. When these errors have been compensated for, the converter, which normally has a perfect relative accuracy, shows a perfect absolute accuracy (Fig. 5.9).

Scale Factor and Gain Error

One specifies with this term the deviation of the real transfer function from the ideal function (Fig. 5.9). This error is stated in terms of the theoretical value; it can be compensated for by adjusting the gain.

Nonlinearity

The nonlinearity is the maximum deviation from the straight line of the ideal transfer function. Normally, it is given in fractions of the least significant bit.

Settling Time

The time interval necessary for a D/A converter to settle to an accuracy of ±1/2 LSB from a full-scale code conversion is called the *settling time* (Fig. 5.10). In this context the slew rate is also of interest. It signifies the maximum gradient by which the output voltage of

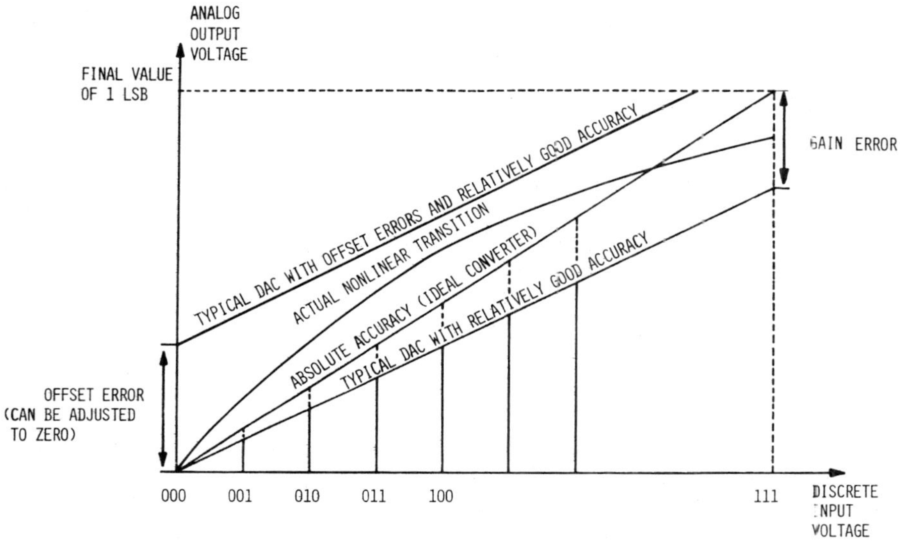

FIG. 5.9 Scaling and offset errors.

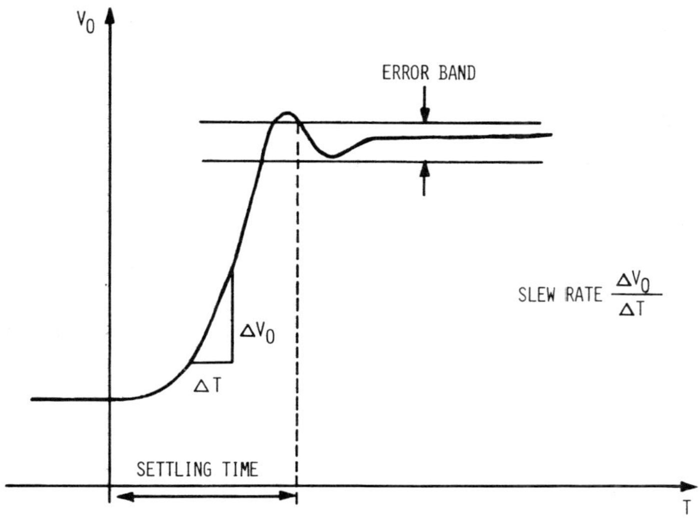

FIG. 5.10 Step response with settling time of a D/A converter.

the D/A converter is able to follow a change of the digital input value. In Fig. 5.10 this parameter is denoted by

$$\frac{\Delta V_O}{\Delta T}$$

Conversion Time of an A/D Converter

This time indicates how long the A/D converter needs to digitize completely one analog value once. With A/D converters operating on the principle of successive approximation the conversion time is independent of the input amplitude. With integrating converters it is a function of the amplitude of the input signal.

5.3.3 Operation of Converters

The binary information normally is brought as a parallel word consisting of several bits to the D/A converter. With these signals bipolar switches for current or voltage sources are controlled. They are weighted with the aid of a resistor network. If Fig. 5.11 a binary D/A converter is shown as a block diagram. The digital value to be converted is presented

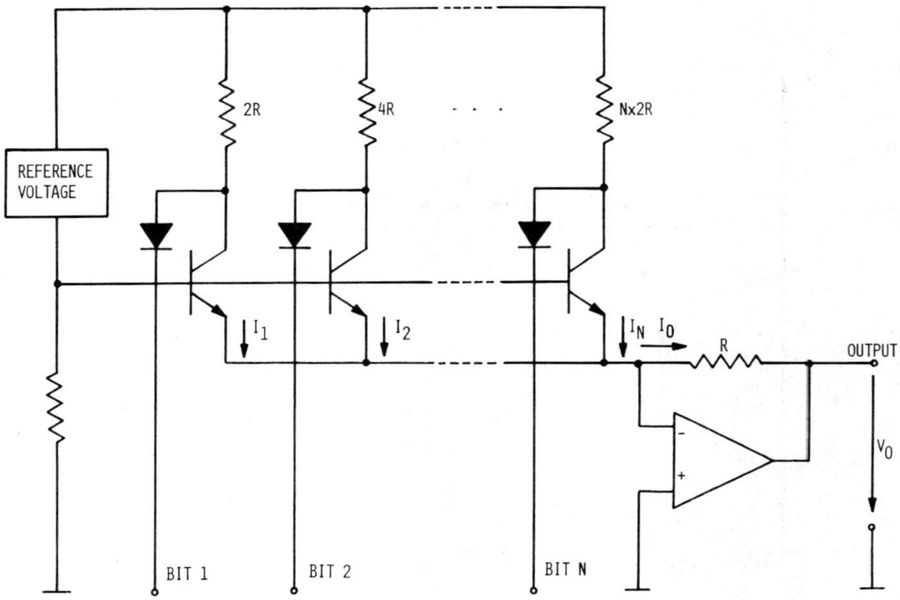

FIG. 5.11 D/A converter with ladder resistor network.

at TTL voltage level to the parallel input transistors. These transistors control the current flow to an operational amplifier. The summed-up currents are converted by the operational amplifier into an analog output voltage. A Zener diode insensitive to temperature changes is used in the power supply to provide a stable reference voltage.

Transformation of an analog signal to a digital signal require more effort in the design of conversion circuits. A/D converters integrated on a signal chip usually operate on the staircase or successive approximation principle. Here the conversion time is kept constant, making it independent of the amplitude of the momentary input signal. With this method the input voltage signal which represents the analog parameter is compared with the sum of a series of voltage increments, where each increment has the same known standard value. As soon as the sum of the increments is equal to the voltage level of the signal to be digitized, a voltage comparator (Sec. 5.2.4) ends the conversion process. The number of increments needed to reach this voltage level represents the desired digitized value. The quantatized analog signal is supplied by a D/A converter in this digital servo system. Its output voltage is compared with the voltage to be digitalized. This procedure is similar to weighing with a balancing scale which is equipped with a set of binary weights. For example, for the unknown value in the range of 0 to 1 kg, weights of 0.5 kg, 0.25 kg, and 0.125 kg, etc., must be supplied.

Successive Approximation A/D Converter

Figure 5.12 shows the block diagram of a A/D converter using the successive approximation principle. In this design a D/A converter is located in the feedback loop of the comparator. It is linked to a register with n bit positions. When an analog voltage is to be converted, first bit 2^{n-1} is set. The corresponding output voltage V_O of the DAC is then compared with the unknown analog voltage V_M coming from the transducer. If V_O is greater than V_M, the control logic resets bit 2^{n-1} in the register. If V_O is less than V_M, the bit remains in the register, and the next lower significant bit 2^{n-2} is set. The new reference voltage V_O generated by the new bit pattern in the register again is compared with V_M. After n comparison steps, V_O approaches V_M and the digital value remaining in the register represents the binary-coded analog value. A timing diagram for a simple 3-bit successive approximation converter is shown in Fig. 5.12. This method is inexpensive; however, noise may cause problems.

Integrating A/D Converter

Another conversion method uses the integration principle. It is used when noise must be reduced and the time necessary for conversion is less important. Especially in digital voltmeters the dual-slope

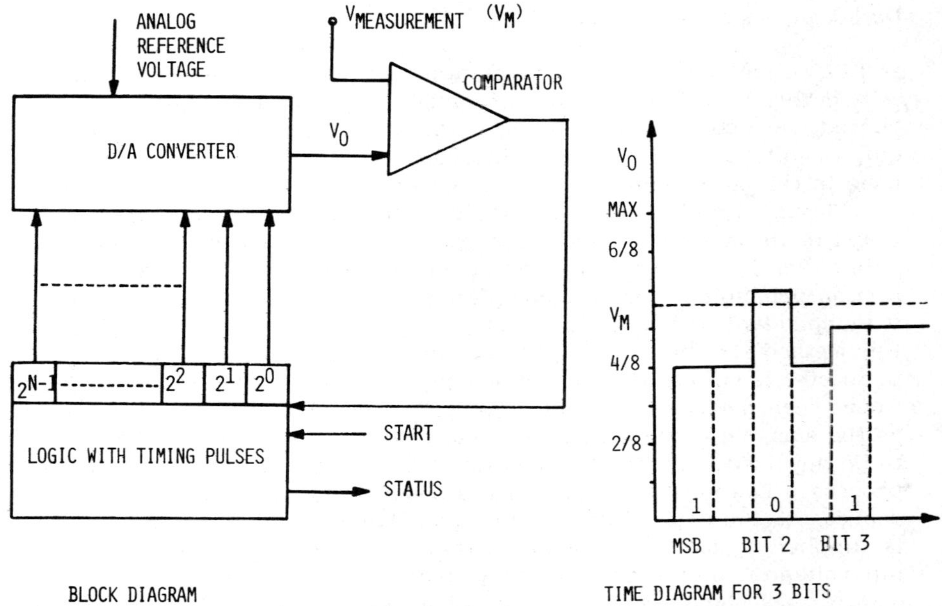

FIG. 5.12 Successive approximation A/D converter: (a) block diagram and (b) time diagram for 3 bits.

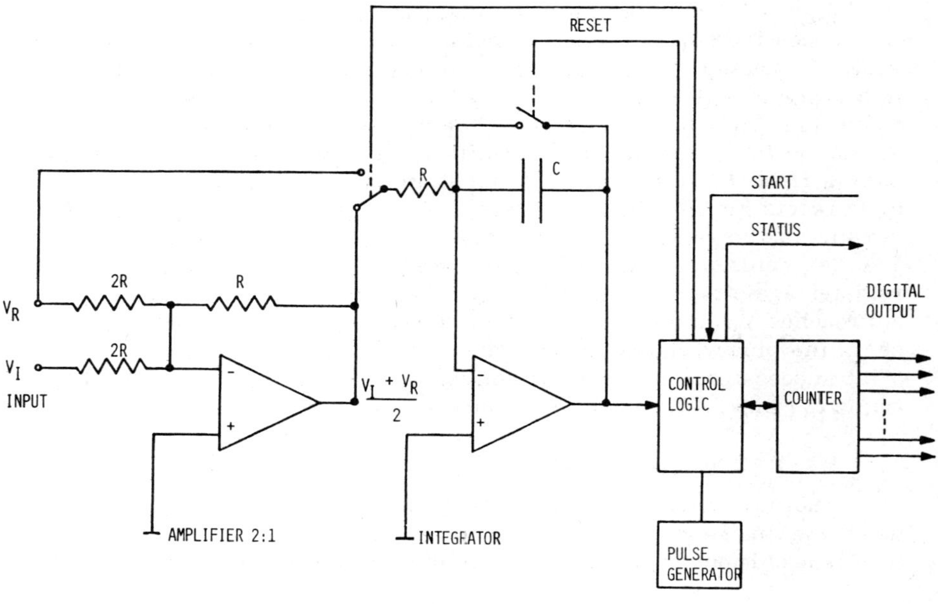

FIG. 5.13 Dual-slope integrating A/D converter.

integration principle is popular. A capacitor is charged by the voltage to be measured. It is then disconnected after a fixed time period. Now in place of the unknown voltage a defined reference voltage with inverse polarity is connected to the capacitor. The time period necessary for the discharge of the capacitor is proportional to the unknown analog voltage.

Figure 5.13 shows a block diagram of an integrating A/D converter which converts positive or negative signals. The input signal is sent to the integrator via an impedance transformer. At the same time a counter begins to register pulses generated by a clock. After a predetermined pulse sequence which corresponds to the time interval T_1, the input signal is disconnected, the reference voltage is switched to the capacitor, and the counter is reset (Fig. 5.14). At this instant the capacitor has been charged to the following voltage level:

$$V_{C,max} = \frac{1}{C}\int_0^{T_1} i_C \, dt = \frac{1}{C}\int_0^{T_1} \frac{V_I + V_R}{2R} \, dt$$

$$= \frac{T_1}{2RC}(V_I + V_R) \tag{30}$$

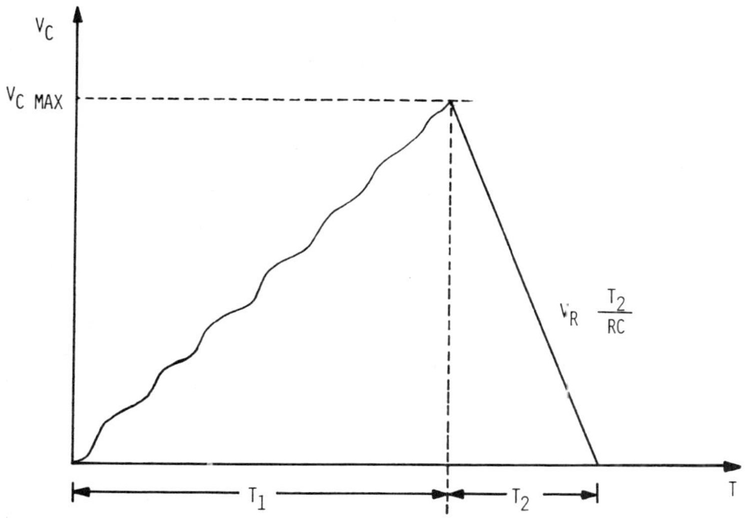

FIG. 5.14 Time diagram of the dual-slope integrator A/D converter.

At the moment of switch-over, the counter again starts to register the pulses, while the capacitor is discharged by the reference voltage according to the following equation:

$$V_C = \frac{T_1}{2RC}(V_R + V_I) - \frac{T_2}{RC} V_R \qquad (31)$$

When the voltage reaches $V_C = 0$, the capacitor is entirely discharged. With the help of the control logic the comparator then stops the counter. By setting the left side of Eq. (31) equal to zero, V_I can be determined:

$$\frac{T_1}{2RC}(V_R + V_I) = \frac{T_2}{RC} V_R \qquad (32)$$

$$V_I = T_2 \frac{2V_R}{T_1} - V_R = K\,T_2 - V_R \qquad (33)$$

If the counter is of binary design its digital value represents the analog voltage V_C offset by the value of the reference voltage V_R. This allows measurement of bipolar voltages in the range of $-V_r < V_I + V_R$.

This mode of operation offers many advantages. Neither the tolerance of capacitors nor variations of the clock frequency influence the conversion accuracy. The integration method compensates for the manufacturing tolerances of components and high-frequency noise. The integration period T_1 is selected in such a manner that noise superimposed over the input signal is reduced as much as possible or is completely eliminated. For example, to avoid noise from line frequency, a minimal integration period of 40 msec should be used, which means 10 measurements per second. In addition one has to observe that only approximately 40 percent of the total measuring cycle can be used for strobing in data. Consequently the converter is not suitable for applications in which analog data are to be sampled at high speed. This conversion method is most often used to convert signals from inductive and thermal sensors having long time constants.

Parallel A/D Converter

The fastest method to convert analog to digital signals is shown in Fig. 5.15. For this parallel conversion $2^n - 1$ comparators are needed. Here n denotes the resolution in bits. The reference voltage is subdivided into multiple voltage increments by a voltage-divider network. Each leg of the voltage divider is connected to one input of the individual comparators. The analog input voltage from the transducer is placed in parallel on the remaining inputs. All comparators whose reference normal is below the unknown voltage are turned on, and the others remain turned off. The $2^n - 1$ output signals are led to a coder

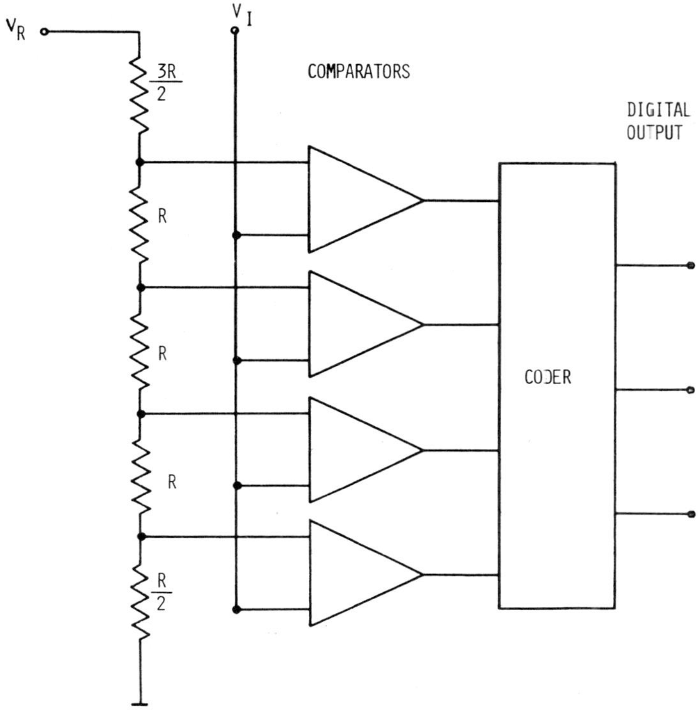

FIG. 5.15 Parallel 3-bit A/D converter.

which converts the information to a binary-coded value representing the analog signal. The signal can be read by the computer.

This parallel conversion method offers considerable advantages. The conversion time is limited only by the switching time of the comparators and that of the coding logic. However, the number of required comparators increases exponentially with increasing resolution n to $N = 2^n - 1$. An improved converter can be obtained by combining the parallel with the successive approximations methods. The parallel converter with its limited resolution generates the least significant bit very quickly. The approximation method determines the remaining more significant bits within a few cycles.

For all three methods discussed so far it should be mentioned that the input voltage from the transducer must remain constant during conversion time. This task is achieved by sample-and-hold components which are explained in Section 5.3.4. Before actual conversion begins the sample element strobes in the analog signal from the input. This sig-

TABLE 5.1 Representative A/D Converters

Resolution (bit)	Conversion time (μsec)	Technology conversion type	Comments
8	0.06	Discrete ECL logic, parallel conversion principle	Ultrafast 17-MHz rate
16	400	Discrete, successive approximation	High resolution, low drift
14	230,000	Discrete, integration principle	High resolution with noise immunity
8	0.9	Hybrid, approximation principle	Fast
8	40	Integrated circuit, approximation principle	Requires external clock and reference
12	40,000	Integrated circuit, integration principle	CMOS autozero circuit requires external components

nal is then held constant during conversion time by the hold element. Temporary buffering of analog data is not necessary for low-frequency or quasistatic signals free of superimposed noise. They hardly change their analog values during the conversion phase.

Comparison of Various Technologies

The original converters consisted of discrete components. Today this technology is still in use for special purpose applications; however, the modules are now cast into plastic material. The discrete components can be optimized to obtain a higher degree of precision and higher speeds. Table 5.1 shows a list of typical performance data.

Although converters of hybrid design can have the same degree of flexibility as discrete circuits, there remains one essential limitation. The individual miniaturized components (semiconductors, resistors, capacitors) are attached to a ceramic wafer and interconnected by hair-thin wires. To minimize the source of errors and for economic manufacturing reasons the number of connections by which the individual components are soldered together must be as low as possible. This also limits the number of elements integrated in a hybrid circuit.

The development of integrated circuits with complex functions can reduce the designing effort for hybrid converters. There are already current sources integrated on one chip which have binary-coded selectors as well as forward/backward counters. They are specifically designed for the approximation conversion method. Resistors with low temperature coefficients whose resistance value is adjusted by laser manufacturing technology, improve the accuracy of hybrid converters. Ordinarily, these units are completely integrated in a plastic module and only need external elements for compensation. Converters of monolithic design are one degree less complex than discrete or hybrid elements. To these large-scale integrated circuits (LSI), external components for fine tuning as well as a reference voltage source must be added. Even with the additional circuitry this integrated device is of relatively low cost.

Monolithic converters operating on the integration principle are very popular. There are already many inexpensive converters available with a resolution of up to 12 bits in which the analog parts as well as the digital logic are integrated on one chip. Table 5.1 displays a list of several converters which are currently available on the market and describes their specifications.

5.3.4 Sample-and-Hold Elements

The sample-and-hold element is a device which takes a snapshot of the analog input voltage during a predefined time interval and renders during a hold cycle to its output a steady-state signal which can be used for further processing. For example, in an A/D converter the

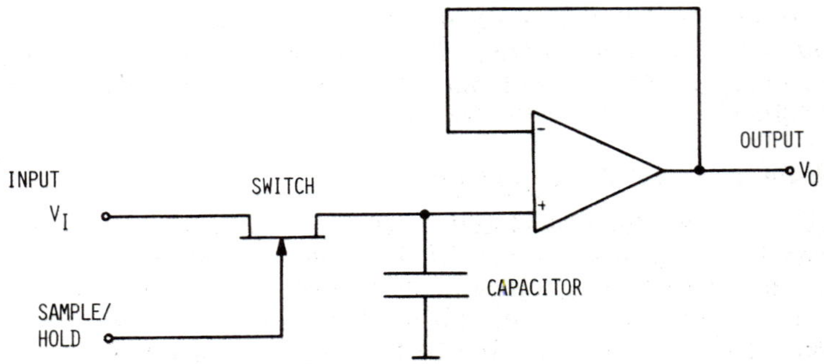

FIG. 5.16 Inverting sample and hold circuit.

momentary value of the input voltage is held and stored from initiation to completion of the conversion. This function is necessary to prevent possible alteration of the analog input signal during conversion time. Otherwise, this could lead to substantial measuring errors.

The use of a sample-and-hold element becomes necessary when the analog input voltage at the input of the A/D converter changes by more than 1/2 LSB (least significant bit) during conversion time. This often is the case with analog voltages of relatively low frequency. It is a well-known fact, for example, that a sine-wave signal during zero crossing generates a considerable voltage change per time unit.

The actual design of a sample-and-hold element (Fig. 5.16) usually consists of a capacitor as a charge element and a switch. The input voltage V is connected to the capacitor via the semiconductor switch. It in turn is controlled by a logic signal. The succeeding operational amplifier functions here as an impedance transformer and prevents deterioration of the stored charge.

To demonstrate the performance data given in data sheets the fundamental parameters of sample-and-hold elements will be described using Fig. 5.17.

Acquisition Time

The time elapsing between the triggering edge of a sample pulse and the settling state of the output voltage defines the *acquisition time*.

Aperture Time

The time delay between the hold pulse and the actual opening of the switch defines the *aperture time*. Sometimes "turn-off time" is used for this definition (Fig. 5.17).

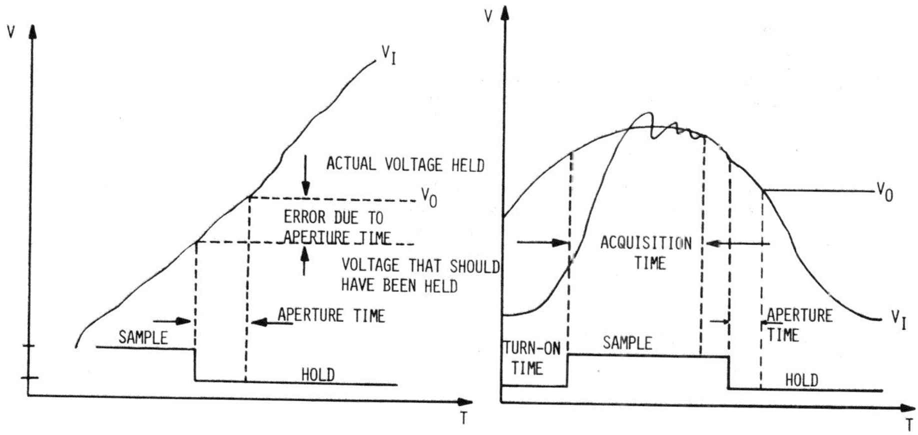

FIG. 5.17 Parameters of sample and hold circuits.

Decay Rate

The decay rate indicates the gradient of the output voltage during the "hold" state.

Bandwidth

The bandwidth denotes the frequency range within which the sample-and-hold element works with the stated accuracy. At higher frequencies errors are caused by distortion and phase shifting.

5.4 PULSE GENERATION AND DETECTION

This section is concerned with the generation, storage, and synchronization of pulses. Problems of this type are present in all digital information systems in many different forms. Here some typical applications are discussed.

5.4.1 Oscillators

In digital-signal processing oscillators have the task of generating a voltage with a well-defined frequency, shape, and amplitude. In most cases it is also required to maintain an exact frequency and to transmit a well-shaped square wave with fast rise and fall times, as well as an amplitude of predescribed height (Fig. 5.18). The parameters of the pulse must comply with the technology used in the circuits of the data processing system.

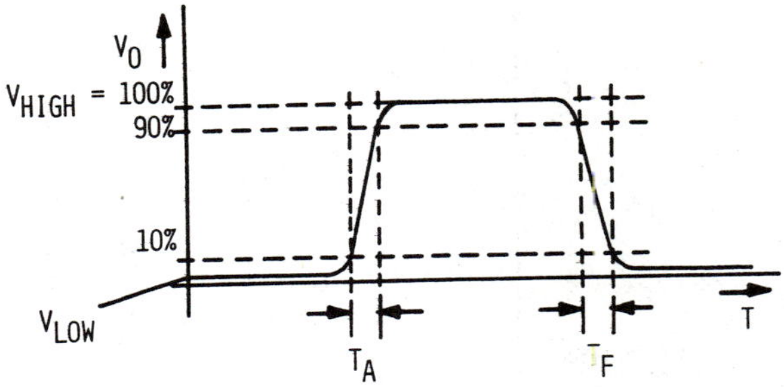

FIG. 5.18 Square wave.

Astable Multivibrator

The astable multivibrator presents the basic form of a digital oscillator. The function of this important circuit for creating a square wave will be demonstrated with the help of a transistor circuit. Although this circuit, built from discrete elements, is seldom used today, it is well suited for explaining the principle of pulse generation. Figure 5.19 shows the circuit of an astable multivibrator and Fig. 5.20 the corresponding wave forms. Since the circuit is self-oscillating, it changes automatically from one state to the other. For our discussion, therefore, it is of no importance with which state we begin.

When the output voltage $V_O{}'$ at the collector of transistor T_1 decreases (point 1 in the pulse diagram, Fig. 5.21), the voltage change is fed back through capacitor C_1 to the base of transistor T_2.

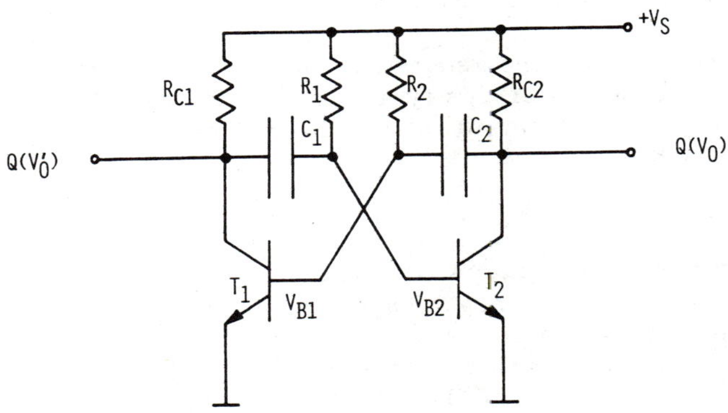

FIG. 5.19 Astable multivibrator.

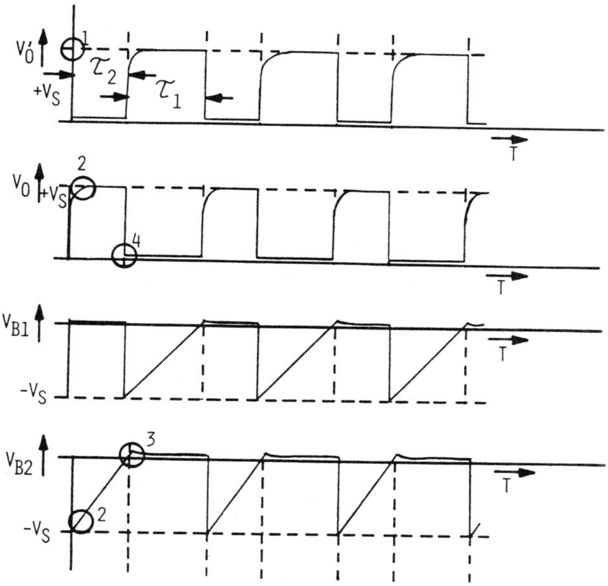

FIG. 5.20 Waveforms of an astable multivibrator: $\tau_1 = R_1C_1 \ln 2$, $\tau_2 = R_2C_2 \ln 2$, V_O = output voltage, V_S = supply voltage, and $V_{B1} = T_1$ base O voltage.

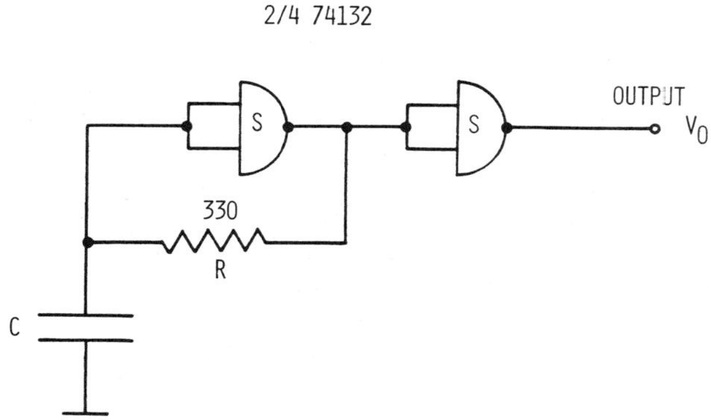

FIG. 5.21 TTL-oscillator with two Schmitt triggers.

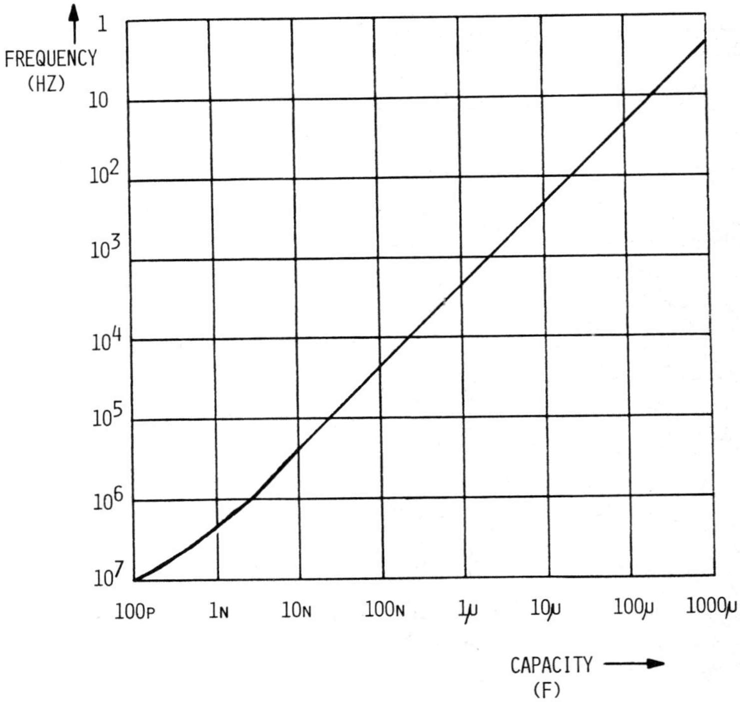

FIG. 5.22 Diagram to determine the value of the capacitor C of the oscillator shown in Fig. 5.21.

This causes T_2 to turn off, thus V_O increases at the collector of T_2. This increases the voltage at the base of T_1 via the feedback connection through capacitor C_2. Thus the started process is accelerated. V_O' drops quickly to the level of the collector emitter voltage which is nearly 0 V. Finally T_2 turns off completely. Now the capacitor C_1 charges up via resistor R_1. This means that the voltage at the base of T_2 increases during the charging time t_1. When V_{B2} reaches the voltage level at which the transistor begins to turn on, V_O falls off because of the flow of the collector current T_2. The feedback connection via C_2 transmits this voltage change to the base of T_1. Thereby V_O' and V_{B2} increase, and T_2 turns on completely. The charging cycle of capacitor C_2 via resistor R_2 then initiates the next change of state. The operation of this astable multivibrator depends on the RC time constants of the resistor capacitor combinations R_1C_1 and R_2C_2.

Unfortunately the output pulse does not have an exact square shape. It depends on the values of the resistors R_{C1} and R_{C2} which are connected to the collectors of the transistors. These resistors generate time constants with their corresponding capacitors. This phenomenon is responsible for the relatively slow rise time of the output voltage. However, there are numerous special circuits which have very short rise times.

Oscillators Built from Integrated Circuits

One often used oscillator consists of two so-called Schmitt triggers. The circuit is shown in Fig. 5.21. The function of a Schmitt trigger will be explained later. Here only the fact will be stated that the Schmitt trigger switches its output signal very fast from one state to the other, as soon as the input signal has reached an upper threshold or falls below a lower threshold. Here a very slow change of the input voltage level which is equivalent to a long rise time of the triggering edge can be tolerated: thereby a fast change of the output voltage is still assured. The Schmitt triggers described here are the TTL-series type 74 N. They are manufactured today by almost all companies producing integrated circuits. The circuit operates in a frequency range of 0.1 Hz to 10 MHz. Its frequency can be altered by changing the capacitor C (Fig. 5.22).

The frequency, however, is not very stable. In this circuit the feedback resistor of 330 Ω may not be altered. If a logic HIGH is present at the first output of the Schmitt trigger, the capacitor charges via resistor R. Thereby the voltage level at the input of the second Schmitt trigger increases. As soon as the threshold is reached, the logic state of LOW appears at the output. Thus the capacitor is charged anew via R and the voltage at the input decreases. Upon reaching the turn-off threshold, the trigger changes state and the entire process starts over again. The second Schmitt trigger only shapes the signal. At the output of this stage a well-shaped square signal is available.

Crystal Oscillators

If a high frequency stability is required the more expensive crystal oscillators must be used. This circuit, for example, is necessary to generate pulse signals for synchronous data transfer. Such circuits can only oscillate at a fixed preset frequency, which is crystal dependent. The stability factor obtainable with a crystal oscillator is approximately

$$\frac{\Delta f}{f} \approx 10^{-6} \div 10^{-10} \qquad (34)$$

Figure 5.23 shows a crystal-controlled oscillator built from two simple inverters [1]. They generate the required amplification and phase shift for the circuit. Both inductances L_1 and L_2 generate a reverse-current feedback. They have a resistance value of approximately

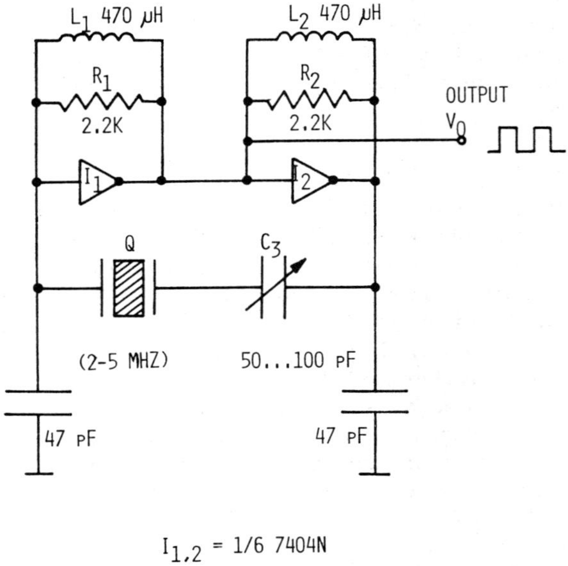

FIG. 5.23 Crystal controlled oscillator. (From Ref. 3.)

100 Ω. Thus the inverter has a sufficient linear amplification. The resistors R_1 and R_2 prevent the system from oscillating at the natural frequency of the inductances. During operation of the circuit the crystal Q has its lowest impedance at its resonance frequency. Thus a positive feedback dominates and the circuit is locked in to the resonance frequency of the crystal. Fine tuning of the feedback path to the resonance frequency is done with the adjustable capacitor C_3. The 47-pF capacitors prevent oscillation of the crystal below and above the fundamental frequency. The circuit shown has the advantage that it is self-starting and can operate with crystals of any impedance. Further examples of crystal oscillators can be found in the literature.

Utilizing the Power-Line Frequency

In many cases it is useful to utilize the frequency of the power line to generate a square wave of the desired control frequency. By using inexpensive components a good square wave with a frequency stability of approximately 1 to 2 percent can be obtained. Depending on the local power supply, the basic frequency as well as the stability of the wave can vary. In the United States, for example, the frequency is 60 Hz, whereas in Europe it is usually 50 Hz. Figures

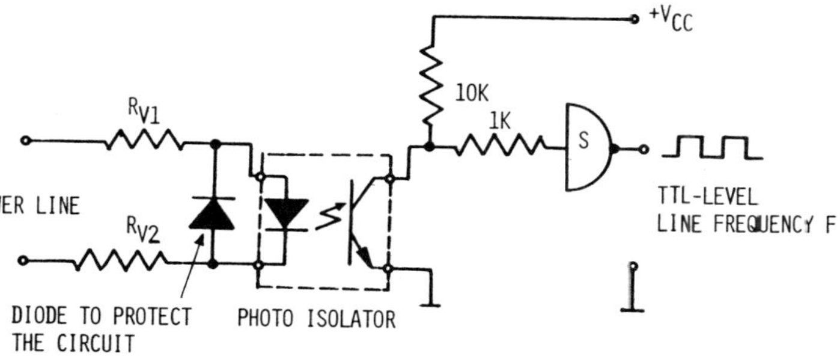

FIG. 5.24 Circuit to generate clock pulses from power-line frequency.

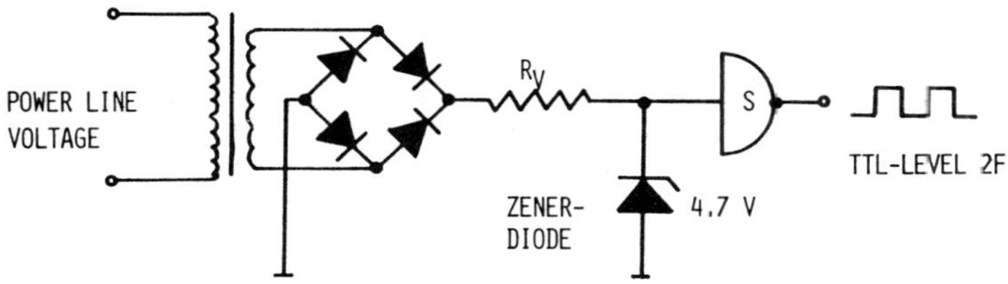

FIG. 5.25 Circuit to generate clock pulses from power-line frequency.

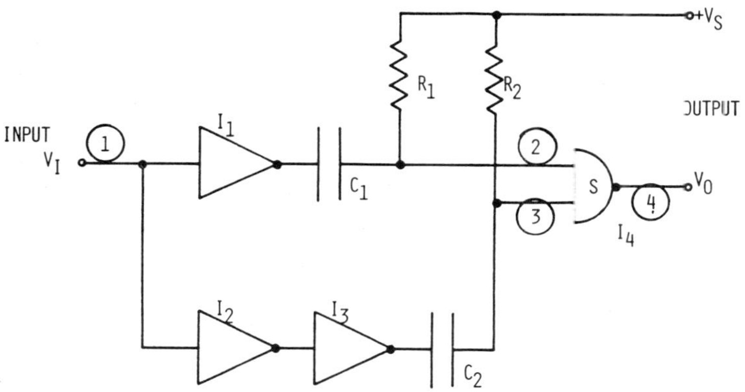

FIG. 5.26 Frequency-doubler circuit.

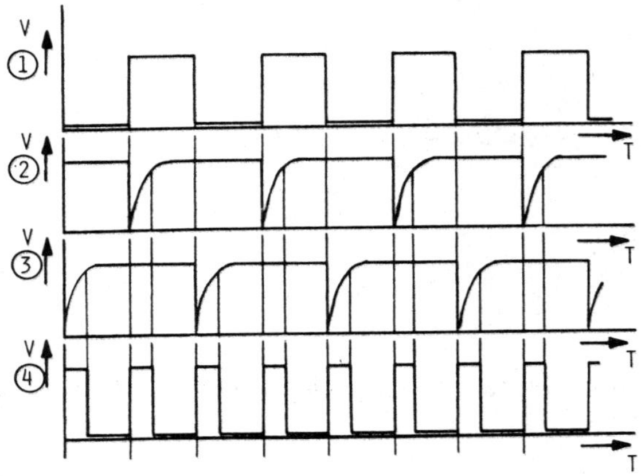

FIG. 5.27 Circuit waveforms: frequency doubler.

5.24 and 5.25 show examples for circuits using the power-line frequency. In the first case the signal is obtained with the aid of a photoisolator, in the second case by an inductive method. By using a single-phase bridge rectifier, the frequency is doubled.

Frequency Doubler

Some applications require clock frequencies which are a multitude of those generated by a crystal oscillator or an other device. In this case the fundamental frequency can be increased to an even number of harmonic frequencies. A very simple method of doubling the frequency is that of differentiating the positive and negative triggering edges of the clock signal. The pulse train generated thereby has twice the frequency of the input signal. By proper selection of the differentiation constants R_1C_1 and R_2C_2, respectively, the duty cycle can be maintained within close tolerances. Figure 5.26 shows the circuit, and Fig. 5.27 the wave forms. Both of the inverters I_1 and I_2 have the task of preventing feedback from the capacitor to effect the function of the oscillator. The NAND gate I_4 must have a Schmitt trigger input so that the slow edges of the differentiated input signal can be processed properly.

5.4.2 Storage of Digital Signals

For storing of digital signals bistable circuits are used. These circuits have two stable output stages. By applying input pulses of short duration, one of the two possible output stages will be assumed and maintained until another input pulse introduces a change. For

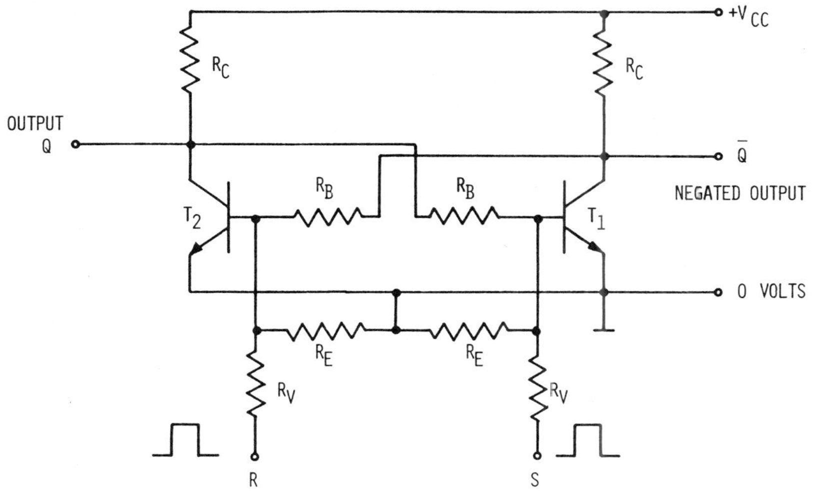

FIG. 5.28 RS flip-flop with transistors.

any length of time between two succeeding input pulses, the element acts as an information storage device. In interface technology various types of flip-flops are used for storage of digital signals.

RS Flip-Flop

The basic circuit of the digital storage elements is that of the RS flip-flop. More complex flip-flops can be derived from this building block. The function of the RS flip-flop will be explained with the help of the transistor circuit of Fig. 5.28.

A positive pulse at input S sets the flip-flop by turning transistor T_1 on. Thereby transistor T_2 is turned off, and the output Q assumes a HIGH state, and correspondingly, the output $\bar{Q}$ is LOW. With the help of a feedback connection via resistor R_B this state remains stable for an unlimited time, even after the pulse at input S has been removed.

If the input R receives a positive pulse, the flip-flop will be reset correspondingly. Transistor T_2 is turned on, and transistor T_1 is turned off. The output at Q becomes LOW, and the output of $\bar{Q}$ HIGH. If positive pulses appear simultaneously at both inputs, the states of both outputs are undefined. The states of the outputs depend on very small time differences between the arrival of the input pulses. This combination of two arriving input pulses is therefore note allowed. The truth table for a RS flip-flop shows the relationship between the states of the input signals and output signals (lower part of Fig. 5.29).

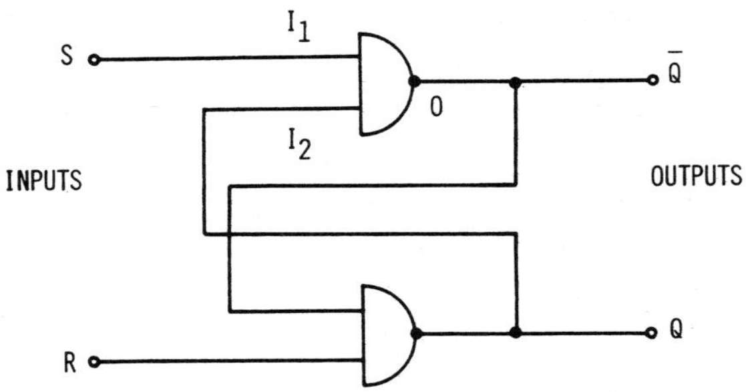

INPUT		OUTPUT
R	S	Q
L	L	Q_0
H	L	L
L	H	H
H	H	INDETERMINATE

FIG. 5.29 RS flip-flop with NOR gate.

RS flip-flops are obtained by connecting the output of one gate with the input of a second gate, and vice versa (NOR or NAND gates). Figure 5.29 shows the equivalent of the transistorized circuit of Fig. 5.28, however, made from two NOR gates, and Fig. 5.30 shows the symbol for the flip-flop. In many cases it is necessary to set a flip-flop synchronously with a clock pulse. For this purpose the RS inputs are AND gated with the clock input T.

D Flip-Flop

A D flip-flop is obtained when the R and S inputs are connected to the common input D, as shown in Fig. 5.31. By inverting the R input, it is possible to eliminate the not-permitted state HIGH-HIGH and LOW-LOW of the RS flip-flop. If a HIGH state is present at D,

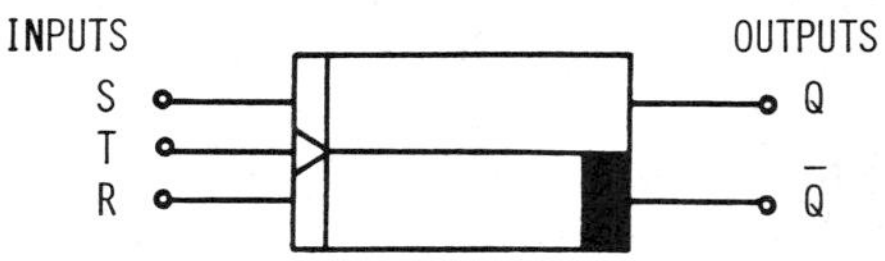

FIG. 5.30 Logic symbol of a clock-controlled RS flip-flop.

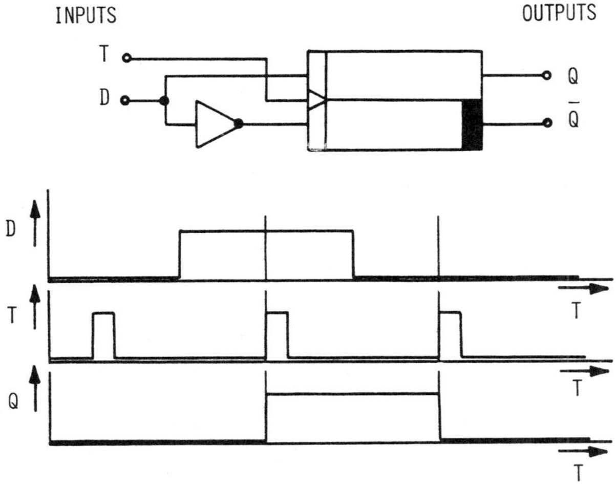

FIG. 5.31 D flip-flop and pulse diagram.

then upon arrival of a positive pulse edge at the T input, this condition is transferred to the Q output. There it is stored until the arrival of the next pulse at T. At this time the state of the D input is interrogated again. If it is at the state HIGH, nothing changes at the outputs. If D has returned to LOW, the outputs are reversed synchronously with the arrival of the triggering edge at T.

JK Flip-Flop

A JK flip-flop is obtained by connecting the Q output of a flip-flop to the K input and the $\bar{Q}$ output to the J input (Fig. 5.32). From

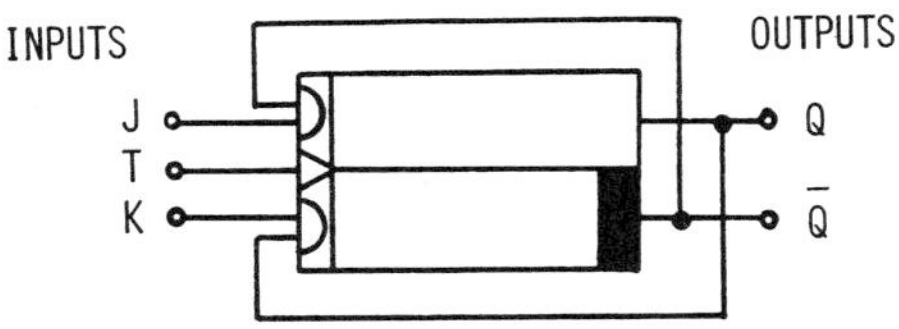

J	K	Q_{T-1}	Q
L	L	L	L
L	L	H	H
L	H	L	L
L	H	H	L
H	L	L	H
H	L	H	H
H	H	L	H
H	H	H	L

FIG. 5.32 JK flip-flop and function table.

the function table one can see that the JK flip-flop behaves exactly like the RS flip-flop, except when both the state of input S and the state of input R are HIGH. If both of the states of R and of S are HIGH, there will be no indeterminate state. This combination, however, will invert the states of the outputs of the flip-flop. Changes in the states as shown in Fig. 5.32 only take place upon arrival of a positive triggering edge at the input T. Here we speak of a positive-triggered JK flip-flop. There are also many other types of flip-flops, which can be derived from the basic designs shown here. Details can be found in the literature, for example, in Refs. 2 and 3.

5.4.3 Shaping of Pulses

In this section problems will be discussed which occur when pulses have to be reshaped. In order to switch digital circuit components, correctly shaped input trigger signals have to be provided. This is done by using pulses with a short rise time and an appropriate voltage level. In case a signal does not have the correct triggering properties, the requirements can be met by reshaping the signal before it is used for switching.

Schmitt Trigger

One of the important functions for the Schmitt trigger is to shape or reshape an irregular signal, which is unsuitable to perform a switching function, into a square wave. A Schmitt trigger consists of a two-stage amplifier interconnected with a feedback line (Fig. 6.33). Figure 5.34 displays the function of the Schmitt trigger when an irregular wave form arrives at its input.

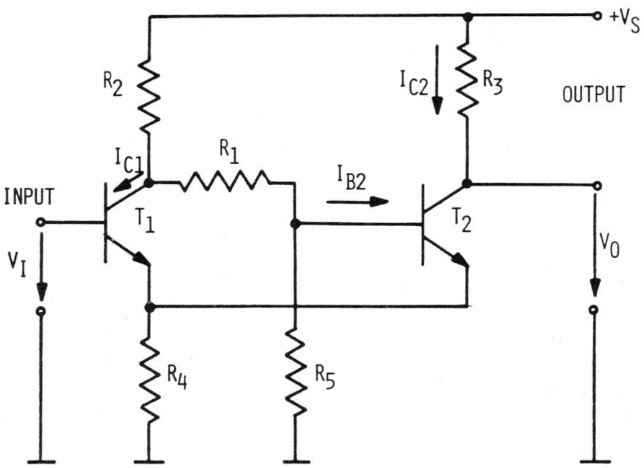

FIG. 5.33 Schmitt trigger with emitter-coupled transistors.

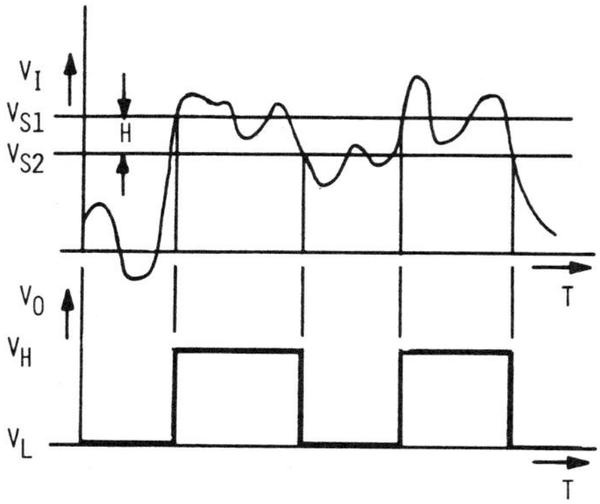

FIG. 5.34 Schmitt trigger waveform. V_I = input signal, V_{S1} = turn-on threshold, V_{S2} = turn-off threshold, H = V_{S1}-V_{S2} = hysteresis, V_O = output signal of Schmitt trigger, V_H = logic HIGH, V_L = logic LOW.

If the input voltage applied to the Schmitt trigger reaches the turn-on threshold V_{S1}, a logic HIGH will appear at the output. Only after the input voltage has fallen below the turn-off threshold V_{S2} will the output fall back to a logic LOW potential. This change of state occurs quickly. Thus, a good square wave is assured. A Schmitt trigger has a hysteresis between the turn-on level and the turn-off level (Fig. 5.35). For this reason, it produces very clean switching operations even when the input signal changes only very slowly in the vicinity of the threshold voltage.

The exact function of a Schmitt trigger will be explained with the help of a transistor circuit (Fig. 5.33 and 5.34). This circuit shows

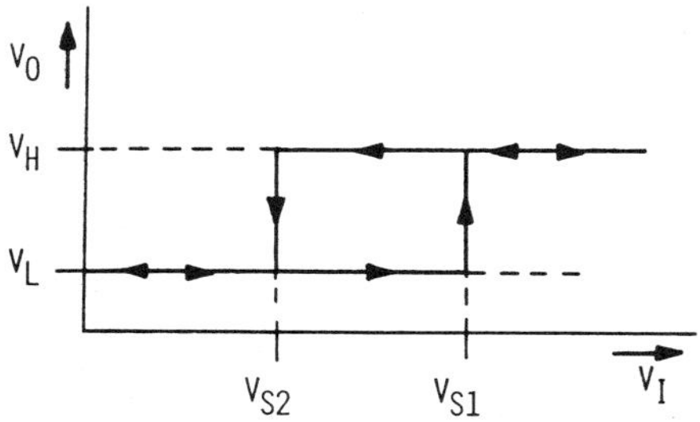

FIG. 5.35 Hysteresis of a Schmitt trigger.

a two-stage amplifier with a feedback loop. The feedback occurs via the common emitter resistor R_4. This causes transistor T_2 to be completely turned on or completely turned off. Transition states last only for a very short time. When the input voltage V_I is close to 0 V, transistor T_1 is turned off. A relatively high base current flows across the voltage divider R_2, R_1, R_5 and to the transistor T_2 and turns it on. Thus, a collector current I_{C2} flows across R_3 and R_4. This current causes a voltage drop across R_4, and the emitter of transistor T_1 becomes more positive than its base. T_1 is entirely turned off. This condition is stable. If the input voltage V_I exceeds the magnitude of the voltage at R_4, transistor T_1 begins to conduct. The collector current which now begins to flow increases the voltage across R_4 and simultaneously reduces I_{B2}. Since prior to this stage T_2 was fully turned on, the change of the base current does not effect the current I_{C2} immediately. When V_I further increases, the voltage at R_4 increases, and because of the feedback the base current I_{B2} decreases. At one point I_{B2} becomes so small that the transistor T_2 is no longer oversaturated, and I_{C2} begins to decrease. The input voltage V_I has then reached the turn-on threshold V_{S1}. From here on the events proceed very fast. Since I_{C2} decreases faster than I_{C1} increases, the voltage across R_4 decreases. As a result of the increasing voltage difference between this voltage and the input voltage, the input current to transistor T_1 increases independently, which in turn increases I_{C1}. The increase in I_{C1} results in an additional reduction of I_{C2} so that the described operation is further reinforced. This unstable state is ended when the base current I_{B2} is close to zero so that T_2 turns off. T_1 now operates in saturation. During the described very short unstable phase, also called toggle operation, the output voltage V_O changes from V_L, turned-off state, to V_H' turned-on state. When the input voltage V_I decreases, the reverse operation takes place. While I_{C1} decreases, the turn-off threshold V_{S2} is reached, and the base current I_{B2} begins to flow. With the help of the feedback across R_2, I_{C2} rises very fast. Thus, the output potential drops off from V_H to V_L very quickly.

The Schmitt trigger has a stable ON and a stable OFF state as well as an unstable transition state, limited by the threshold voltages V_{S1} and V_{S2}. The transition through this range is very fast during the turn-on and turn-off operations. A typical Schmitt trigger in the TTL family is the 74N132 integrated circuit. This element can handle very slow rising times of 1 V sec^{-1}.

Monostable Multivibrator

The characteristic feature of a monostable multivibrator is that it only has one stable output condition. When an input event occurs, for example, a trigger pulse arrives, this stable state will be maintained for an externally adjustable time period. Thereafter, the circuit flips

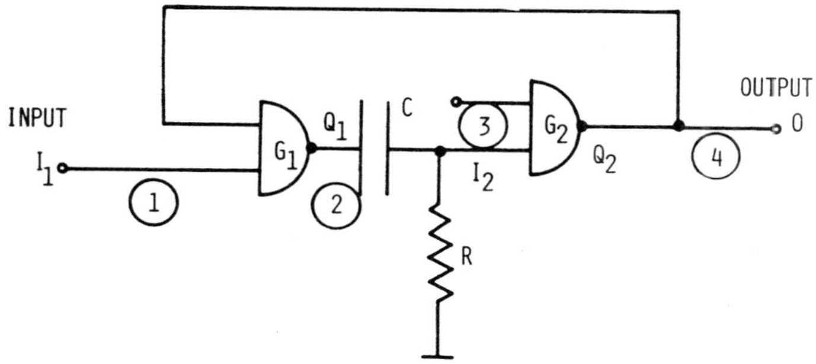

FIG. 5.36 Monostable multivibrator with NAND gate.

back to its stable condition. One can say that a monostable multivibrator maintains the input condition for an exact defined time. Such circuits are used when a signal of a defined form and length must be generated. Figure 5.36 shows a monostable multivibrator designed from two NAND gates. In the stable condition a HIGH signal prevails at the input I_1 (Fig. 5.37). Via resistor R one input of gate G_2 is grounded. The HIGH potential at the output of gate G_2 is fed back to gate G_1. Since both inputs of G_1 are at HIGH potential, the output Q_1 is at LOW potential. In this condition the capacitor C carries no charge. If a negative edge appears at input I_1, the toggle operation begins. The output of Q_1 becomes HIGH. This sudden change in voltage level is transmitted via C to G_2. Thus the output Q_2 of G_2

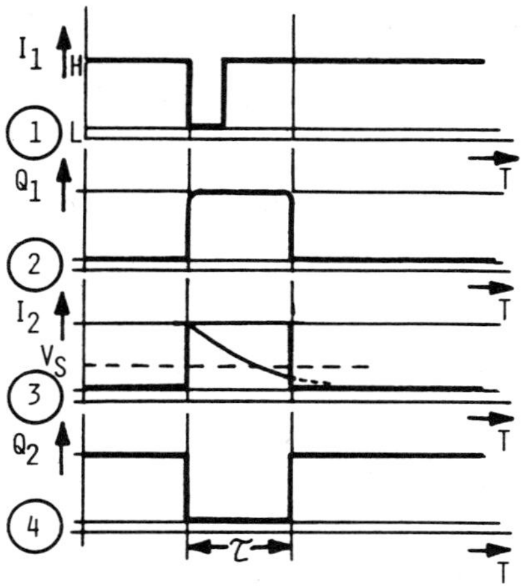

FIG. 5.37 Circuit waveforms of a monoflop.

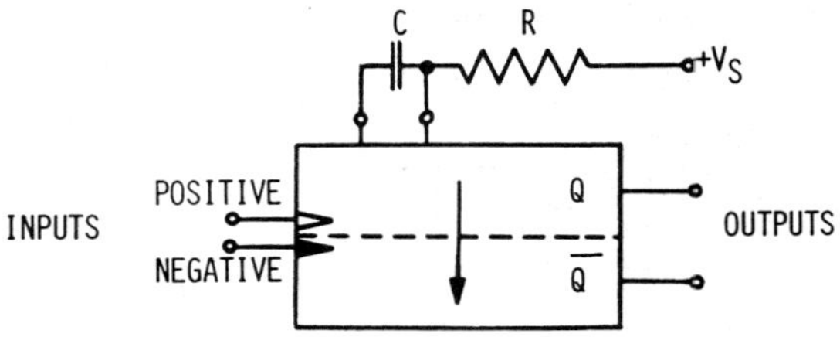

FIG. 5.38 Logic symbol of a monoflop.

becomes LOW. Now capacitor C will be charged via R. The voltage at I_2 drops off exponentially. As soon as this voltage has fallen off to the level of the threshold voltage V_S of G_2, this gate G_2 begins to switch again and the output Q_2 becomes high. The transition time depends on the dimension of the RC components. The active turn-on time τ can be described with the help of the following relation:

$$\tau = RC \ln 2 \tag{35}$$

Since the toggle operation is initiated by the trailing edge of the signal, this circuit is called the *trailing-edge-triggered monostable multivibrator*. Electronic elements which are suitable to perform this function are of the integrated-circuit series 74 N. Figure 5.38 shows their logic symbol. Integrated circuits differ from circuits of discrete design. There are specific types of monostable multivibrator which can be retriggered and other which cannot be retriggered. In the case of a retriggerable system, each new clock signal occurring during the unstable phase leads to a new trigger time τ. If such an element is provided cyclically with a short pulse whose length is shorter than the time τ, the output Q remains turned on. Only when the input pulse train stops will Q return to the stable condition. In the nontriggerable monostable multivibrator, pulses arriving during the unstable phase are ignored. Only after Q has returned to the stable condition can the element be retriggered by a new input pulse. Figure 5.39 shows this relationship.

Time-Delay Circuits

When controlling a technical process by computer often the switch-on or switch-off point of an event has to be delayed for a predefined time period. Monostable multivibrators are suitable for this task. Figure 5.40 shows a circuit having a variable turn-on delay, in this case realized with the integrated-circuit TTL 74123. An output signal at Q_2 can only be initiated when a positive edge appears at $\bar{Q}_1$. The delay time can be adjusted with the help of resistor R_1. In Fig. 5.41 a variable turn-off delay is shown. Here the monostable

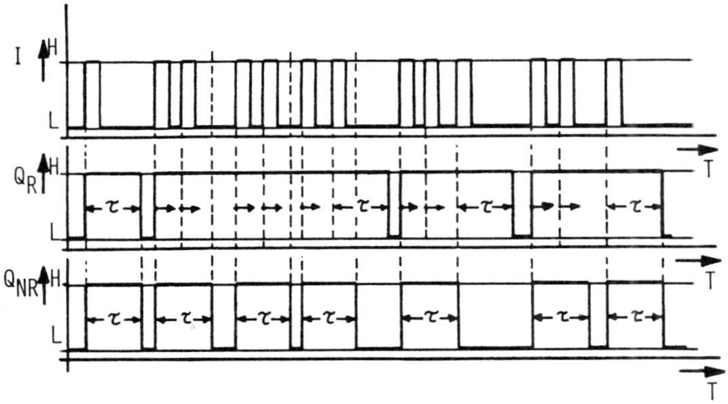

FIG. 5.39 Effect of an input pulse train on a retriggerable and nonretriggerable multivibrator: I = input pulse train, Q_R = output of retriggerable monostable multivibrator, and Q_{NR} = output of nonretriggerable monostable multivibrator.

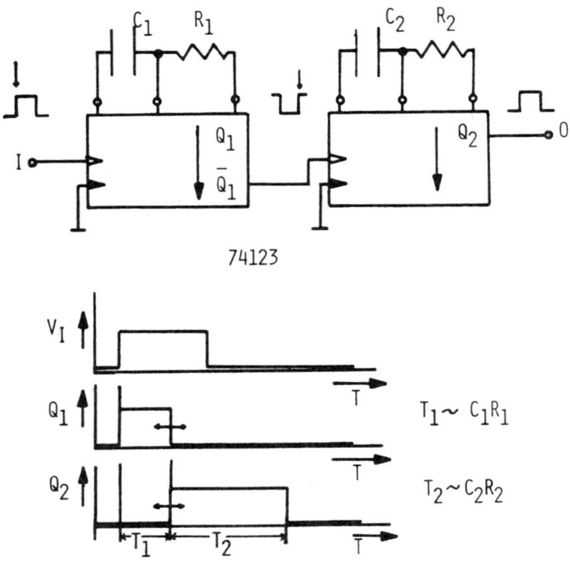

FIG. 5.40 Turn-on delay.

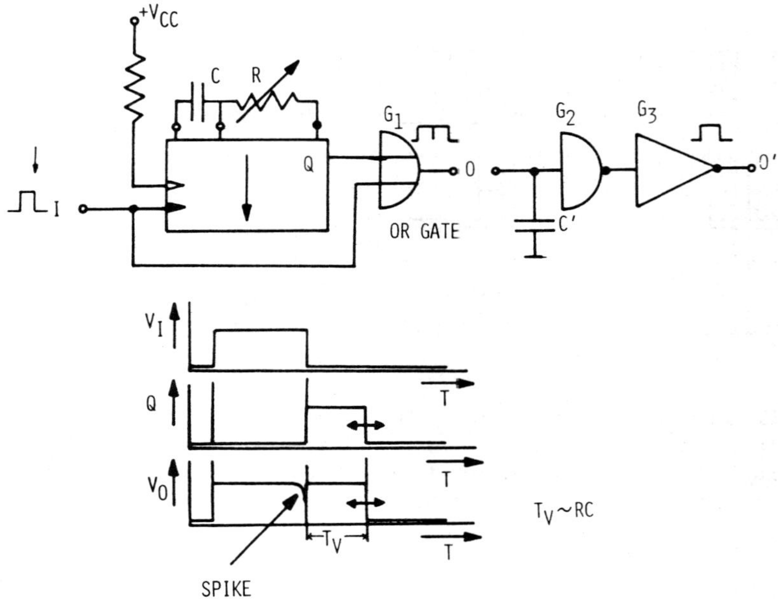

FIG. 5.41 Variable turn-off delay.

multivibrator is triggered by the trailing edge of the input signal V_I. With the help of an OR gate, both pulses are combined so that their length is added.

During turn-off of the input signal, the output signal must stay at the LOW state for approximately 50 nsec. The output signal of the monoflop has a delay of approximately 50 nsec, which induces noise in the OR gate. If the output pulse is not to be processed further by edge triggering, this short voltage spike can be disregarded. In the other case possible interference must be filtered out with a capacitor and a Schmitt trigger.

Debouncing of Contacts

One important use of pulse shaping circuits is for debouncing of contacts. In technical processes transducers for digital information are very commonly designed with mechanical reed contacts. They are typically contained in hand-operated switches, contactors, data input keys, microswitches, limit switches, relays, thermoswitches, etc. Inherent to all these elements is that they have no exactly defined turn-off and turn-on time. The change of state between a static off and static on signal cannot be obtained instantaneously, however; in general, sev-

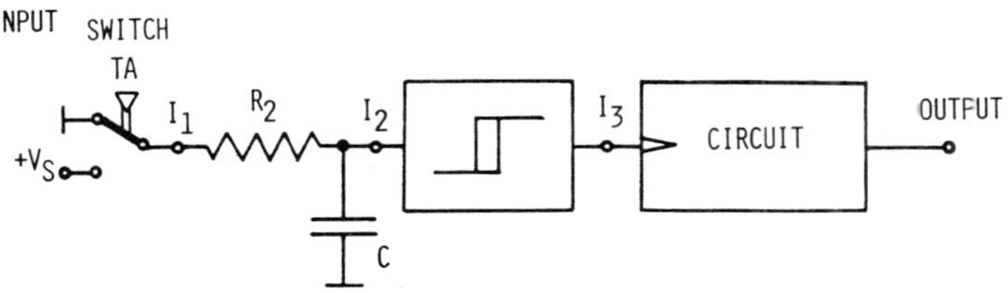

FIG. 5.42 Debouncing with the help of a capacitor and Schmitt trigger.

eral intermediate on and off states occur before a new stable state is reached. This contact bouncing is caused by mechanical vibration initiated by the transient behavior of the contact reed. However, when connecting a process to the computer care must be taken that all information available to the computer is defined exactly. This means that all information transferred from mechanical transducers must be converted into well-defined pulses. The simplest method to debounce a mechanical contact is obtained by using a capacitor (Fig. 5.42 and 5.43).

In the quiescent state the capacitor has no charge. When the key is depressed, the capacitor is charged via resistor R_2 by the transient pulses caused by bouncing. With the help of a correctly dimensioned time constant R_2, the time T_C becomes greater than the

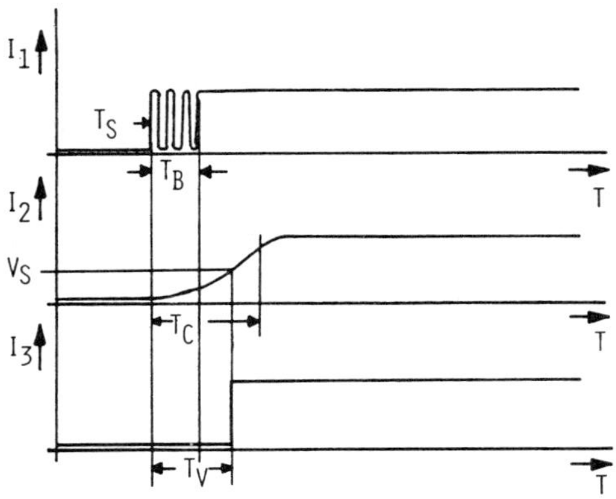

FIG. 5.43 Debouncing waveforms: T_S = turn-on point, T_B = bouncing time, T_C = time constant of RC_2, and T_V = delay time to activate Schmitt trigger.

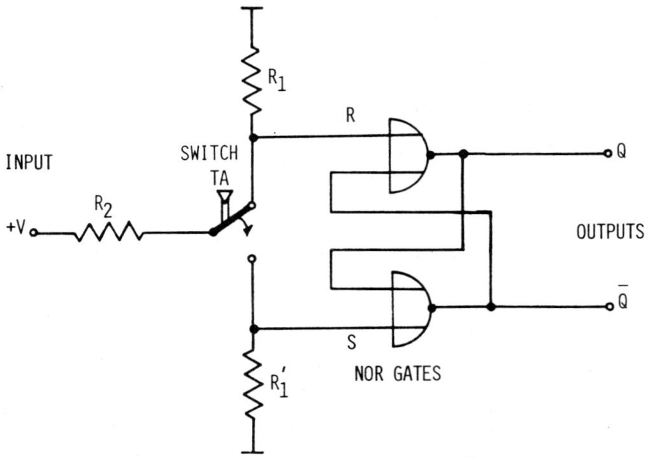

FIG. 5.44 Debouncing with RS flip-flop.

bouncing time T_B. The slow increase of the voltage at the capacitor (point I_2) prevents the unwanted pulses during the bouncing period from affecting the connected circuit. The Schmitt trigger shortens the rise time of the signal. Thus a well-defined input pulse is obtained for switching. When a debouncing circuit is designed one must take the delay time T_V to activate the Schmitt trigger into consideration.

Another commonly used circuit for debouncing a switch is shown in Fig. 5.44. Here a well-defined edge is obtained with the aid of a RS flip-flop. Resistor R_1 and R_2 are dimensioned such that the voltage dividers consisting of R_2 and R_1 or R_2 and R_1', respectively, at the input of each gate generates a HIGH signal. When the switch is open and R_2 is not connected to the input of the gate, the input signal is LOW. In the initial state shown, Q is LOW. When the key TA is depressed, both input gates go LOW during the switching period.

When the reed of the switch touches the lower contact for the first time, a HIGH potential will be at the S input and the output Q will assume a defined state HIGH. During bouncing of the reed contact only one input condition can exist. The RS flip-flop does not respond to different input combinations. When the key of the switch is released and the reed returns to the initial position, the reverse process occurs. Here also the output variables Q and Q are defined exactly at any time. The interrupted signal delivered by the reed (caused by bouncing) during the switching process will be replaced by a very clean LOW-HIGH-LOW transition.

If a pulse is to be generated whose transition period is independent of the hold time of the key of the switch, the already introduced mono-

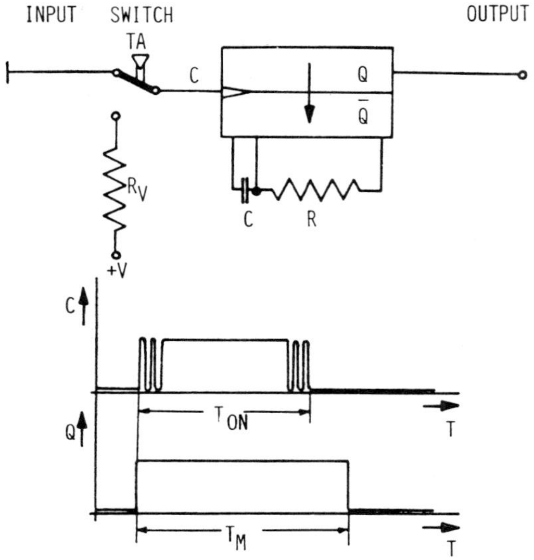

FIG. 5.45 Debouncing with monoflop.

stable multivibrator can be used (Fig. 5.45). When the first triggering edge arises, the output Q turns to HIGH potential before bouncing begins. Variations of the input signals are no longer of importance. After the time T_M has expired, which is defined by the RC components, the monostable multivibrator switches independently to LOW. In order to use this circuit for debouncing, the time T_{ON} during which the key is depressed must be shorter than the turn-on time of the monostable multivibrator. Otherwise, upon release of the key the bouncing pulse again turns on the flip-flop.

5.4.4 Synchronization of Pulses

External process signals often have to be synchronized with the operation of the computer. Here there must be distinguished first between a constant or quasi-static process signal for which the characteristic switching time is substantially greater than the clock cycle of the system and second between an event represented by a single edge of a pulse. In the first case, the event is usually characterized by a defined signal level. Here we speak of signal-state control. In the second case, we speak of signal-edge control.

For transfer of state-controlled signals the D flip-flops may be used as it was described in Fig. 5.31. When the process signal is available at the D input, it can be strobed into the flip-flop by each

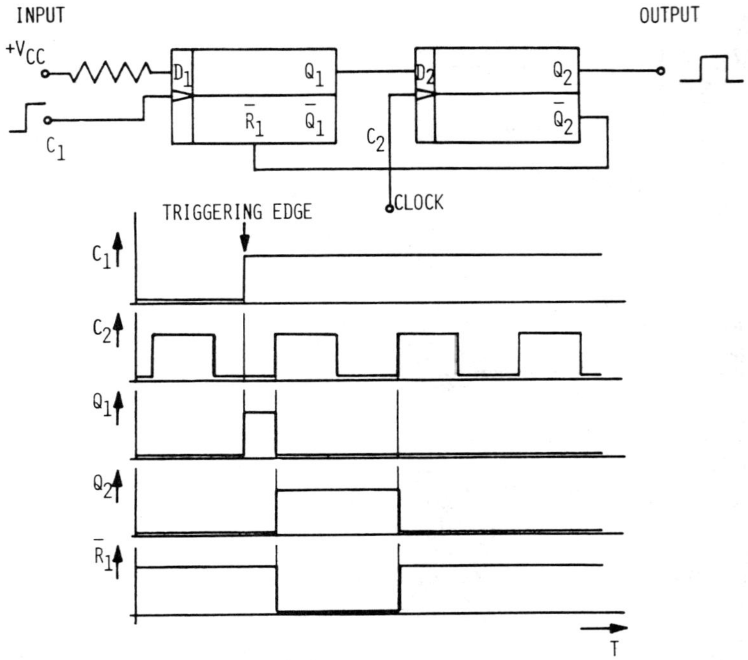

FIG. 5.46 Pulse synchronization of asynchronous signals. I_B = base current, I_C = collector current, V_{CE} = collector-emitter voltage, R_L = load.

arriving pulse edge of the clock signal. An asynchronously arriving trigger edge can, for example, be synchronized with the help of the circuit SN7474, which has two D flip-flops (Fig. 5.46). The D_1 input of the first flip-flop is kept HIGH at all times. The positive signal edge at C_1 obtained from the process puts Q_1 on HIGH potential. The next leading edge of the system clock pulse transfers the HIGH signal at input D_2 of the second flip-flop to Q_2. At the same time Q_2 turns LOW and resets the first flip-flop via the R_1 input. Upon arrival of the next leading edge, Q_2 is reset to LOW once again. Thereby upon arrival of a positive triggering edge at C_1, a synchronous HIGH pulse appears at Q_2.

The circuit used for these flip-flops must also be provided with a D input and a reset input. In this case we speak of combined D and RS flip-flops.

5.5 INTEGRATED-CIRCUIT FAMILIES

5.5.1 Introduction

Integrated-circuit families is the name given to a series of integrated-circuit elements which all meet defined specifications and conceptually are based on the same design principles. The integrated-circuit families are distinguished by their

Logic level
Clock rate
Noise margin
Power dissipation

During the evolution of this technology over the last 10 years, manufacturers introduced many different circuit concepts, not all of which have gained practical importance. Only the most important groups of integrated circuits will be explained in this chapter.

5.5.2 Bipolar Technology

This group includes all integrated-circuit families which have bipolar transistors as active elements (Fig. 5.47). General characteristics of this technology are

High clock rate
Relatively high power dissipation
Large space requirements for integration
Relatively high costs per active element

Direct-Coupled Transistor Logic (DCTL)

In this logic family the input signal is directly introduced into the base of the transistor (Fig. 5.48). The disadvantages of the family lies in the fact that the transistor has a relatively high base current. This can pull down the available voltage level at the output to approximately 0.9 V (HIGH) when several stages are connected to it in parallel. The static noise margin is then reduced considerable, 0.5 V when HIGH, 0.2 V when LOW. The *noise margin* is defined as the difference between the voltage level which represents the static output state (HIGH or LOW) of a preceding gate and the voltage level at the input of the following gate, at which its output changes state (Fig. 5.49).

If the transistors in a DCTL gate are not completely identical, the transistor with the lowest base emitter potential draws the main current, which leads to current hogging. For very high potential differences between the base and emitter, the gate is no longer functional. This family is being replaced by newer circuit technologies.

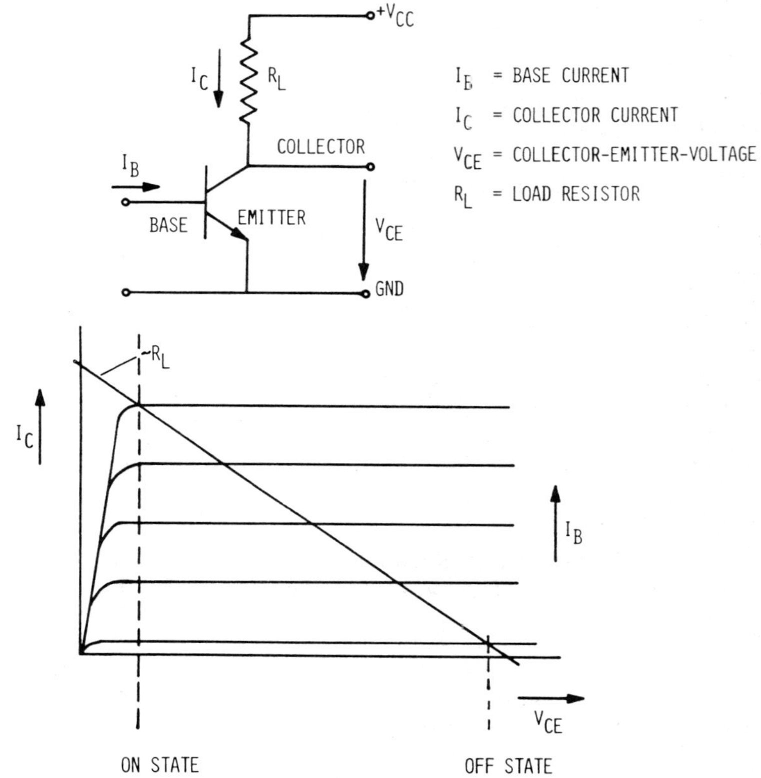

FIG. 5.47 Bipolar transistor working as a switch.

Resistor-Transistor Logic (RTL)

The problem of current hogging can be reduced in the RTL circuit with the help of base resistors (Fig. 5.50). In this circuit the number of gates which can be connected to the output is much higher than in the DCTL family (fan out). However, the clock rate with which the circuit can be triggered is limited by the time constant, which is determined by the values of the base resistors and the input capacity of the circuit. The disadvantages of the small voltage swing which can be tolerated between the HIGH and LOW output level, as well as the small noise margin, remain the same as with the DCTL family. Today RTL gates have no further significance.

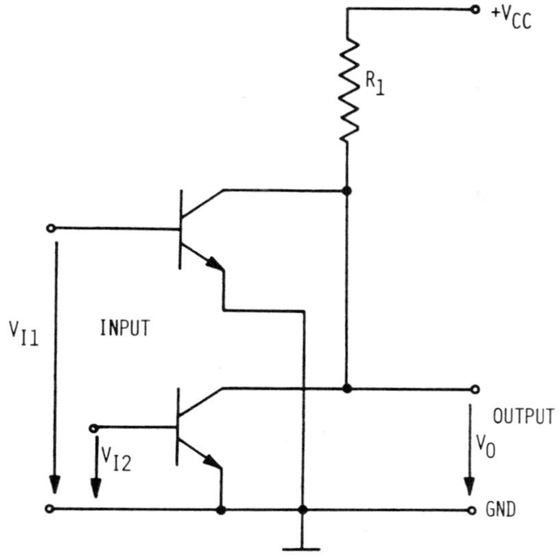

FIG. 5.48 DCTL-NOR gate.

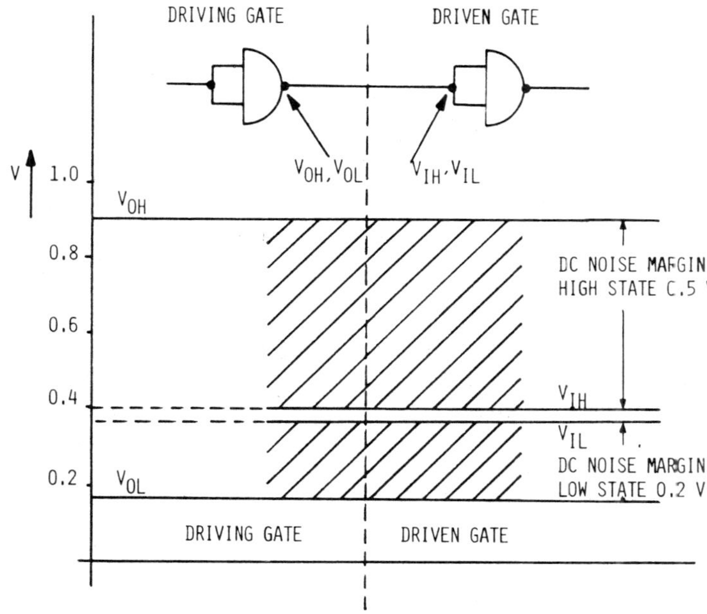

FIG. 5.49 DC noise margins of a DCTL gate.

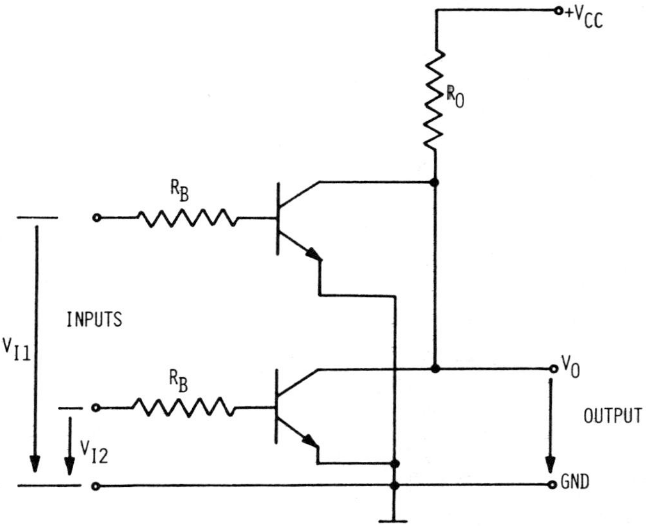

FIG. 5.50 RTL-NOR gate.

Diode-Transistor Logic (DTL)

The diode-transistor logic is the forerunner of the TTL technology commonly used today. The supply voltage $+V_{cc}$ is 5 V. Since the logic level of both the DTL and TTL families are the same, pin-compatible circuit components can be directly interchanged. The input connections to the DTL circuits are made via diodes (Fig. 5.51). For open inputs or when V_{I1} and V_{I2} are at HIGH potential, there practically is no input current flowing through the diodes D_1 and D_2. Current flows across the resistor R_S via both the diodes D_{S1} and D_{S2} to resistor R_1 and into the base of the transistor. Thereby the transistor is turned on. At both anodes of the input diodes D_1 and D_2 lies a potential which is equal to the sum of the forward voltages of both diodes D_{S1} and D_{S2} plus the base emitter voltage of the conducting transistor. This value is approximately 2.1 V and that of the output voltage V_O is approximately 0.2 V (LOW). If the input voltage drops below the threshold voltage of a gate, the diode connected to this gate conducts. The voltage at the anode drops to 0.7 V. The voltage drop across both diodes D_{S1} and D_{S2} reduces the flow of the transistor base current to a level at which the transistor turns off. The output level jumps to V_{cc} (HIGH).

It can easily be seen that the threshold of the DTL gate is 1.4 V with the two diodes D_{S1} and D_{S2} in the circuit. The output voltage

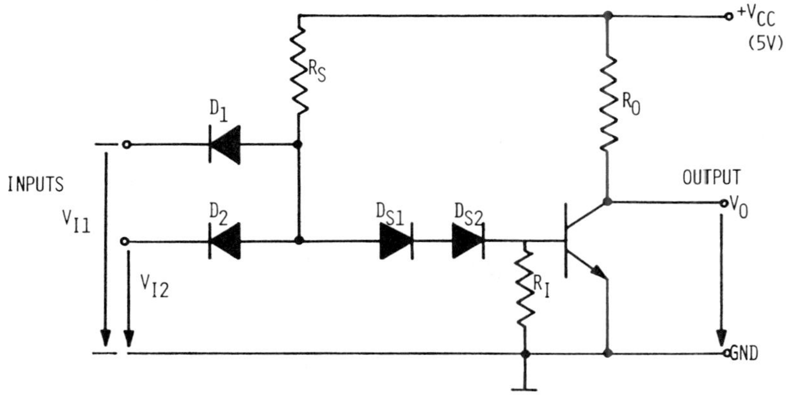

FIG. 5.51 DTL-NAND gate.

swing lies between 0.2 (LOW) and 4 V (HIGH). Thereby a substantially improved noise margin is obtained in comparison to the prior discussed families. The progagation delay of the DTL gate is approximately twice as high as that of the RTL gates (typically 25 nsec), which is due to the ohmic resistance of the diodes D_{S1} and D_{S2} located in the connection line leading to the base of the transistor. Today the DTL technology is almost completely compatible with TTL logic and has therefore been replace by it when used for interface circuits.

High-Threshold Logic (HTL)

For applications under extremely unfavorable environmental conditions the HTL logic family can be employed. This family is characterized by its very high static and dynamic noise margin. This advantage, however, is offset by the large propagation delay time of approximately 150 nsec. Figure 5.52 shows the fundamental circuit of the family, and Fig. 5.53 shows the transition diagram. The high noise immunity is achieved with the help of a zener diode at the gate input and a supply voltage of 15 V.

Transistor-Transistor Logic (TTL)

Presently the TTL circuit family is the most commonly used family and there is a large variety of different circuits available. Figure 5.54 shows the fundamental design of the TTL-NAND circuit. The supply voltage ($+V_{CC}$) here equals 5 V. The functional concept of the NAND gates shown here is essentially the same as that of the DTL circuit. In place of the discrete diodes, however, several diodes of the multiemitter transistor type are used. This has the advantage that many

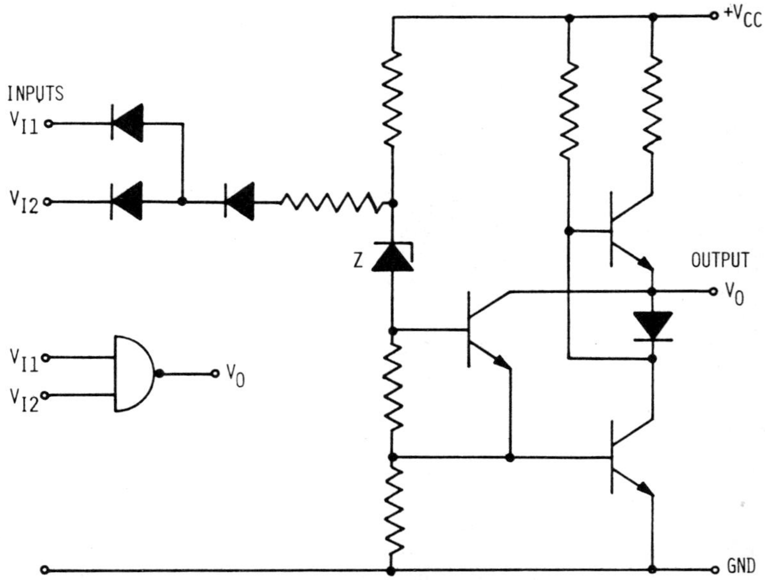

FIG. 5.52 HTL-NAND gate (+V_{CC} = 15 V).

components can be integrated on one semiconductor crystal. Figure 5.55 shows the transition diagram of this basic TTL circuit. The threshold lies at approximately 1.4 V, and the guaranteed noise margin at approximately 0.4 V.

Commercially available TTL gate circuits contain an output amplification stage in order to obtain the highest possible fan-out. The lowest guaranteed fan-out value is 10 TTL gates. Special-purpose drivers, however, allow the connection of as many as 50 gate inputs or even more. Figure 5.56 shows the design of a typical TTL-NOR gate.

The TTL family has a relatively short delay time (typically 5 to 30 nsec). This time depends on the saturation properties of the output transistors. The advantages of the TTL technology are its relatively high fan-out as well as a high noise margin. The commercial availability of a large number of different circuits makes the family of interest for many applications. The disadvantage is the high power consumption (for example, in comparison to MOS logic) and the large amount of space needed to integrate the circuit on one semiconductor chip. This problem is mainly caused by the large dimensions of the resistors in the gate circuits; they occupy a considerable area on the chip.

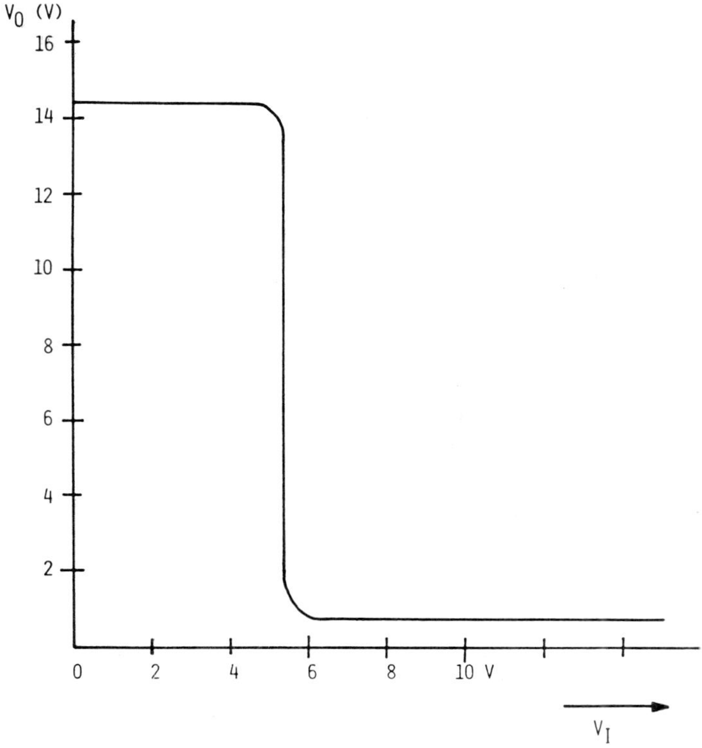

FIG. 5.53 Transition diagram of a HTL gate.

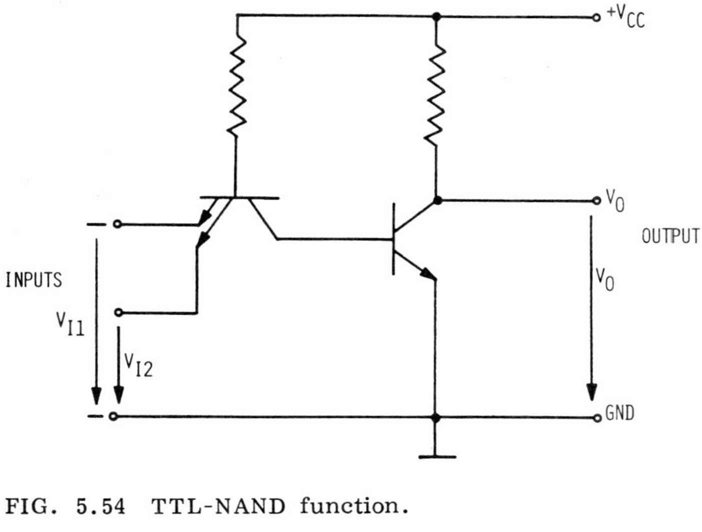

FIG. 5.54 TTL-NAND function.

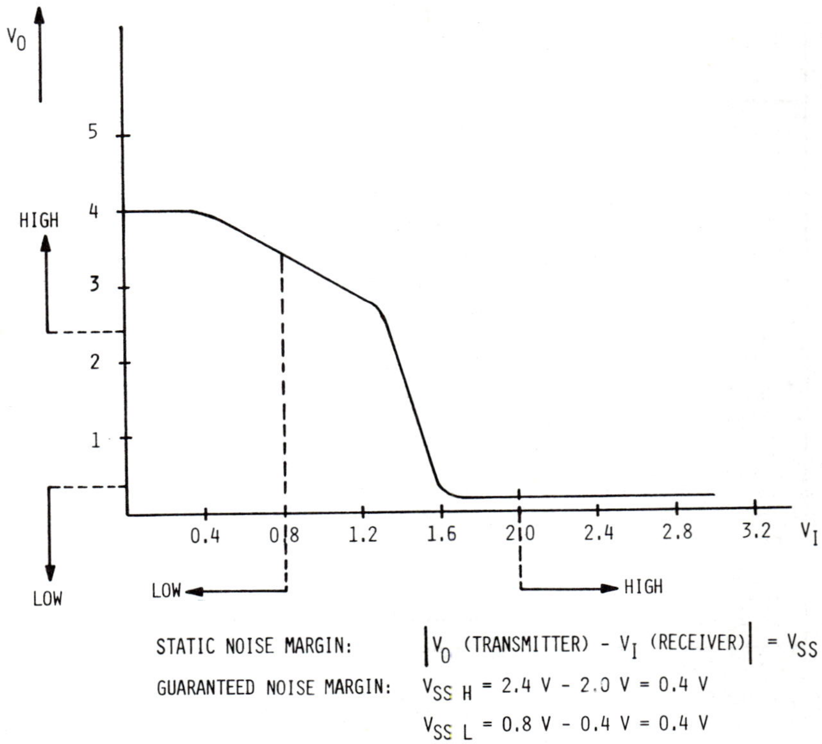

FIG. 5.55 Transition diagram of a TTL gate.

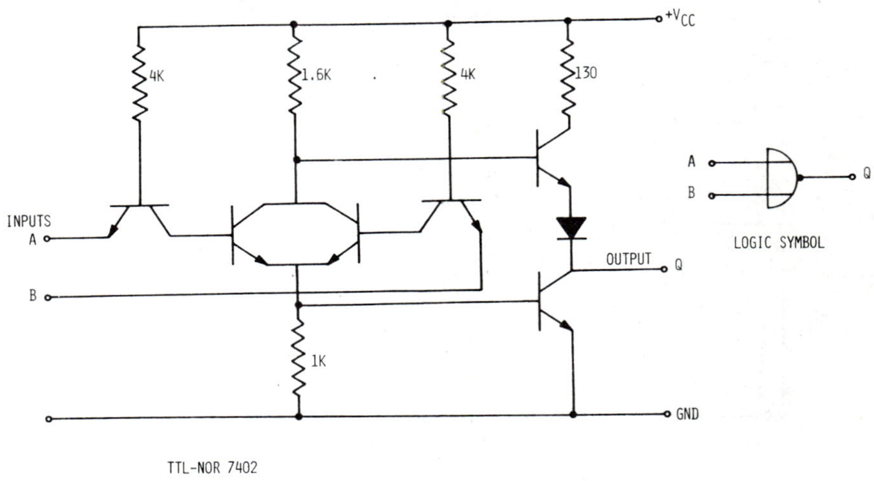

FIG. 5.56 Typical TTL-NOR gate.

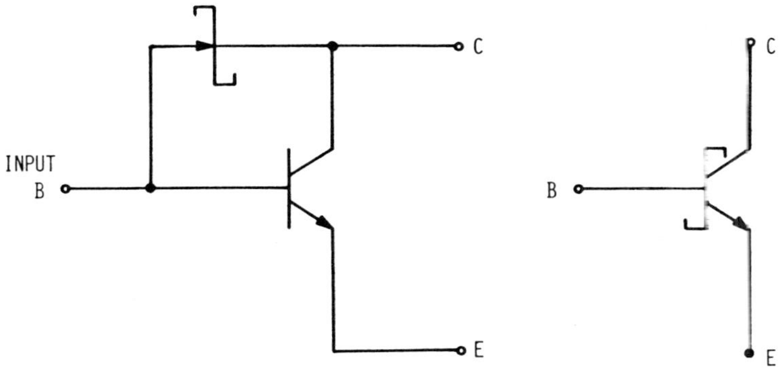

FIG. 5.57 Schottky transistor and circuit symbol.

Schottky TTL Logic

A newer development of the standard TTL series is the Schottky TTL series. In this design some selected transistors of the TTL circuit are built as the so-called Schottky transistors. In a Schottky transistor a Schottky diode lies between the base and the collector. It limits the current flow into the base to a predefined value by holding the base collector voltage constant when it has reached the for-

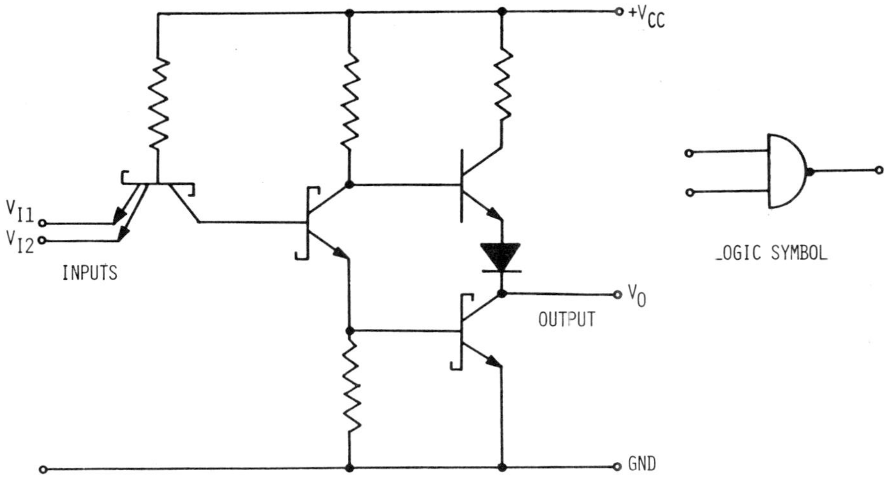

FIG. 5.58 Schottky-TTL-NAND gate.

ward voltage. Thus the transistor will not go into saturation. By this method a propagation delay time of 9 nsec is obtained for the LS series and 3 nsec for the extremely fast S series of the TTL family. Figure 5.57 shows a Schottky transistor and Fig. 5.58 a Schottky gate.

Emitter-Coupled Logic (ECL)

The ECL technology was developed for extremely fast switching operations, as they occur in central processing units of large computers and in special interface components. Here the obtainable propagation delay time per gate is about 0.1 nsec. Figure 5.59 shows the fundamental circuit.

The emitters of the input stages T_1 and T_2 are connected to the reference stage T_3; whereby the resistor R_3 has the function of a constant current source. The reference potential for T_3 is supplied by the two diodes D_1 and D_2, the transistor T_4 and the corresponding resistors. This reference potential also determines the gate threshold. The transistors T_5 and T_6 and the corresponding emitter resis-

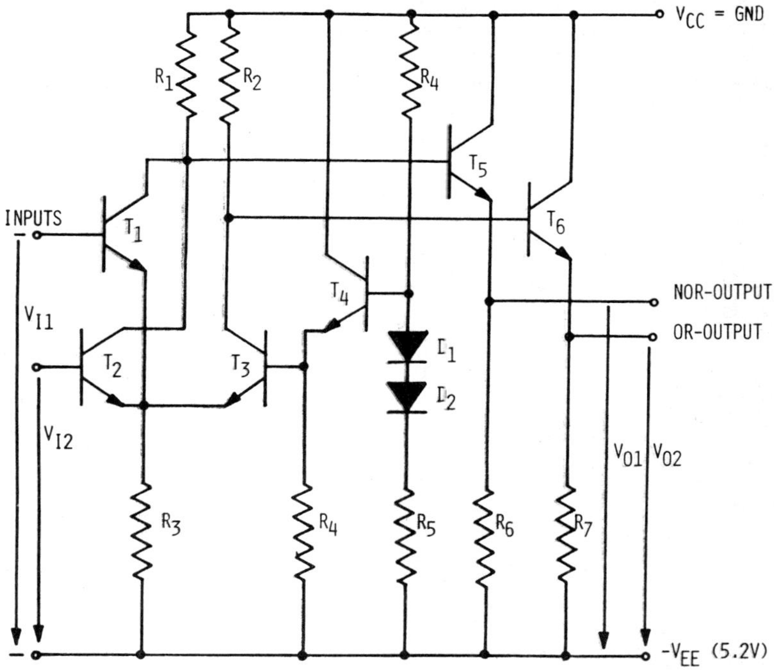

FIG. 5.59 ECL-NOR-OR circuit.

tors operate as an emitter follower. Thus, the outputs have a very low impedance.

If V_{I1} and V_{I2} are near 0 V (LOW), the transistors T_1 and T_2 are turned off, so that T_5 conducts and the NOR output is HIGH. If one of the input voltages becomes more positive than the reference voltage at T_3, then the corresponding input transistor turns on; whereby the collector potential decreases and T_5 turns off. The NOR output becomes LOW. The reason for the low propagation delay time of this type of gate lies in the fact that saturation of all transistors is prevented with the help of the emitter coupling. Because of the low impedance output, ECL gates can drive up to 30-gate inputs. However, the low ohmic resistances cause higher power dissipation. Typical levels for ECL gate are

$V_{CC} = 0\ V \quad V_{EE} = 5.2\ V$

Threshold: -1.175 V
HIGH: -0.7 to -1.025 V
LOW: -1.325 V to V_{EE}

Integrated Injection Logic (I^2L)

A modern European development is the bipolar logic family I^2L, also called merged transistor logic (MTL). It is characterized by extremely low space requirements when integrated on a chip and, additionally, by the small power dissipation it has a propagation delay of approximately 2 nsec.

Figure 5.60 shows a NAND gate designed in I^2L technology. The gate consists solely of two complementary transistors. The absence of

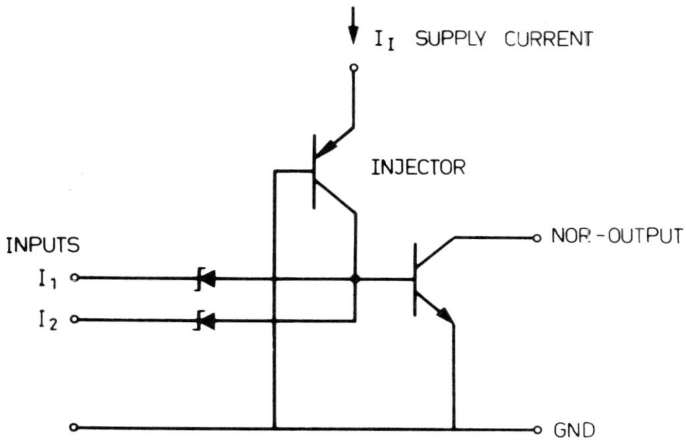

FIG. 5.60 I^2L(MTL) NOR gate.

load resistors, which are responsible for the large space requirement and high power consumption in the TTL family, is apparent. In place of a resistor the lateral PNP transistor is used, whose emitter is called an *injector*. By means of the supply current, which flows into the injector, the effective load resistance, and also the propagation delay time can be controlled within wide limits. The I^2L process is suitable for large-scale integrations because of its small space requirement and power dissipation. Several microcomputer chips using this technology are already on the market (such as the SBP400 and SBP9900).

5.5.3 MOS Technology

This group of integrated-circuit families is characterized by the use of MOS field-effect transistors as active elements. A distinction is made between P-channel MOS and N-channel MOS transistors, as well as between enhancement and depletion types. For physical details, see Delholm [4].

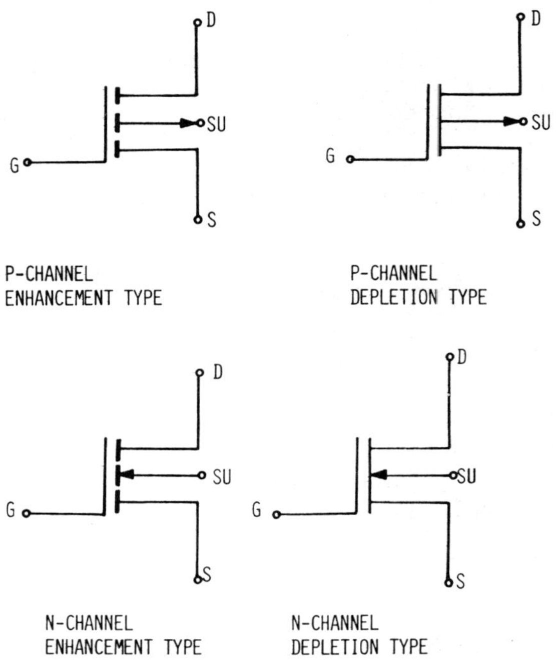

FIG. 5.61 MOS field-effect transistor: G, gate; D, drain, S, source; SU, substrate.

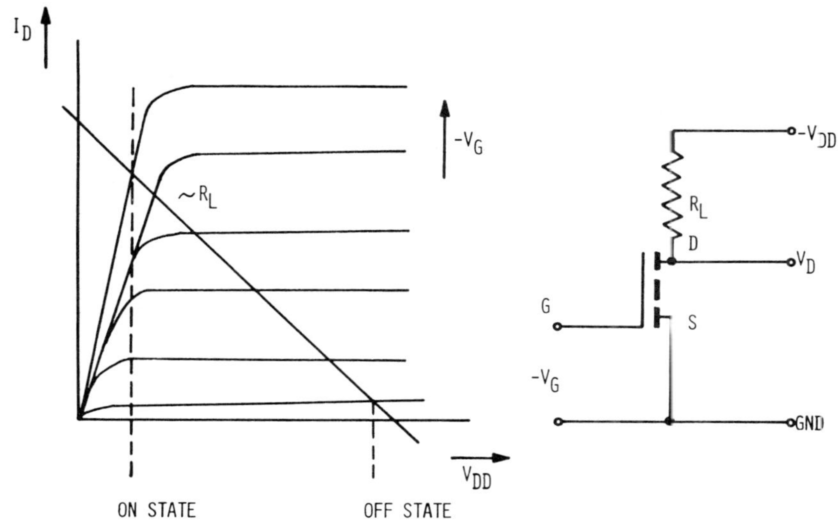

FIG. 5.62 MOS transistor operating as a switch. I_D = drain current, V_G = gate voltage, V_{DD} = drain voltage source, R_L = load resistor.

Figure 5.61 shows the four most common MOS transistor types. In modern integrated circuits usually N-MOS transistors of the depletion type are used. Figure 5.62 shows the voltage-current characteristics of a MOS transistor.

The main characteristics of logic circuits designed with MOS transistors are

Average propagation delay times
Very low input currents
Extremely high input impedance
Low power dissipation
Small space requirement
Low cost per integrated element

Fundamental Circuits in P-MOS and N-MOS Technology

The original integrated circuits based on MOS technology consisted of the so-called P-MOS enhancement type. When producing the circuit a N-doped silicon substrate is taken, in which a P-doped drain source diffussion is embedded. The control potential at the gate of the P-MOS transistor has to be more negative than the potential of the substrate. The current-carrying capability of the P-MOS transistor

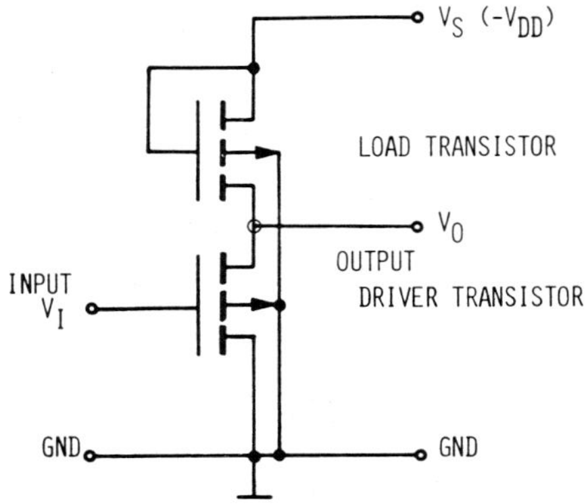

FIG. 5.63 MOS-inverter base circuit.

is obtained by doping it with positively charged current carriers (holes). The holes have less mobility in the semiconductor material than the negative charge carriers (electrons). The propagation delay time of the P-MOS elements is relatively long.

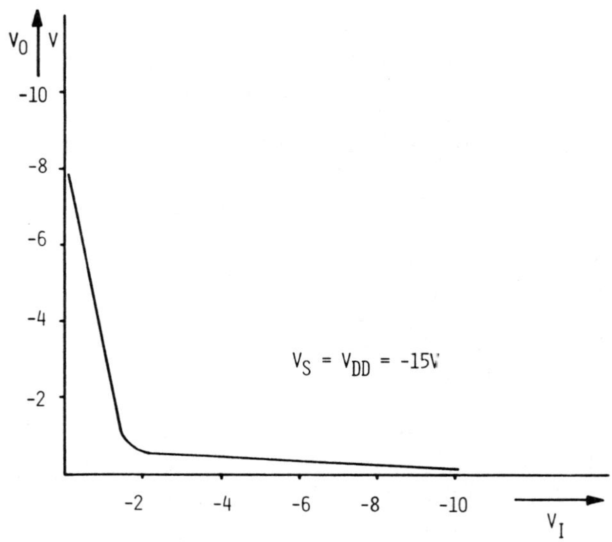

FIG. 5.64 Static transition diagram.

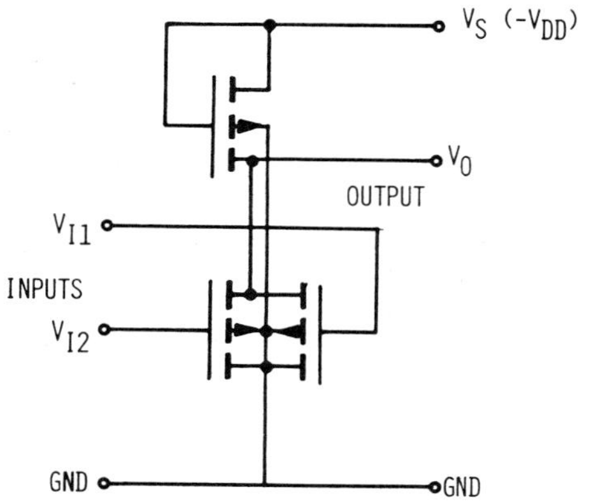

FIG. 5.65 MOS-NOR gate.

For N-MOS transistors the substrate is P doped. The current carrying capability is provided by electrons. Modern MOS processors are an advanced development of the N-MOS technology.

Figure 5.63 shows a MOS transistor used as a switch. As previously mentioned, load resistors occupy a large amount of space when integrated on a chip. Therefore, the MOS inverter circuit (Fig. 5.63) has its load resistor replaced by an additional MOS transistor. When this load transistor is carefully laid out, an accurate resistance value is obtained. It lies within the range of 10 K. A typical transition curve for this circuit is shown in Fig. 5.64. By adding a second driver transistor to the circuit, the inverter can easily be changed to a NOR gate. To obtain the same resistance value as in the inverter circuit, a change must also be made to the load transistor (Fig. 5.65). The P-MOS and N-MOS technologies are currently used in many circuits especially in large-scale integrated digital memories. Integrated components containing only single gates are seldom designed.

Static MOS Registers

Large-scale integrated circuits such as registers, memories, decoders, etc., designed in MOS technology, can in principle be manufactured to any size by repeatedly using the basic elements shown in Fig. 5.61. For each inverting stage the output voltage is a function of the input voltage specified by the static transition curve. The function of such circuits is called *static*. The memory function of flip-flops designed in this manner can be maintained as long as the supply voltage is connected to it. This is similar in the bipolar technology.

Dynamic MOS Registers

Static MOS registers, broadly viewed, have the same properties or behavior as bipolar circuits. The gate current to actuate one MOS stage, however, is smaller than the current needed to operate a bipolar element. In the dynamic MOS technology the gate current does not have to flow constantly (Fig. 5.66). One utilizes the fact that MOS transistors have a relatively high gate capacitance, which makes it possible to store a charge for a certain time period. To actuate a MOS transistor for a longer period of time, clock pulses are applied cyclicly to the gate. A load current flows only during the active clock phase. The charge in the gate maintains the state of the gate during inactive intervals. Thus it is possible to reduce the power consumption to operate the circuit. The amount of power reduction depends on the duration of the on-and-off cycle of the clock signal. Figure 5.66 shows one register stage designed in the so-called dynamic two-phase technology. The area inside the rectangle indicated by the broken lines shows both inverter stages. The transistor T3 and T4 are used as resistors and actuated by the two clock phases ϕ_1 or ϕ_2.

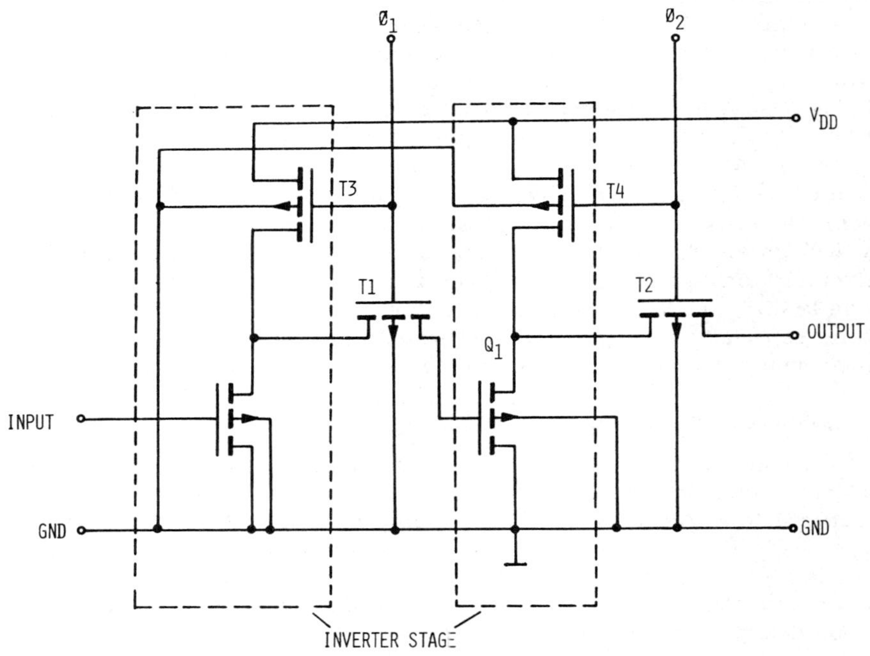

FIG. 5.66 Dynamic two-phase MOS register stage.

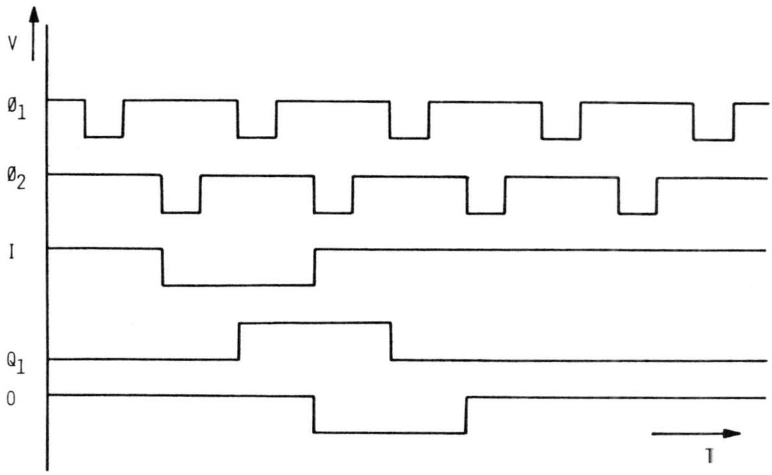

FIG. 5.67 Two-phase MOS storage.

There are two switching transistors T1 and T2 which transfer the information at the correct time to the next stage. Figure 5.67 shows the pulse diagram of a two-phase MOS register stage.

An important characteristics of the dynamic MOS technology is the lower cutoff frequency limit for the clock pulse which must be maintained. In order to make sure that the potential of the gate capacitance does not fall below the switching threshold, the delay time between two clock pulses may not exceed a predefined period. This sets the lower cutoff frequency limit to 5 and 10 kHz.

A disadvantage of the dynamic two-phase technology is that for high integration the resistance value of the load transistors must be exactly maintained. This disadvantage is overcome in the dynamic four-phase technology, in which all transistors are designed as switching transistors. Thus all transistors can be laid out uniformly to have the smallest possible dimension. Figure 5.68 shows one register stage built from four-phase MOS technology. There is no dc-voltage supply necessary. The low load currents are taken from the clock generator. In Fig. 5.69 the time events are indicated by arrows at which a logic level is interrogated for its transfer to the next stage. It can be seen that such circuits can only operate together with elements of the same type. For other logic levels they need a converter at the input and output. This technology is thus only suitable for large-scale integrations. Four-phase MOS stages are the fundamental building blocks of modern large-scale integrated circuits, for example, microprocessors.

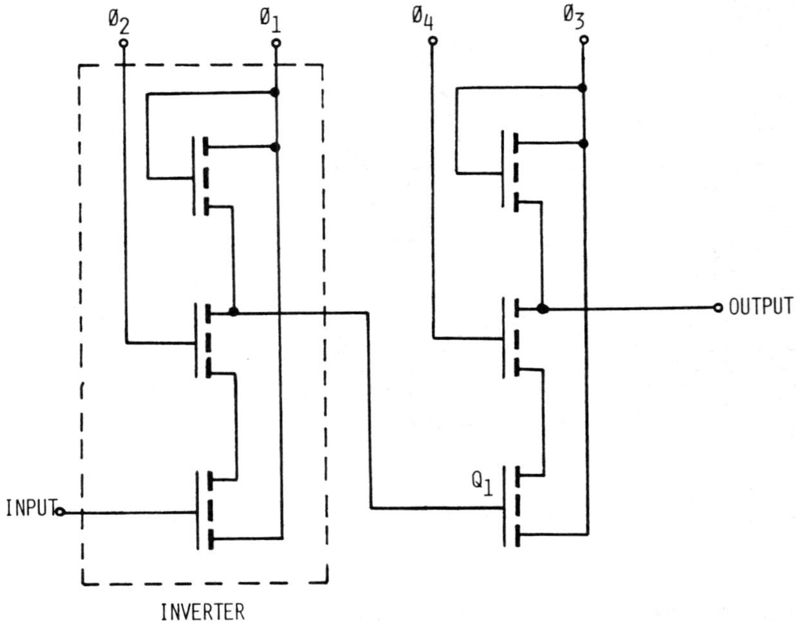

FIG. 5.68 Four-phase MOS register stage.

Usually all driver stages, as well as the elements for generating the four clock phases are integrated on the same semiconductor chip. Memories built in this manner today already have a capacity of 64,000 bits per integrated circuit. The shape of the signal between two active clock pulses is not very critical. Therefore, cyclical interference occurring at the low state of the signals between ϕ_1 and ϕ_2 has no influence on the function of the register. The interference is generated at regular intervals during which the gate capacities are charged. It must be suppressed when the information is transferred to external logic circuits.

C-MOS Logic

In C-MOS technology (complementary metal-oxide semiconductor) one P-channel transistor and one N-channel transistor are connected to form a complementary circuit. Thereby low power consumption, high noise immunity, and insensitivity to temperature changes are obtained. The supply voltage can be in the range of 3 to 16 V. The voltage levels needed for operating TTL elements also falls within this range. Therefore C-MOS logic to a great extent is TTL compatible,

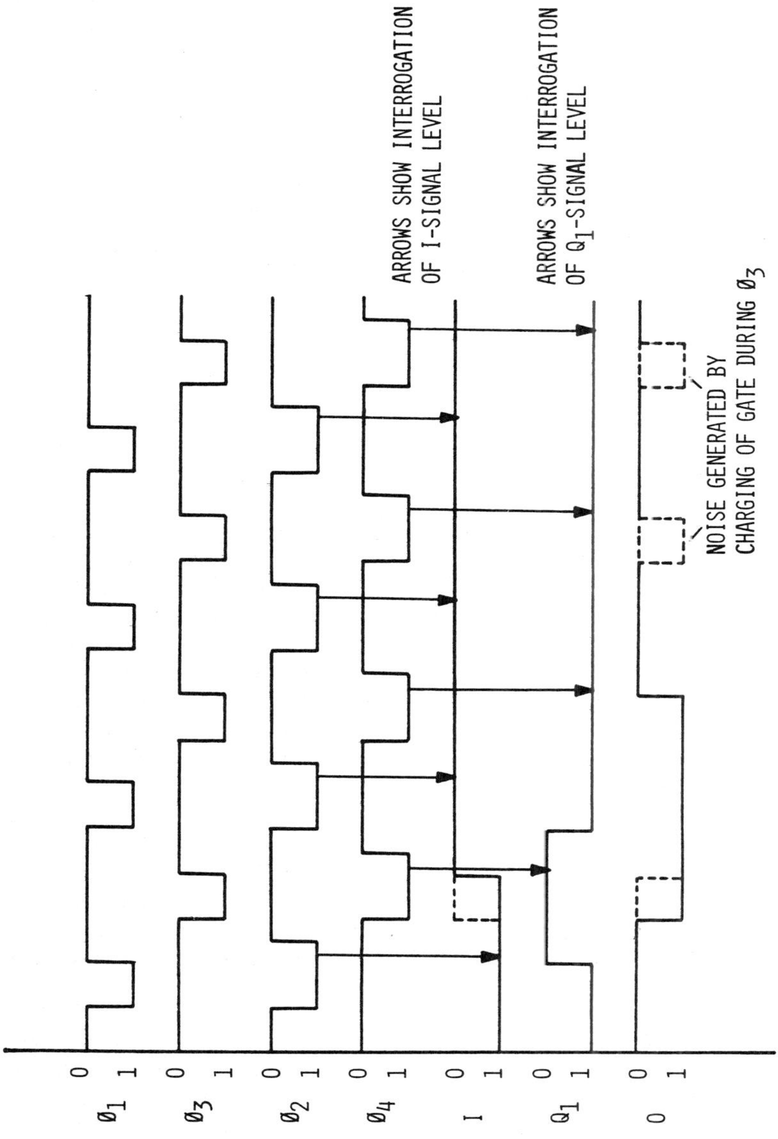

FIG. 5.69 Waveforms of dynamic four-phase MOS register.

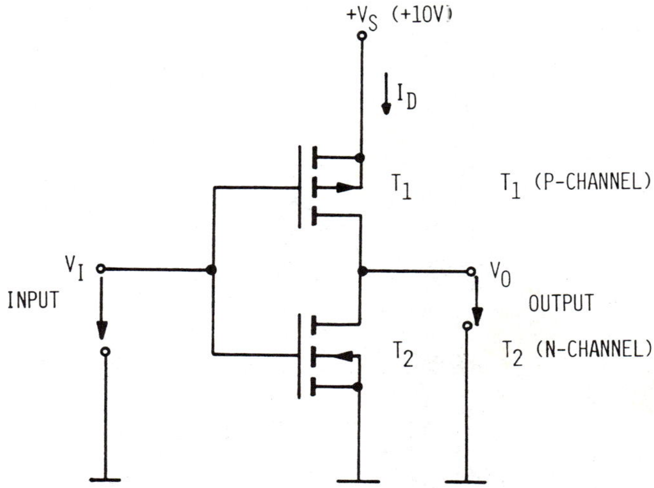

FIG. 5.70 CMOS base circuit (inverter).

although it has a much lower power consumption. For this reason this logic family replaces the TTL technology in many applications.

Figure 5.70 shows an inverter designed in C-MOS technology. In order to understand the function of this circuit we will divide the input voltage into three ranges. Also, the potential V_{th} (threshold voltage) must be defined between the gate electrode and the substrate, which is characteristic for the switching behavior of the field-effect transistor (FET). At a supply voltage of $V_S = +10$ V and a typical threshold voltage of 2 V we obtain the following relations:

Case a. $0 \leq V_I \leq 2$ V. In this case the potential of gate T_1 is much more negative than that of the substrate. The voltage between the substrate and the gate is higher than V_{th}. This causes T_1 to conduct. The voltage at the gate of T_2 is lower than V_{th}, so that T_2 is turned off. The turn-off resistance of a FET circuit typically is 1000 MΩ. At this state the output of the inverter is at HIGH potential (+10 V).

Case b. $8\ \text{V} \leq V_I \leq 10$ V. Now the potential of gate T_2 is more positive than the threshold voltage V_{th}, and T_2 conducts. Since the voltage difference between the gate and the substrate of T_1 is lower than V_{th}, T_1 is turned off. The output of the inverter now is at LOW potential (0.2 V).

Case c. $+2\ V < V_I < 8\ V$. In this range both transistors conduct. It is a transition range passed through quickly during the switching operations of the gates.

In each of the first two cases, one of the FETs always was turned off ($R_{out} \sim 1000\ M\Omega$). This means that only a very small supply current flows through this high ohmic path. The power consumption of this device is low since a gate, except during transition times always is in either of the two functional states. Figure 5.71 shows a simple transition curve. In addition, the current consumption for the signal-switching phases is indicated.

These favorable properties of the C-MOS circuit hold true mainly for low operating frequencies. At higher frequencies the current consumption is predominantly determined by the capacitive load, which the succeeding stage imposes on the preceding stage of interconnected C-MOS inverters. The current necessary for charging the gate capacities is much higher than the supply current needed to operate the gate. This supply current flows between the supply voltage source and the ground across the activated circuits paths.

The disadvantage of the C-MOS technology lies mainly in the complexity of its design. Since two different transistor types are needed, so-called isolation boxes are necessary on the crystal. This prevents short circuits. The complete circuit needs a great deal of space and is therefore expensive.

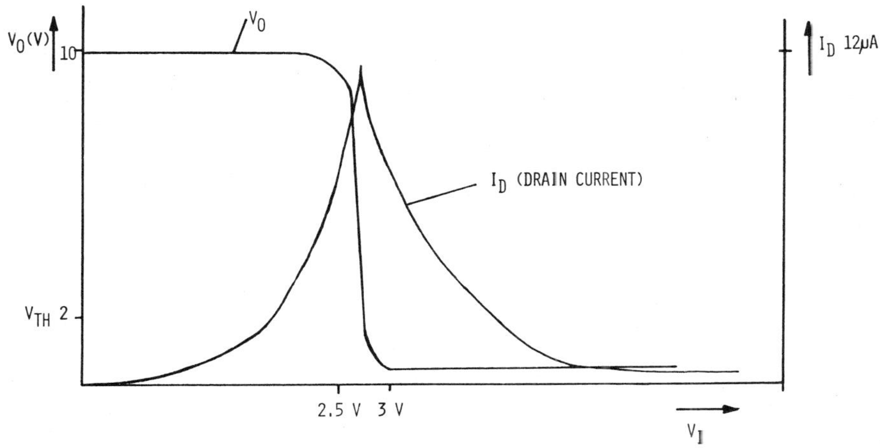

FIG. 5.71 Transition diagram and current consumption.

TABLE 5.2 Most Important Properties of the Integrated-Circuit Families

	Bipolar		MOS	
Criteria	TTL, ECL, etc.	I^2L	P-MOS, N-MOS	C-MOS
Size	∿ 150 μ × 150 μm	∿ 30 × 30 μm	∿ 50 × 50 μm	∿ 100 × 100 μm
Complexity of the integration process	140 fabrication stages	25 fabrication stages	25 fabrication stages	50 fabrication stages
Isolation boxes	Needed	Not needed	Not needed	Needed
Resistance characteristics	Dependent on the size; diffused resistors; relatively low ohmic	Dependent on injector current; lateral PNP transistor as variable resistor	Dependent on channel geometry; MOS transistor used as high ohmic resistor	Complementary MOS transistors used as switches; relatively low ohmic
Upper frequency limit	1000 MHz	Dependent on current; max. 100 MHz	1-5 MHz	5 MHz
Power consumption	High	Dependent on upper frequency limit	Low	Very low
Price	Expensive	Inexpensive	Inexpensive	Expensive
Application (preferable)	Individual logic components; MSI; LSI; large-scale circuits for extremely fast applications	Inexpensive large-scale circuits; microprocessors, memories, etc.	Inexpensive large-scale circuits with frequency limits less than those of I^2L; microprocessors, memories, etc.	Individual logic components; large-scale circuits with extremely small power consumption

5.5.4 Comparison of Different Logic Families

After having discussed briefly the most important logic families and their characteristics, the typical properties of these families are shown in Table 5.2.

5.6 DESCRIPTION OF SEVERAL DIGITAL CIRCUIT COMPONENTS

5.6.1 Code Converters

A process parameter deriving from a mechanical transducer, for example, for one to measure linear displacement, angular displacement, pressure, etc., in general is directly converted to digital voltage signals by the transducer or a converter. In a succeeding transformation these parameters are converted to binary code suitable for further processing by the computer. This, however, is not the only kind of conversion required in an installation. Whenever a control computer is connected to the technical process via an A/D or D/A converter, code converters are necessary.

One coding and transformation mechanism is briefly explained with the help of a displacement transducer using the gray code prin-

TABLE 5.3 Conversion of Gray Code to Binary Code

Gray code				Binary code			
G_3	G_2	G_1	G_0	D_3	D_2	D_1	D_0
L	L	L	L	L	L	L	L
L	L	L	H	L	L	L	H
L	L	H	H	L	L	H	L
L	L	H	L	L	L	H	H
L	H	H	L	L	H	L	L
L	H	H	H	L	H	L	H
L	H	L	H	L	H	H	L
L	H	L	L	L	H	H	H
H	H	L	L	H	L	L	L
H	H	L	H	H	L	L	H

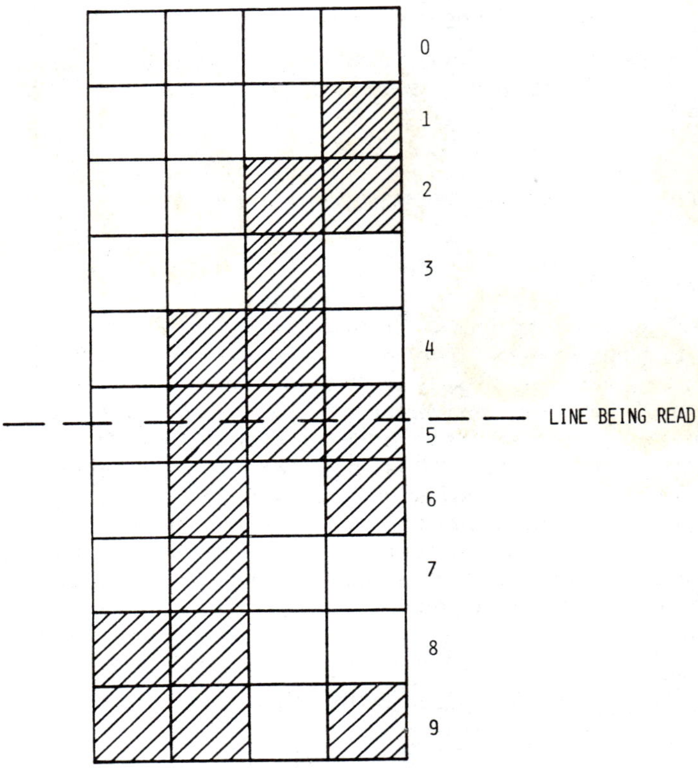

FIG. 5.72 Gray-code scale.

ciple (Fig. 5.72). A measuring scale which is fastened to the moving object is scanned by photodiodes. These convert the analog parameters of the displacement to the discrete value of the code. In this code only one bit position changes when moving from one displacement increment to the next one. An error occurring during conversion thus only can be of 1 bit. By using optical scanning in this design it is possible to achieve a resolution of a few micrometers. Since, generally, rotary measuring systems are simpler and cheaper than linear systems, an altered gray-code is used for decoding angular positions. In this case the linear displacement is transferred by a mechanism consisting of a pinion gear and a rack to the rotary transducer of the instrument.

The value obtained in gray code must be converted to binary code before further processing by the computer. It is possible to derive the following rules from the truth table of Table 5.3. The digit

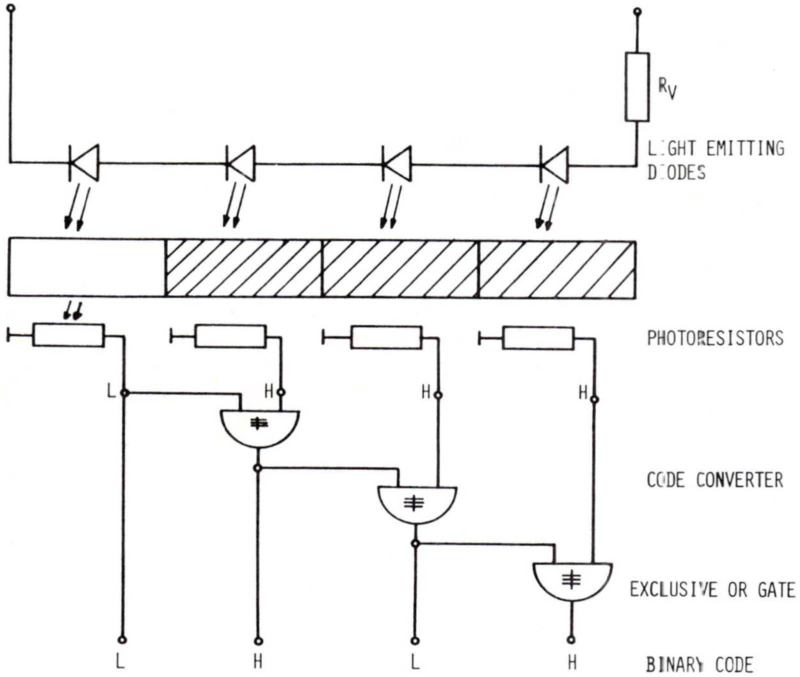

FIG. 5.73 Conversion of gray code to binary code.

D_m of the binary code of the power m is obtained from the output signal of an inclusive OR when the two input signals of this gate are

1. The position m of the gray code G_m
2. The position m + 1 of the binary code D_{m+1}

Only the highest position of the gray code is directly transferred to the most significant position of the binary code. Figure 5.73 shows the circuit of a gray code converter, briefly termed *decoder*. It can be purchased as an integrated circuit.

In this example, the measured value which represents the traveled distance or the angular position can be represented as a decimal number by using a seven-segment LED display. Since the computer stores the information in binary form, a code conversion to BCD code must first be performed to display a decimal number with the help of a 4-bit word. The code converter for this purpose can be purchased as an integrated-circuit module. This circuit also contains the drivers for activating the segments of the LED digits.

5.6.2 Multiplexer

The multiplexer can be visualized as a special type of code converter. It converts information from n input channels to one output channel. In digital circuit technology where signals are represented in binary form, the function of the multiplexer can be compared to the "1 out of n" decoder.

The general principle of a multiplexer is to switch digital as well as analog signals and is shown in Fig. 5.74. A rotary switch scans the input lines I at a constant angular velocity. During a defined time period, the so-called sample time, it connects each input with the common output O. Thus the input signals from several channels are brought to the multiplexer, where they are converted and transmitted in time segments over a single line. When multiplexing analog signals the Shannon theorem must be observed. This implies that the sampling frequency to poll the input channel must be at least twice as high as the highest frequency of the signal sampled.

If the rotary switch of the device shown in Fig. 5.74 is operated in reverse direction, it is called a *pulse distributor*, or *demultiplexer*. It distributes successively arriving discrete signals to the individual output lines. A synchronization device handles the distribution of signals at a predescribed sequence. Where information must be concentrated to reduce the overall cost of a transmission system, a multiplexer can be used together with a demultiplexer. These so-called DEMUX devices are suitable for serial transmission of originally parallel binary information over long distances. Also, in analog technology the number of A/D converters normally needed can be reduced by using a multiplexer.

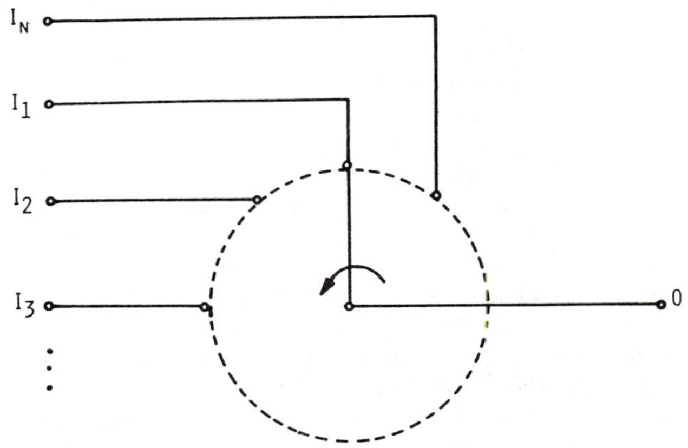

FIG. 5.74 Mechanical model of a multiplexer.

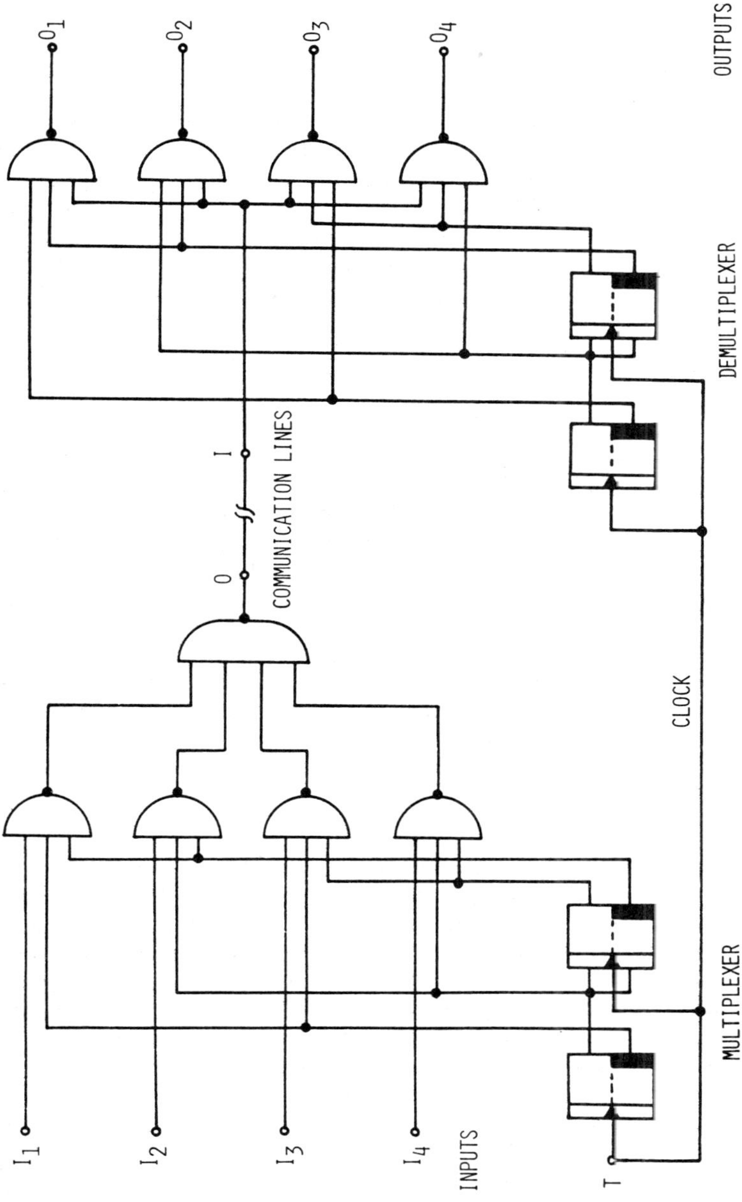

FIG. 5.75 Digital signal transfer with the help of a multiplexer.

Figure 5.75 shows the principle of digital serial data transmission. Here the information of four parallel output channels is concentrated via a multiplexer to one communication line. In the receiver station which may be located a longer distance away a demultiplexer distributes the information to four outputs. A common clock pulse controls the data distribution to the various channels.

The rotary switch of our mechanical model (Fig. 5.74) represents a synchronous counter. It can be duplicated with an integrated circuit which switches the signals from the input to the output with the help of gates. These logic functions of the gates in the multiplexer or demultiplexer are already integrated in the circuit of the control and selector inputs. There are today devices available for handling up to 16 binary channels. They can be controlled by control and selector signals. Some of them possess a second control input for connecting the output to a high ohmic or tristate level. This makes it possible to bring the outputs of several multiplexers to one common line.

In an analog multiplexer the function of the rotary switch is realized with a semiconductor circuit made from complementary metal-oxide semiconductors or from field-effect transistors. With the help of input gates which are actuated by a binary-coded control word individual channels can be selected. Within certain limits they can also transfer bipolar signals. For selecting a channel the decoder may be activated by a TTL signal. The addressed channel is then connected to the output via a low ohmic resistance of 100 Ω. All other analog signals meet an input impedance of approximately 10^9 Ω.

Typical applications of these analog switches or multiplexers are for data acquisition devices where several analog signals are brought over one channel to a programmable amplifier, a complex filter, or an A/D converter.

5.6.3 Counters

In the logic circuits discussed so far (coders, multiplexers, etc.) the binary output variables are directly generated from binary signals of the input variable. In this case the logic operations are not a function of time. In the data sheets they are shown with the aid of truth tables, whereby the different combination of the input variables and the corresponding output variables are presented. These logic circuits do not contain memory and are known as *combinational networks*.

A circuit which maintains the state set by the preceding logic operation is a sequential network. This preceding state is maintained in memories called *flip-flops*. In the simplest case the combination of the internal state variable with the input variables is performed sequentially at discrete time events. A circuit working in this mode is called a *synchronous* network operating with a common clock pulse. Asynchronous networks are characterized by the fact that logic combinations take place without a central clock pulse and that the output variables are instantaneously generated by a change of state at the input. A counter, for example, illustrates a circuit which can be either a synchronous or an asynchronous network.

Asynchronous Counters

Asynchronous counters are usually designed by an empirical method. One begins with the timing diagram or the truth table and conceives the sequential circuit according to the stated relationships. The timing diagram and truth table of an asynchronous BCD counter are shown in Fig. 5.76. This counter is automatically reset after 10 clock pulses. The problem of resetting complicates the design of the circuit. A pseudostate following the count "9" in the binary counter is used for resetting. The timing diagram of Fig. 5.76 shows this short reset pulse. It has a length of approximately 50 nsec, depending on the transfer time. The outputs of the first and third flip-flops are connected together with an AND gate. The output of this gate initializes all flip-flops of the counter via an asynchronous reset input.

The preceding discussion leads us to the circuit of Fig. 5.77. It is equipped with a reset (clear) switch to initialize the counter. The NAND gate generates a short pseudostate after each 10th clock pulse. This pseudostate, however, can be eliminated by proper design of the circuit. The asynchronous method of strobing in information is a

CLOCK PULSE NUMBER	BCD CODE 2^3	2^2	2^1	2^0
0	L	L	L	L
1	L	L	L	H
2	L	L	H	L
3	L	L	H	H
4	L	H	L	L
5	L	H	L	H
6	L	H	H	L
7	L	H	H	H
8	H	L	L	L
9	H	L	L	H
10	H	L	H	L
0	L	L	L	L
1	L	L	L	H

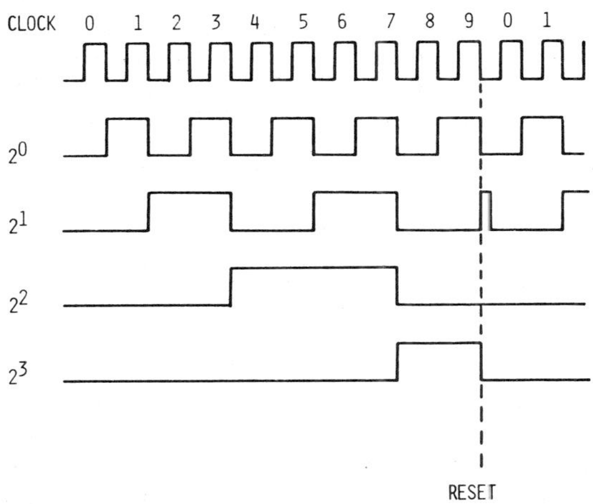

FIG. 5.76 Truth table and pulse diagram of an asynchronous BCD counter.

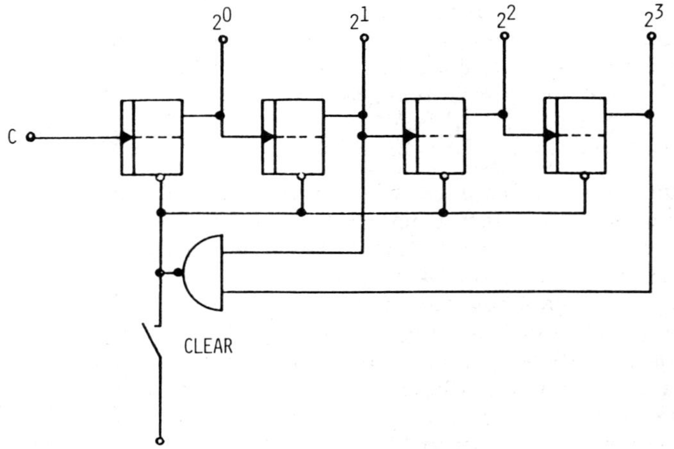

FIG. 5.77 Asynchronous BCD counter.

further disadvantage of this circuit. Each flip-flop strobes in at its input the output information of the preceding flip-flop. Thus the asynchronous counter delay time is the sum of the delay times of the individual flip-flops. This means that information is only available after a certain settling time.

Synchronous Counters

Several systematic methods are available for the design of synchronous counters. In this counter the required number of flip-flops K can be determined from the maximum number of the states M of a cycle. This is done with the following equations:

$$2^K \geq M \tag{36}$$

$$K \geq \ell d\ M \tag{37}$$

The various types of flip-flops can be classified by their ability to perform logic combinations with the input/output variables and the clock pulse. Since logic combinations are a time function, the variables are marked by an index n signifying the sequence number of the logic operation. In our special case, where the circuits are controlled synchronously by a common clock pulse, the indexes n, n + 1, etc., indicate the discrete time period of the active pulse edge.

In data sheets the function of a flip-flop is presented with the help of the following information:

Symbol
Truth table
Characteristic switching functions

In Fig. 5.78 data for the two most common types of flip-flops are shown. When designing sequential circuits, in principle all different types of flip-flops can be used; the JK flip-flop is the most versatile one. With the proper selection of a flip-flop for an application the circuit complexity may be minimized. By starting with the output variable in relation to the preceding input variable and with the internal states of the circuit at the time t_n, the truth table or the Karnaugh diagram can be drawn up for the succeeding time t_{n+1}. The so-called problem function can be obtained in standardized form by an optimization procedure; thereby redundancy is eliminated.

The input function of the flip-flop to be used is obtained from the logic equation derived by setting equal the calculated problem function and the characteristic switching function. The input variables then have to be combined with the help of the results of this equation.

In the following example the use of this method to synthesize a sequential circuit is shown: A binary counter is to be designed to count up to 4 and down to 0 (modulo 4). The input variable X determines whether the counter counts up (X = H) or down (X = L). With the help of Eqs. (36) and (37) it can be shown that in order to obtain the four different states of Fig. 5.79, two flip-flops are needed.

The problem functions can be written individually for each of the two flip-flops. Thereby the output function of the individual flip-flops at the time t_{n+1} is described with the aid of its preceding state. This is the state which the circuit maintained during the time t_n.

The problem function can be derived directly from the truth table of Fig. 5.74. To optimize the logic expressions and to standardize it, the Karnaugh diagram is used. Figure 5.80 shows the resulting problem function.

During the next step the input functions of both flip-flops are determined. These are found by setting the characteristic switching functions of the JK flip-flop (Fig. 5.78) equal to the corresponding problem function. Now it is possible to derive for the input function the following logic combinations:

For the flip-flop Y_0:

$$J_0 = H \quad K_0 = H \tag{38}$$

For the flip-flop Y_1:

$$K_1 = Y_0 \equiv X \tag{39}$$

TYPE OF FLIP FLOP	SYMBOL		TRUTH TABLE	CHARACTERISTIC SWITCHING FUNCTIONS
	DIN* 40700	USA		
DELAY-, LATCH-FLIP-FLOP	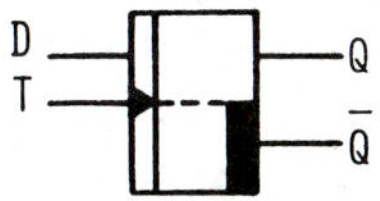	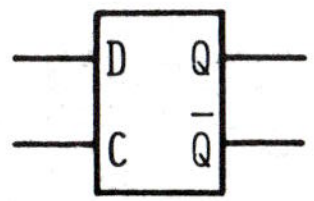	D_N | Q_{N+1}: L | L; H | H	$Q_{N+1} = D_N$
J-K-FLIP-FLOP	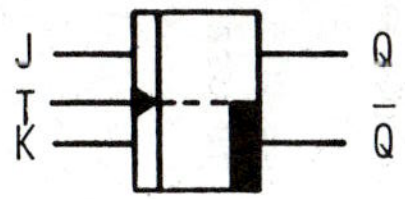	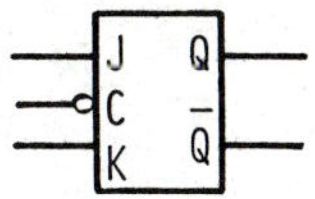 D = DATA C = CLOCK	T_N (J K) | T_{N+1} (Q): L L | Q_N; L H | L; H L | H; H H | $\bar{Q}_N$	$Q_{N+1} = (\bar{K}Q + J\bar{Q})_N$

*GERMAN INDUSTRIAL STANDARD

FIG. 5.78 Graphical symbols and truth tables of different flip-flops. (*German industrial standard.)

STATE AT TIME N			STATE AT TIME N+1	
X	Y_1	Y_0	Y_1	Y_0
L	L	L	H	H
L	L	H	L	L
L	H	L	L	H
L	H	H	H	L
H	L	L	L	H
H	L	H	H	L
H	H	L	H	H
H	H	H	L	L

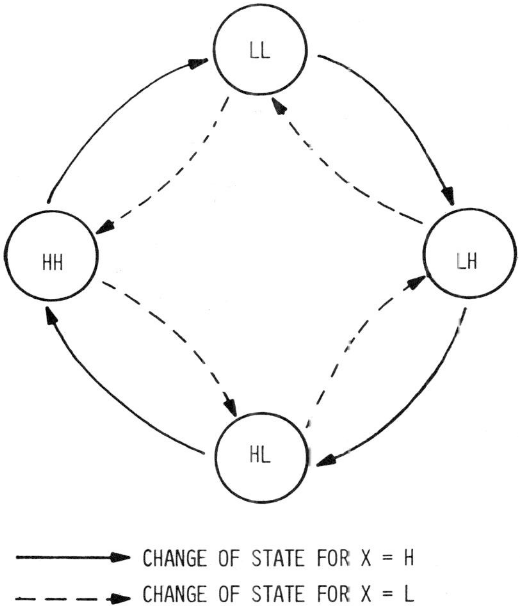

FIG. 5.79 Truth table and Mealy graph for a modulo 4 binary up and down counter.

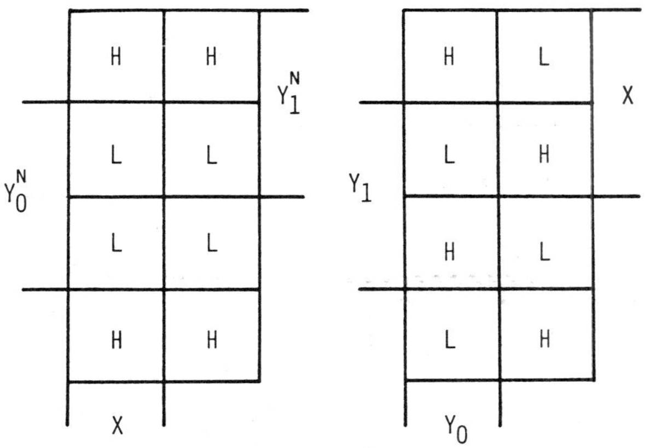

FIG. 5.80 Karnaugh diagram for the stage variables Y_0 and Y_1 at time n + 1. $Y_{0,(n+1)} = \overline{Y}_{0,n}$, $Y_{1,(n+1)} = [(X \not\equiv Y_0)Y_1 + (X \equiv Y_0)\overline{Y}_1]_n$, where $\not\equiv$ exclusive or.

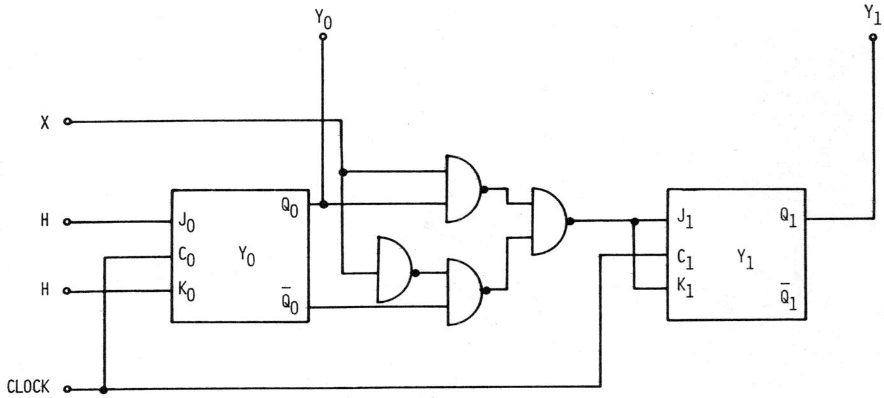

FIG. 5.81 Binary up and down counter.

This last stage completes the synthesis or the design of the sequential circuit. The realization of the two-digit binary up and down counter is shown in Fig. 5.81. The circuit was realized with NAND gates; the switch X determines the counting direction.

5.6.4 Shift Registers

Shift registers are used to store information as well as to control shifting operations of digital signals. The same design methods ex-

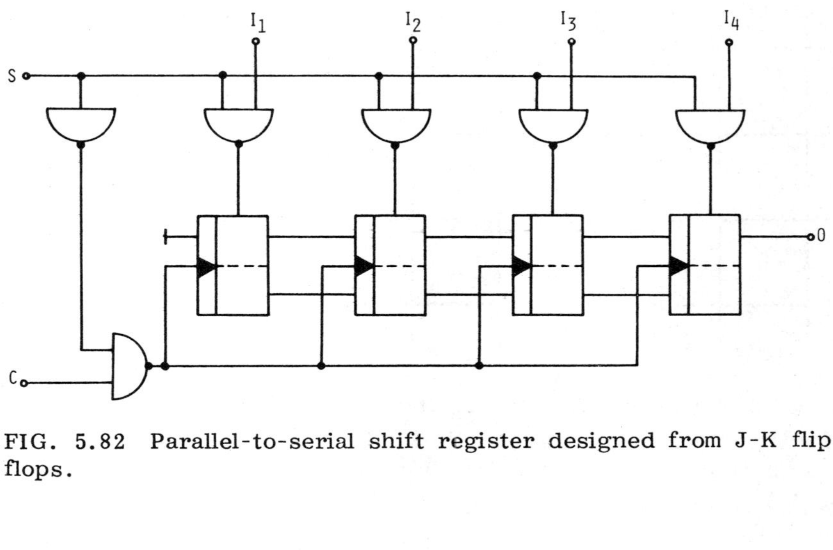

FIG. 5.82 Parallel-to-serial shift register designed from J-K flip-flops.

plained for the synthesis of counters can be employed for the layout of shift register. Discrete shift registers are preferably designed with JK flip-flops. As an example, a parallel/serial shift register is shown in Fig. 5.82. In this device data can be strobed in parallel form via the four inputs I and read out serially at output O. The desired mode of operation is determined by the control signal S. If a HIGH signal prevails at the S input, data can be strobed in parallel at inputs I_1 to I_4 into the flip-flops. At the same time the signal line for the clock is blocked. With a LOW signal, the stored data can be shifted to the right. In addition to discrete shift registers, there are also many similar circuits for various other shift functions. Here we will only briefly discuss a universal shift register and a dynamic shift register.

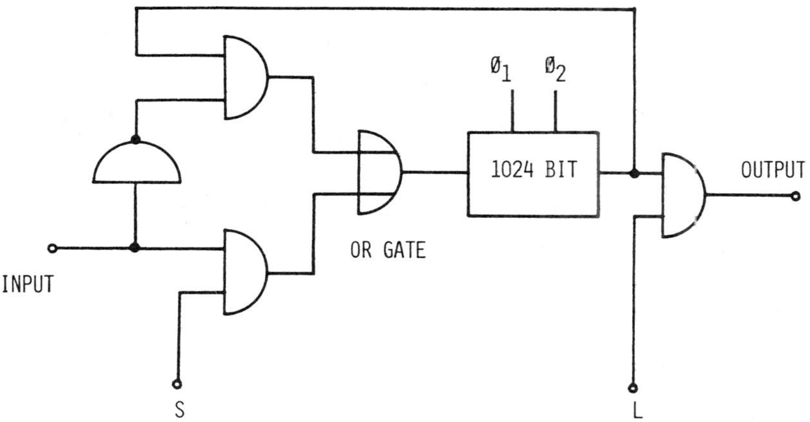

No	INPUT CONTROL SIGNALS S	L	MODE OF OPERATION
0	L	H	DATA ARE ROTATED, OUTPUT ON LOW
1	L	H	DATA ARE ROTATED AND ARE READY FOR OUTPUT
2	H	L	STROBE-IN OF DATA, OUTPUT ON LOW, STORED DATA GET LOST
3	H	H	STROBE-IN OF DATA, STORED DATA ARE PLACED ON OUTPUT

FIG. 5.83 MOS shift register and truth table for control signals.

The universal shift register can strobe-in parallel data and shift these either to the left or to the right. The mode of operation is controlled by a special input.

Recently large-scale integrated MOS shift registers have been introduced. They are characterized by low power consumption and minimal space requirements. Figure 5.83 shows a block diagram of a dynamic MOS shift register. Each memory location consists of a P-channel MOS transistor (of the self-depleting type) and uses the gate capacity as the actual storage device. The very low charge stored is rapidly depleted by leak currents. Therefore the contents of the storage cell must be regenerated at short time intervals. For this reason the shift register must be constantly supplied with two-phase clock pulses ϕ_1 and ϕ_2. They lie in the range of 10 kHz to 5 MHz. Normally digital data is circulated cyclically in a shift register. With the help of the control inputs, however, it is possible to set four different modes of operation (see table on Fig. 5.83).

REFERENCES

1. F. Dokter and J. Steinbauer, *Digitale Elektronik in der Meßtechnik und Datenverarbeitung*, Bd. I, Philips-Fachbücher, 3. Aufl. 1972.
2. B. A. Artwick, *Microcomputer Interfacing*, Prentice-Hall, Englewood Cliffs, N.J., 1980.
3. R. H. Beerbaum, Two Inverters and a Crystal Assure Oscillator Start, *Electronics*, *49* 2 (1976), 115.
4. L. A. Delholm, *Design and Application of Transition Switching Circuits*, Texas Instruments, Inc., McGraw-Hill, New York, 1968.
5. *COS/MOS Integrated Circuits Manual*, RCA Solid State Division, Somerville, Mass., 1972.

BIBLIOGRAPHY

Caughlin, R. F., and Driscoll, F. F., *Operational Amplifiers and Linear Circuits*, Prentice-Hall, Englewood Cliffs, N.J., 1977.

Lenk, J. D., *Logic Designer's Manual*, Reston Publishing Co., Reston, Va., 1977.

Peatman, J. B., *The Design of Digital Systems*, McGraw-Hill, New York, 1972.

Santucci, Demos, Data Acquisition Can Falter Unless Components Are Well Understood, *Electronics*, *48* (November 1975) 114-118.

Siemens Datenbuch, *Digitale Schaltungen*, Siemens AG, Munich, 1974.

The Interface Circuits Data Book, Texas Instruments Learning Center, Munich, 1974.

6

COMMUNICATION BETWEEN THE INTERFACE AND THE CONTROL COMPUTER

6.1 INTRODUCTION

In Sec. 2.6 we briefly discussed the various methods of data input and output for control computers. We distinguished between program-controlled input/output using no interrupt, program-controlled input/output using interrupt, and direct memory access (DMA). The method selected for a particular application depends on the required speed of the data communication between the computer and the peripheral, as well as on the cost of the computing equipment and communication programs. Program-controlled input/output using no interrupt is seldom used, since with this method the computer must wait for the slow peripheral devices until it has data available or is ready to receive data. For this reason this type of input/output in most cases is done with the help of interrupt control. Here data communication is performed simultaneously by hardware and software. Direct memory access is initiated by software and carried out under interrupt control.

Since there is a multitude of different bus architectures used in control computers, it is almost impossible to discuss them all. In this chapter the main emphasis will be placed on the description of the functions of the interface for a two-bus system. The interface design discussed is a simplified structure of that of the Interdata Computer Model 1* [1].

One bus of the input/output architecture of this "model computer" concept is equipped with a multiplexer so that data from several slow peripherals can be exchanged via the CPU with the working memory (Fig. 6.1). The second bus is used for fast peripherals which

*The computer is a product of the Perkin Elmer Corp., Oceanport, N.J.

exchange data autonomously with the memory under DMA and thus bypass the CPU. The second bus architecture to be discussed is that of the Dietz 621 computer (Fig. 6.2). A detailed analysis of this bus, however, will not be given. The universal bus of the PDP 11 and many microcomputers is similar to that of the Dietz computer [2,3].

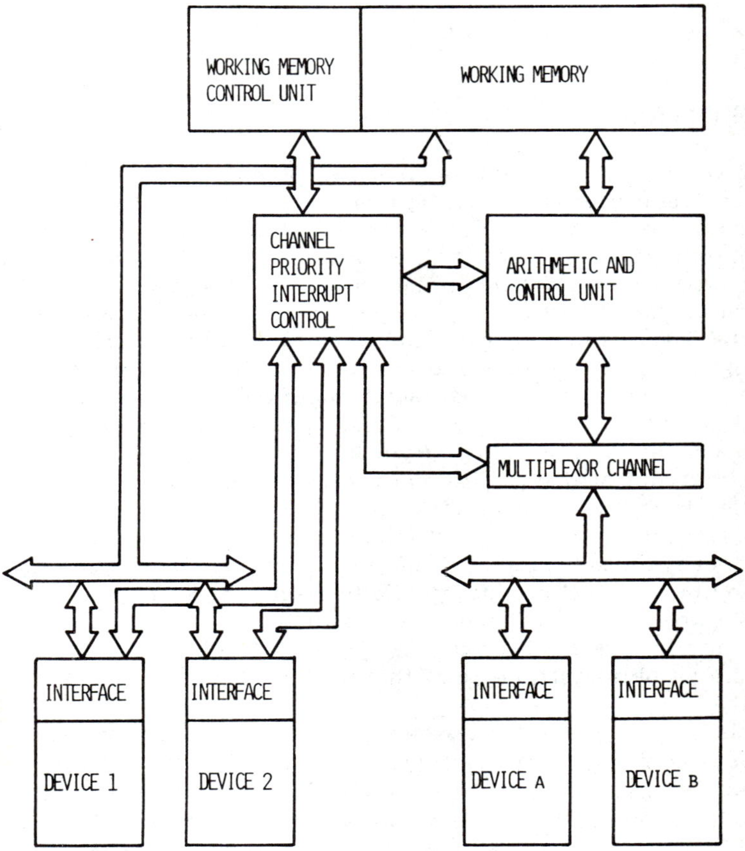

FIG. 6.1 Computer with a two-bus system.

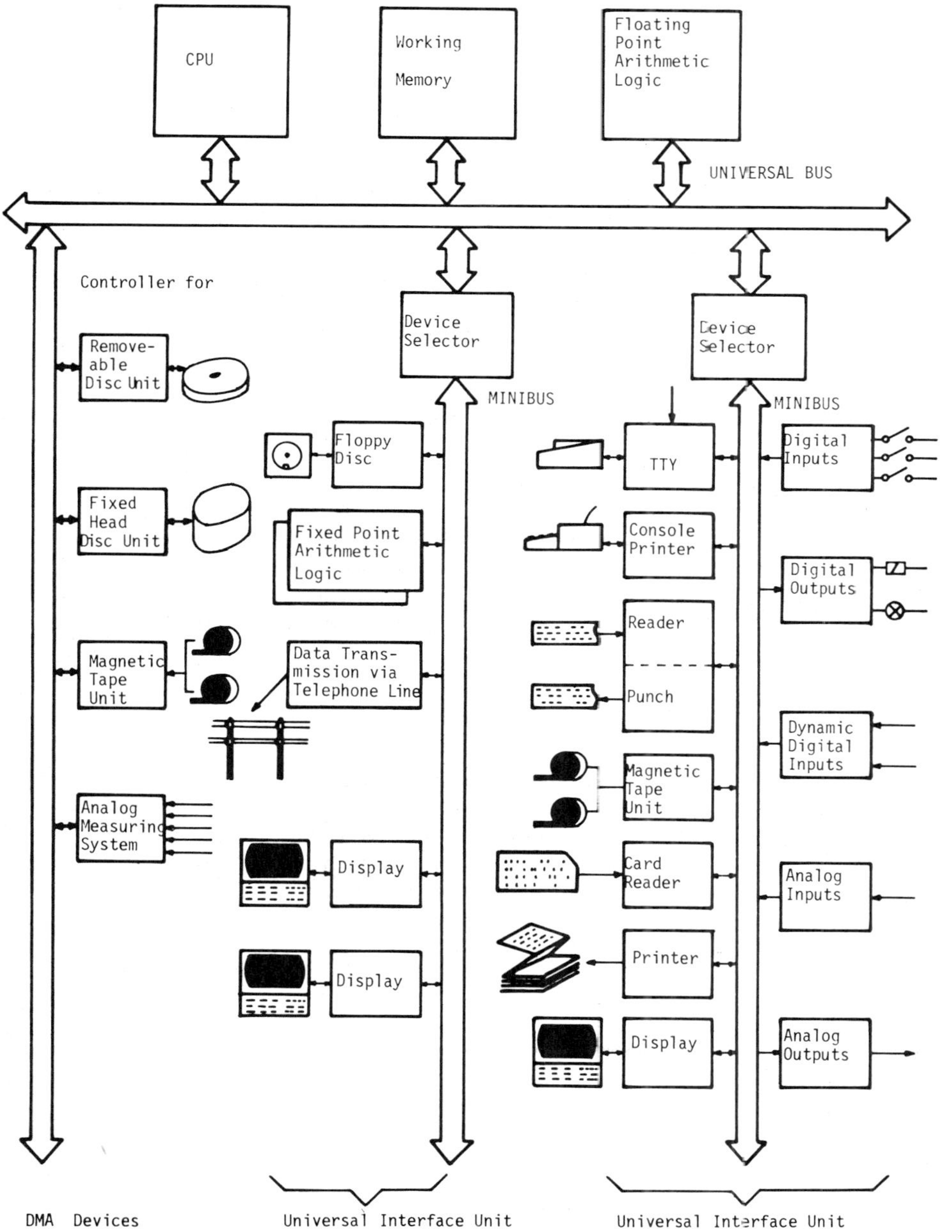

FIG. 6.2 Universal bus of the Dietz computer. (From Ref. 2, courtesy of Dietz Computer Systems, Mülheim a.d. Ruhr, Federal Republic of Germany.)

6.2 BUS COMMUNICATION LINES

6.2.1 Number of Communication Lines

The input/output architecture of a control computer may vary considerably, depending on the model and manufacturer. Thus interfaces are in general not exchangeable between computers. The design of the computer influences the architecture of its input/output system. The number of communication lines in different bus concepts lies between 25 and 100. This number is determined by the complexity of the input/output system, by the byte or word structure of the computer, and by uni- or bidirectional data communication (Sec. 4.3).

6.2.2 Synchronization of the Signal Flow

Transfer of control signals and data takes place under strict communication protocol. For example, it is necessary to address a peripheral before data can be sent to it. Modern computers use two different synchronization methods. These are the synchronous and asynchronous input/output. It is also possible that both methods can be found in one computer. Data transfer can be triggered by the leading or trailing edge of the control pulse. This depends on the technology of the computer and interface.

Synchronous Input/Output

With this method initiation of the various communication phases such as addressing and transfer of data is done by control pulses with constant time intervals (Fig. 6.3). The control logic for synchronous data transfer is relatively simple. During each input/output phase all signals can be transferred correctly when the time between clock pulse has been selected so that it corresponds to the longest phase which is encountered in the data transmission procedure. Therefore,

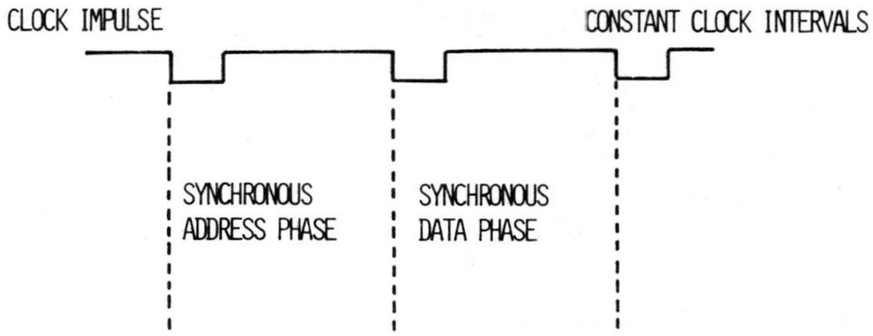

FIG. 6.3 Synchronous I/O operation.

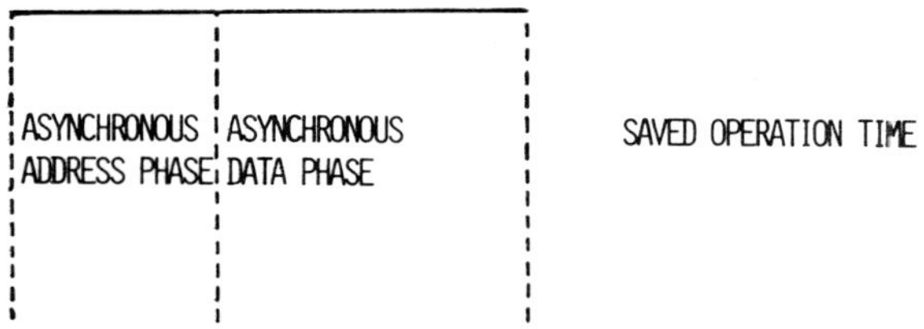

FIG. 6.4 Asynchronous I/O operation.

this method is not suitable for extremely fast processes. When determining the required time period between two succeeding signal pulses, the travel speed in the cable and in the emitter end receiver circuits must be taken into consideration. Likewise, the signal must be stabilized when it is used for triggering so that transient disturbances and noise do not affect the transmission.

Asynchronous Input/Output

By using a more complex control logic the time disadvantage of the synchronous data input/output and synchronous data transmission can be eliminated. With asynchronous data input/output each new phase of the communication procedure is initiated directly after completion of the previous phase (Fig. 6.4 and Sec. 5.4.4). This method will be preferred for time-critical processes.

6.3 INTERFACE FOR PROGRAM-CONTROLLED DATA INPUT/OUTPUT

6.3.1 The Functional Components of an Interface

The interface has various functions to perform for data communication. A circuit design to fit all interfaces cannot be conceived. Depending on the type of computer and peripheral, there are many different circuits used; however, many basic components of different interfaces have the same function. Figure 6.5 shows a simplified block diagram of a typical interface. This interface has been designed with the following components:

1. An address decoder for interpreting the address of the peripheral
2. An instruction decoder for interpreting instructions
3. An instruction register for receiving and interpreting control commands sent to the peripheral
4. A timing and control logic for initiating control functions

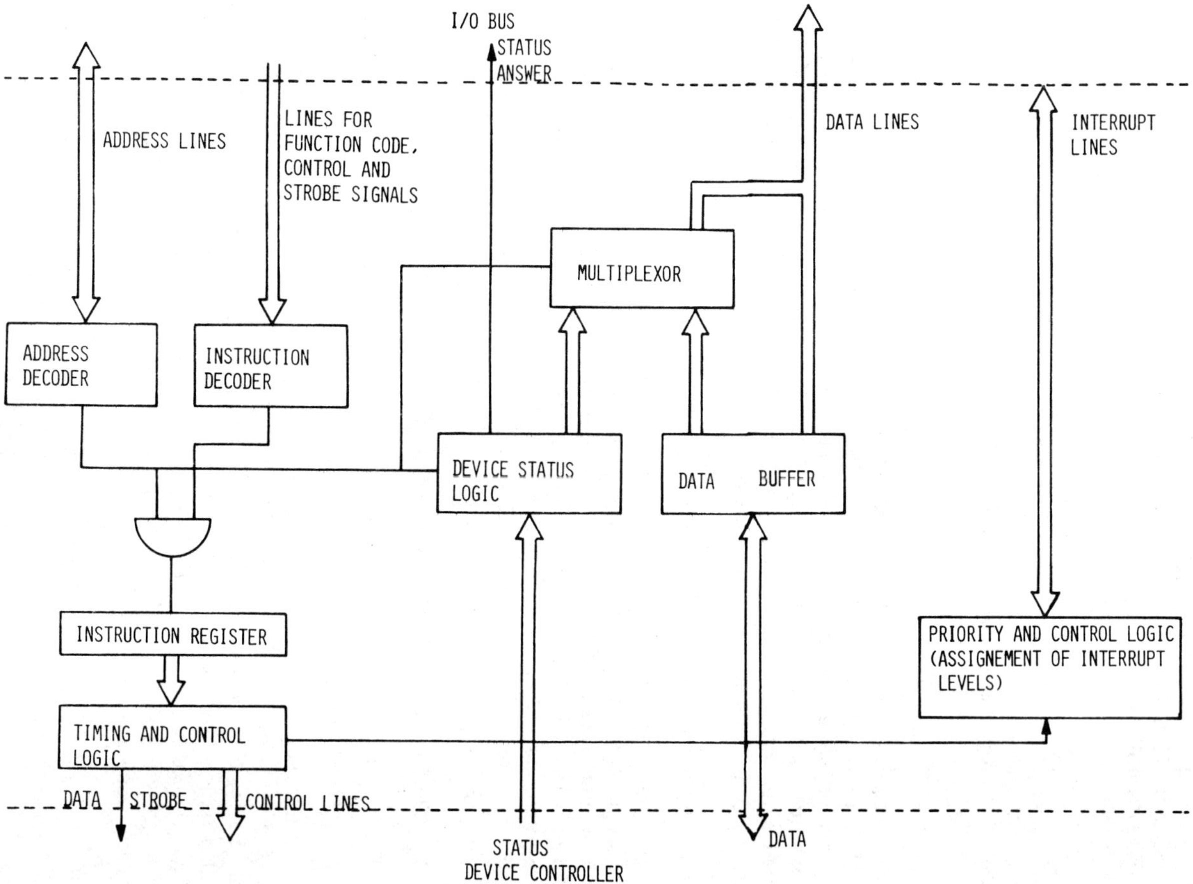

FIG. 6.5 Simplified control logic of a typical interface. (Redrawn from A. A. J. Hoffman et al., IEEE Spectrum, February 1974. Copyright © 1974 IEEE.)

5. A status logic for determining the status of the peripheral
6. A data buffer for temporary storage of data
7. A multiplexer for connecting either the status logic or the data buffer to the data lines
8. A priority control logic to determine the interrupt levels
9. Data lines to receive data from the input/output bus
10. Data lines to send data from the peripheral to the computer

The multiplexer channel for program-controlled data transfer, of the selected model computer is byte oriented (Fig. 6.1). Thus, the 8-bit address field makes it possible to connect 256 peripherals to the channel. The bus consists of eight input, eight output, eight control, and two test lines, as well as a clear line. The input/output lines are used both for addressing and data exchange. Each peripheral connected to the bus has a distinct priority assigned to it which depends on the location of the peripheral on the bus.

The bus can only be occupied at a given time by one peripheral. During data transfer a global mask must be set to exclude other interrupts. Without this mask the computer or the peripheral may be supplied with false or incomplete data. The data transfer cycle is determined by the execution time of the input/output instruction. A computer system having slow peripherals can only operate efficiently under interrupt control; otherwise, the CPU may encounter long waiting times.

6.3.2 Assignment of the Bus Lines

For the model computer described here the individual lines of the bus have the functions shown in Table 6.1. The following is a more detailed description of the tasks of the individual lines:

DALOO, ..., DALO7, Data Available Lines: Transfer one byte of an address or data or control information from the CPU to the interface. Information transfer must be initiated by the CPU with either the ACK, SR, or DR signal.

The reader interested in comparing the assignment of these lines with those described in the interface of Fig. 6.5 will notice some significant differences. For example, for the computer chosen here, the eight data available lines are used for addressing, transfer of the function code, and communication of data to the interface. Thus, this interface can only be used for this specific computer.

DRLOO, ..., DRLO7, Data Request Lines: Transfer one byte of an address or data or status informaton from the interface to the CPU. The information transfer must be initiated by the CPU with either the ACK, SR, or DR signal.

TABLE 6.1 Assignment of Bus Lines

Number of lines	Function	Abbreviation	Direction of data transfer CPU —— peripherals	
8	Data available	DAL00, ..., DAL07	Data	→
8	Data request	DRL00, ..., DRL07		←
1	Status request control	SR		→
1	Data request	DR		→
1	Command	CMD		→
1	Data available	DA	Control	→
1	Address	ADRS		→
1	Acknowledge	ACK		→
1	Clock	CLO6		→
1	Clock	CLO7		→
1	Attention	ATN	Test	←
1	Synchronize	SYN		←
1	System clear	SCLRO	Initialize	→

Here too, the assignment of data lines differs considerably from that of the interface shown in Fig. 6.5.

SR, Status Request: Before information can be transferred via the DRL line the CPU requests information about the status of the peripheral from the interface. Data transfer from the peripheral is accompanied by a delayed synchronization signal SYN.

DR, Data Request Control: Before initiating a data transfer, the CPU sends a data request signal to the interface. After initiation of the transfer the peripheral sends a delayed synchronization signal SYN back to the CPU.

ACK, Acknowledge: During an interrupt request ATN, the peripheral places its address on the DRL line. The CPU responds to the interrupt request with ACK. A delayed SYN signal accompanies the address.

DA, Data Available: The CPU sends data to the interface via the DAL lines and announces the availability of data with a delayed signal DA. The interface answers with a SYN signal.

CMD, Command: The CPU sends a control instruction via the DAL line to the interface and simultaneously issues a delayed CMD signal to announce the availability of this instruction. The interface answers with a SYN signal.

ADRS, Address: The CPU sends an address via the DAL lines to the interface and simultaneously issues an ADRS available signal. The interface answers with a SYN signal.

ATN, Attention: A peripheral announcing an interrupt sends an attention signal ATN to the CPU. After receiving an ACK answer from the CPU this signal is removed. Thereafter the interface places its address on the DRL line and simultaneously sends a delayed SYN signal.

SYN, Synchronize: This signal is issued by the interface and sent to the CPU to respond to all other signals except SCLR. This guarantees the correct transfer of information.

SCLR, System Clear: This signal grounds the peripheral during power failure, power-up, or initialization of the computer.

CLO, Clock Signals: Needed to synchronize communication between the CPU and peripheral.

All lines except the acknowledge line are used in parallel by all peripherals. The ACK line is connected to the peripherals in series (Fig. 4.30). After occurrence of an interrupt request, the ACK signal is placed by the CPU on the line to identify the requesting peripheral. A check is made as to whether the first peripheral requested an interrupt (see also Sec. 5.3.2). If not, the signal is sent to the next peripheral and so on, until the unit is found which requested the interrupt with the signal ATN.

6.3.3 Sequence of Input/Output Operations

Data input and output is initiated and controlled by instructions. In our computer we distinguish between six different types of instructions.

1. Addressing instruction for addressing an input/output peripheral
2. Control instruction for issuing control information
3. Status instruction for determining the status of a peripheral
4. Data output instruction for transferring data to the peripheral
5. Data input instruction for receiving data from the peripheral
6. Acknowledge instruction for confirming the interrupt request

During a data exchange the status, control and input/output instructions must be present.

In our model computer the input/output instructions are byte oriented, which means that they have a length of 8 bits. First a peripheral is addressed (by an address instruction) and activated. Thereafter further input/output instructions can be issued. For computers with word-oriented input/output instructions the control, status, and input/output instructions generally also contain an address field, which allows direct addressing of the peripheral by issuing the instruction.

The following is a more detailed description of the execution of the individual instruction.

Address Instruction

The address instruction consists of 8 bits. It is possible to address 256 peripherals with it (Fig. 6.6).

1. The contents of the accumulator which contains the address of the peripheral is placed on the data available lines DALOO, ..., O7.
2. The CPU activates the address control lines ADRS.
3. The address decoders of all peripherals attempt to interpret simultaneously this address. The peripheral with the correct address is activated by the gate G1 of the interface and sets the address flag of the flip-flop FO via gate G1 to HIGH. In the interfaces of all other peripherals the address status flag is set to LOW.
4. Simultaneously the synchronization line SYN is activated by gates G2 and G4. The CPU receives a signal informing it that the addressed peripheral has recognized its address.
5. After receipt of the SYN signal the CPU removes the address of the peripheral from the line and later the interface removes the SYN signal. The selected peripheral remains active until an other

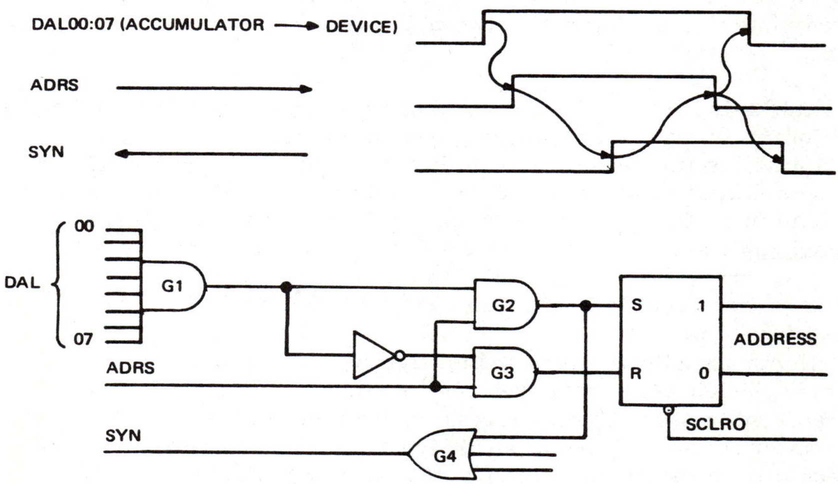

FIG. 6.6 Transmission of the device address to the interface. (Courtesy of Interdata, Perkin-Elmer, Oceanport, N.J.)

TABLE 6.2 Assignment of Control Instruction Lines to Address Thermocouples and to Select Peripheral Functions

Bit	
0	Address thermocouple 1
1	Address thermocouple 2
2	Address thermocouple 3
3	Address thermocouple 4
	Peripheral's function
4	Activate/deactivate peripheral
5	Activate/deactivate thermocouple
6	Activate/deactivate waiting time
7	Activate/deactivate measurement

peripheral is addressed or until the system clear signal SCLRO is issued. As long as the peripheral is active, it responds to all input/output instructions following the address phase.

6. If the SYN signal is not removed after a predefined time, a watchdog timer makes certain that the address phase is terminated. This problem is made known by a status flag to avoid lengthy waiting times for the CPU.

Control Instructions

By using control instructions, it is possible to turn the requested functions of the peripheral on and off. For example, a measurement can be initiated or a peripheral can be advised to turn its motor off. The individual instruction to be issued thus depends on the peripheral's function. Table 6.2 shows the assignment of control instructions for a temperature measuring instrument with four thermocouples. By using bits 0 to 3, the four thermocouples can be individually addressed. With bits 4 to 7 the various instrument functions can be performed. Theoretically it would be possible to code 253 instrument functions with eight control lines. For simplicity, in this example these functions are transferred uncoded via the individual lines. Through the function code the instrument and the thermocouples can be individually turned on and off. It is also possible to allow the measurement to stabilize for a predefined waiting time.

A control instruction is transferred as follows (Fig. 6.7):

1. The CPU outputs the contents of the accumulator, which contains the function code, on the data available lines DALO0, ..., 07.
2. The CPU activates the output command line CMD.

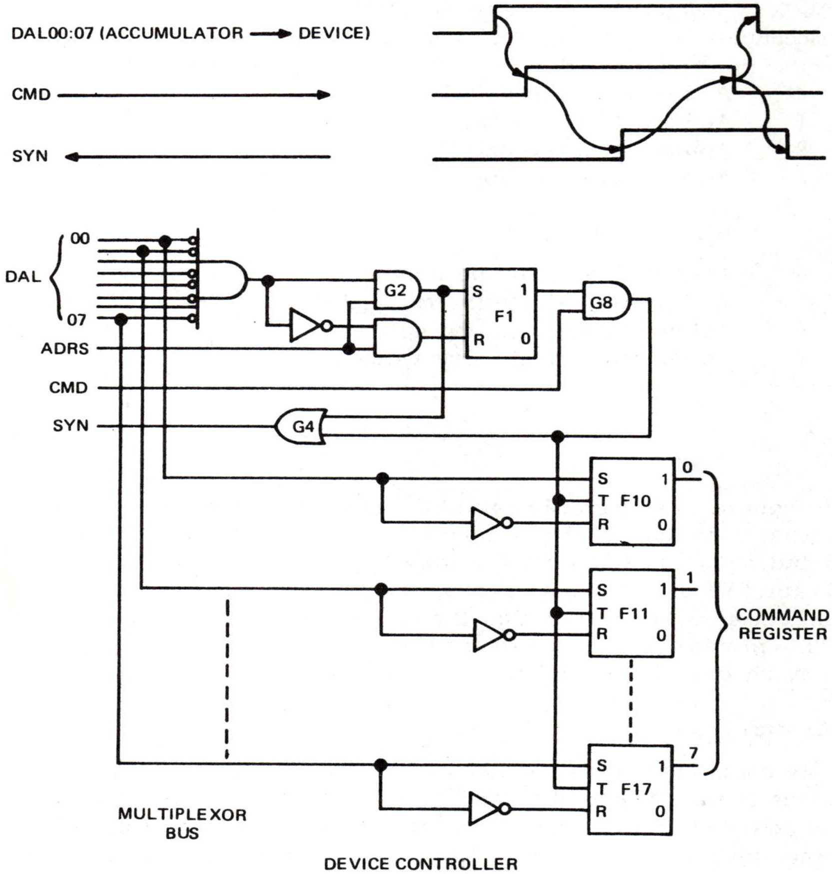

FIG. 6.7 Transmission of the control command to the interface. (Courtesy of Interdata, Perkin-Elmer, Oceanport, N.J.)

3. The previously addressed peripheral strobes the control instruction into its buffer with the help of gate G8. The buffer consists of the flip-flops F10 to F17.
4. Simultaneously a synchronization signal SYN is sent to the computer via the gates G8 and G4.
5. After receiving the SYN signal the CPU removes the function code from the lines and later the interface turns the SYN signal off. The addressed peripheral remains active.

TABLE 6.3 Assignment of Individual Status Lines to Peripheral Functions

Status Lines bit	Peripheral response if true
0	The addressed thermometer is turned on.
1	The addressed thermometer is grounded.
2	The addressed thermometer has an open line.
3	Temperature measurement has been stabilized.
4	Data are available in the buffer.
5	Not used.
6	Not used.
7	Peripheral is busy.

Status Instruction

For faultless data transfer it is necessary that the CPU knows the status of the peripheral. For example, data transfer to a peripheral can be initiated only after the previous transfer is completed. The computer must also be informed of problems with the peripheral. So the interface must be capable of analyzing the status of the peripheral and inform the computer of its status. It is therefore necessary that the status of each peripheral is communicated to the computer individually. Generally, data can be transferred between the peripheral and the computer only if all status signals have been set to LOW. In other words, the peripheral must be operative.

Our computer is provided with eight lines for communicating the status of the peripherals (Table 6.3). Most computers only have three lines for communicating eight individual statuses in coded form. Theoretically it is possible to code 256 statuses with our circuit. In reality though, there is seldom a peripheral that needs to communicate that many types of information on its status.

A status request from the computer proceeds as follows (Fig. 6.8):

1. The CPU turns the status request signal SR on.
2. The currently addressed interface sends its status to the CPU via the eight DRL lines. Gates G7 and G18 to G25 are activated. In addition, via gate G7 the signal SYN is sent to the CPU.

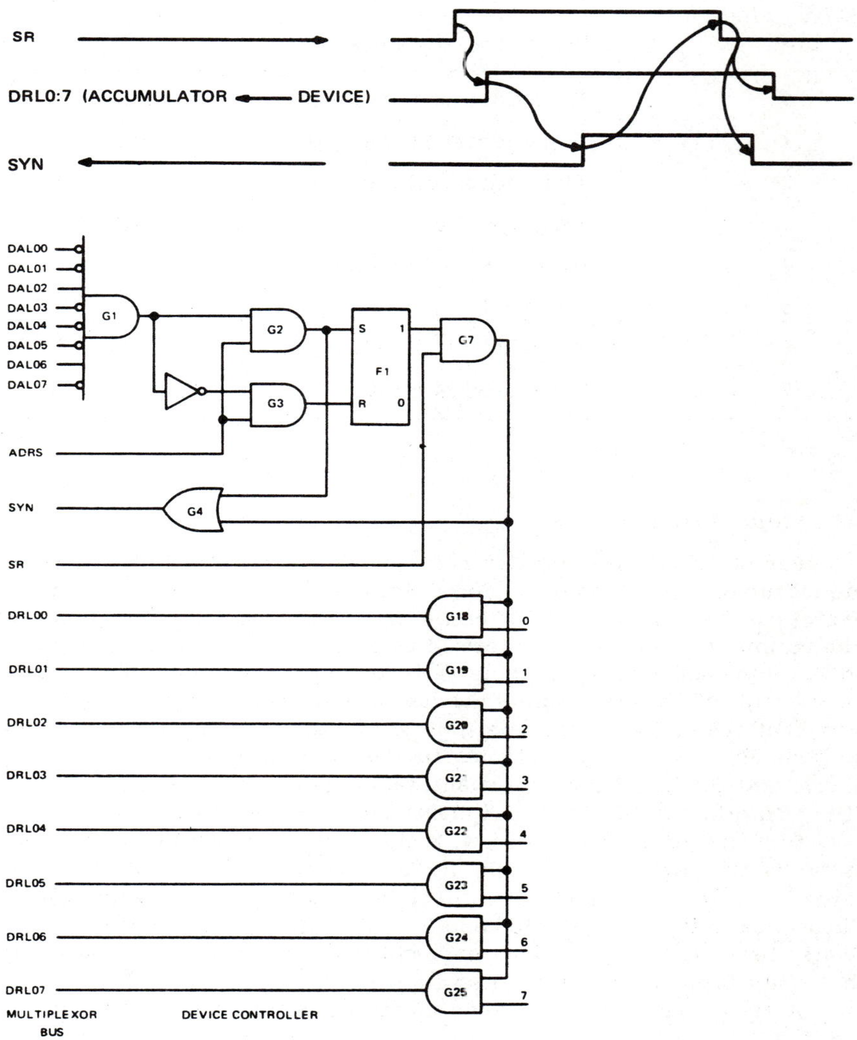

FIG. 6.8 Transmission of the device function from the interface to the CPU. (Courtesy of Interdata, Perkin-Elmer, Oceanport, N.J.)

3. Upon receipt of the SYN signal, the CPU accepts the status and strobes it into the accumulator. The SR line is deactivated.
4. After removal of the SR signal the status transfer is terminated by the interface. The SYN line is deactivated.

Data Output Instruction (Write Instruction)

Data output can only be accomplished when the peripheral is functional and not busy. Before a data transfer the status of the peripheral must be checked. The procedure of status requesting has been described in the preceding section. If the status request indicates that the peripheral is ready to accept data, data transfer will be initiated (Fig. 6.9). When any of the status bits of lines 0 to 6 are set, the peripheral is not functional (Table 6.3). In this case the status byte is placed in the accumulator, and the peripheral's problem can be investigated by a jump into a subroutine. If only bit 7 is up, the peripheral is busy, although functional. Then the program enters a waiting loop and remains there until the peripheral is ready. Afterward data transfer can be initiated.

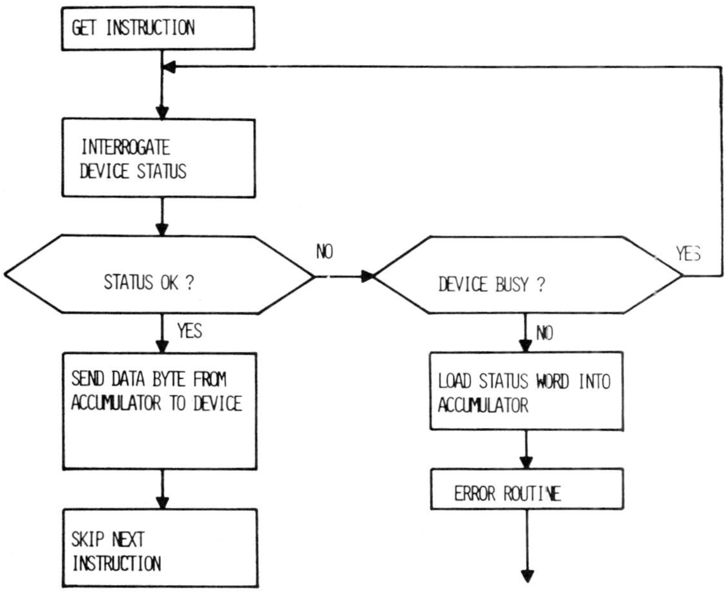

FIG. 6.9 Interrogation of device status.

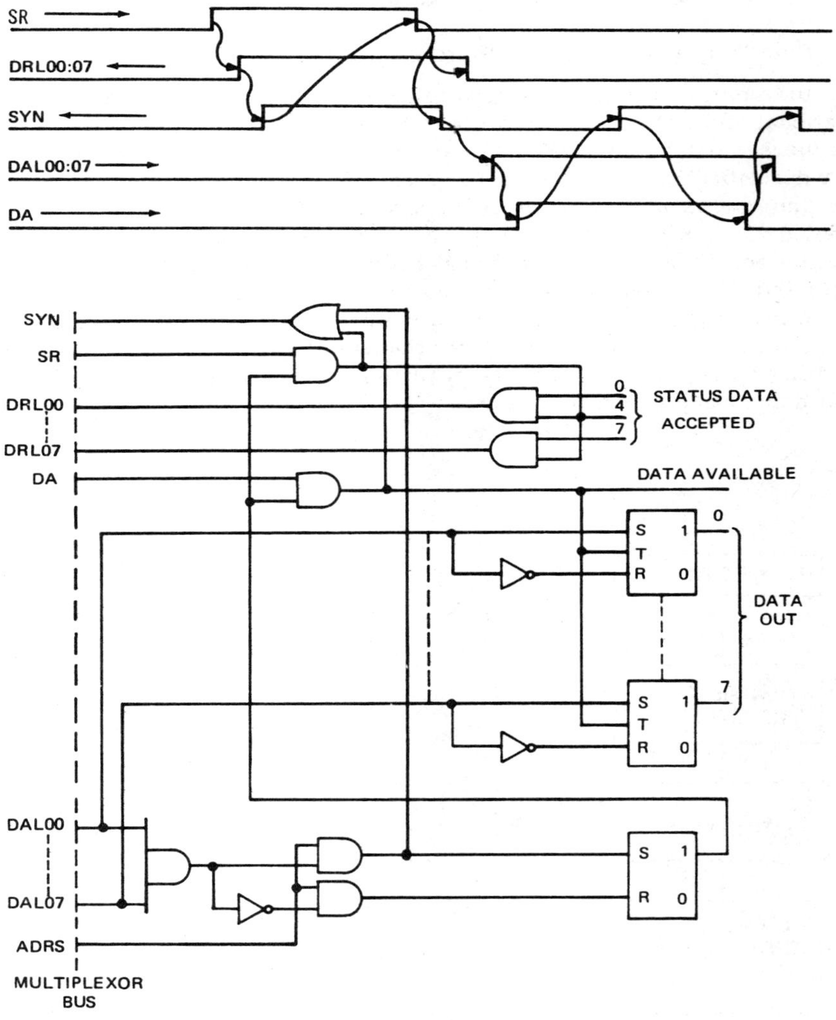

FIG. 6.10 Transmission of a data word with the help of a skip command. (Courtesy of Interdata, Perkin-Elmer, Oceanport, N.J.)

Data transfer takes place as follows (Fig. 6.10):

1. The CPU turns the sense data line SR on. The interface of the addressed peripheral sends a status byte via lines DRLOO, ..., O7 to the CPU, and at the same time the SYN line is activated. (The procedure of checking the status was described under Status Instruction.)
2. The CPU checks the status byte. If any of the status bits 0 to 6 are set, the next instruction of the program is entered which, for example, may cause a jump into an error routine, (Fig. 6.9). If only bit 7 is set, it indicates that the peripheral is busy. Then the CPU leaves the SR signal turned on until the peripheral is ready for data communication.
3. Before initiation of the data transfer, the SYN line is set on LOW. The data to be transferred is put on the DALOO, ..., O7 lines by the accumulator. Simultaneously the data available signal DA is issued.
4. The data available signal DA strobes the data to the flip-flops F2 to F9 with the help of gate G6. At the same time gate G6 activates gate G4, which in turn issues the SYN signal.
5. Upon receipt of the SYN signal, the DA line is turned off and data is removed from lines DALOO, ..., O7. The interface of the peripheral turns the SYN signal off.
6. If the data transfer was successful, the next instruction (the jump to the error routine) is bypassed and the program continues.

Data Input Instruction (Read Instruction)

Data input to the CPU can only take place if the peripheral is functional and not busy. Status control of the peripheral is done similarly as in the read mode (Fig. 6.11).

1. The CPU turns the sense status line SR on. The interface of the peripheral currently addressed sends the status byte to the CPU. (This procedure has been described in the section Status Instruction.)
2. The CPU checks the status byte. If any of the status bits 0 to 6 are set, the next instruction of the program is executed, which, for example, could be a jump to an error routine (Fig. 6.9). If only bit 7 is set, the peripheral is busy. Then the CPU keeps the SR signal HIGH until the peripheral is ready for data exchange.
3. To initiate data transfer, the SYN line is turned off. The CPU activates the data request line DR.
4. With the help of gates G10 to G17 data is sent to lines DRLOO, ..., O7. Simultaneously a SYN signal is issued via gate G4 and data is entered into the accumulator.

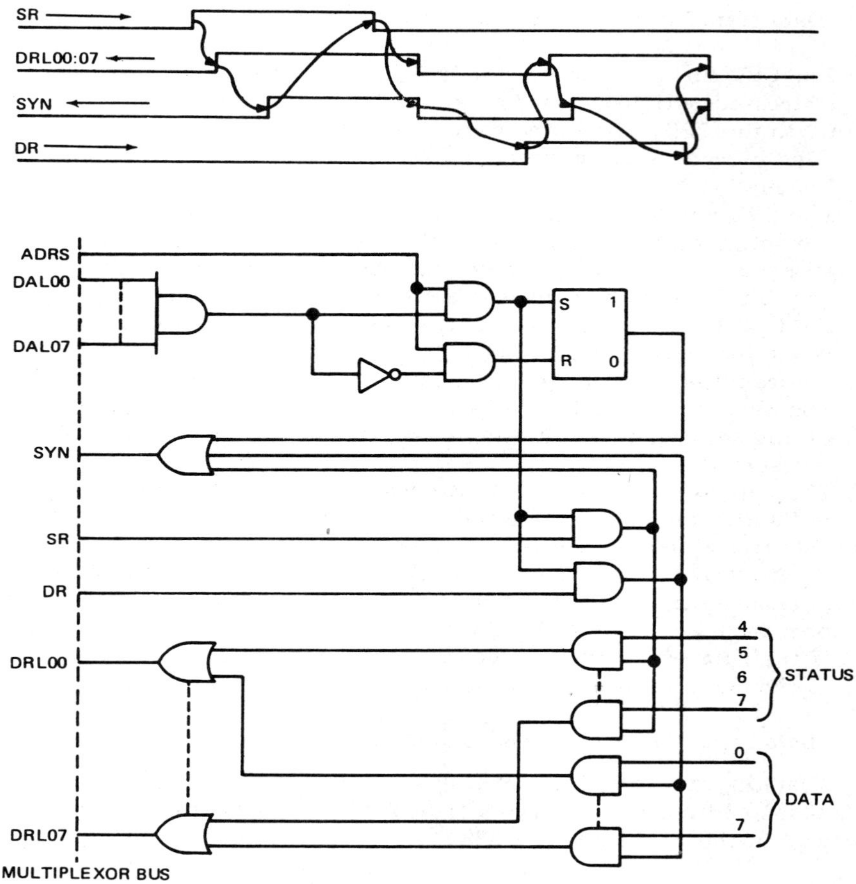

FIG. 6.11 Reading a data word with the help of a skip command. (Courtesy of Interdata, Perkin-Elmer, Oceanport, N.J.)

5. The CPU deactivates the DR line, and the interface removes the data from lines DRLOO, ..., O7. The data transfer has been completed.

Acknowledge Instruction (Interrupt Processing)

This instruction is used to respond to an interrupt, and to determine which of the 256 possible peripherals needs to be serviced (Fig. 6.12).

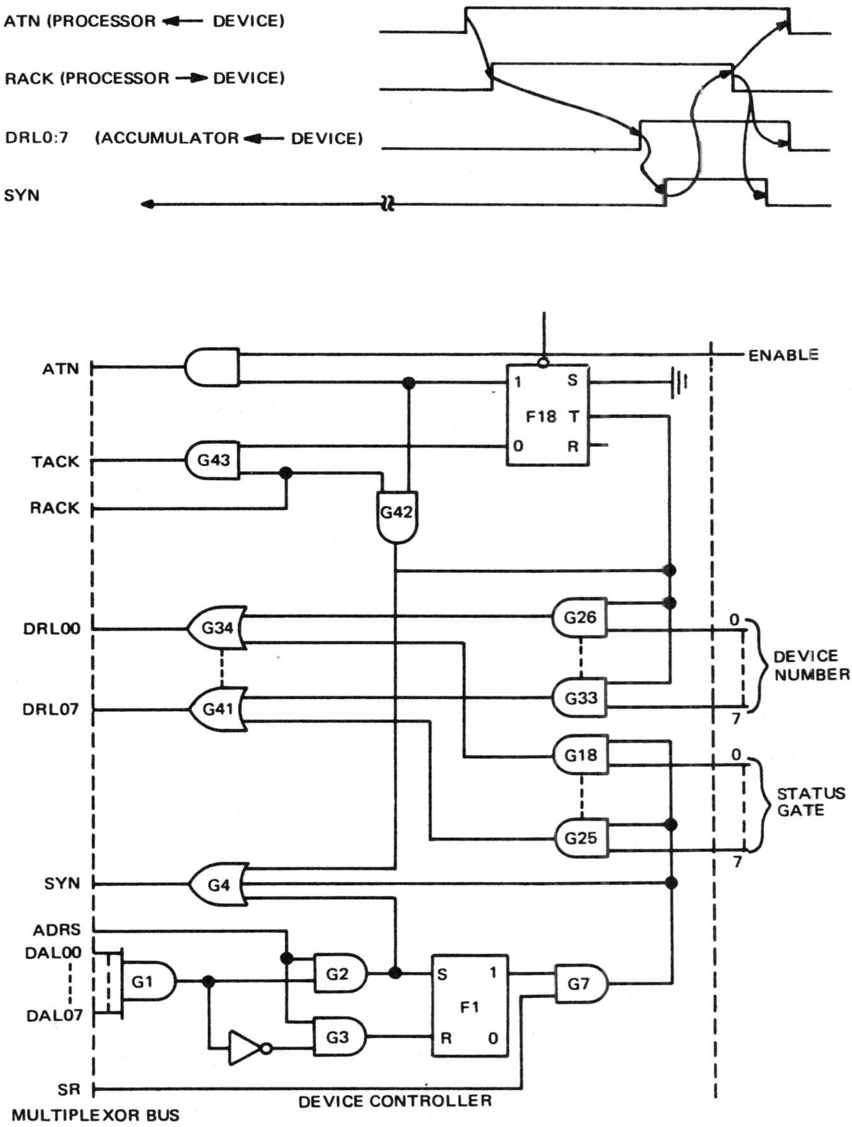

FIG. 6.12 Interrupt logic. (Courtesy of Interdata, Perkin-Elmer, Oceanport, N.J.)

1. The peripheral requesting an interrupt puts its interrupt flip-flop F18 into a HIGH state, initiating the transfer of a signal via the ATN line.
2. The program responds to the interrupt with an acknowledge instruction.
3. The CPU activates the acknowledge line RACK.
4. The first peripheral connected to the bus receives the acknowledge signal RACK. If flip-flop F18 of this peripheral is not set, the acknowledge signal is passed on to the next peripheral as an acknowledge signal RACK. This device in turn receives the signal as RACK. This interrogation repeats itself until the peripheral has been found which requested the service. The located peripheral sets the lower output of flip-flop F18 to LOW, which is connected to gate G43. All peripherals of lower priority are disconnected from the interrupt chain.
5. The signal RACK and the HIGH signal of the flip-flop F18 activate the address circuits of flip-flops G26 to G33 and send the SYN signal to the CPU. Likewise, flip-flop 18 is reset, and the ATN signal is turned off.
6. The CPU strobes the peripheral's address into the accumulator.
7. The ACK signal is removed, and the interface puts the SYN line on LOW.

6.3.4 Address Assignment to an Interface

Usually the design of an interface is such that its address can be hard wired with the help of wire jumpers (Fig. 6.13). Once a priority has been assigned to an input/output peripheral, it will be connected to the I/O bus according to its place in the priority chain. In this example the address can be selected arbitrarily from 256 combinations. However, one must take caution that each address only occurs once. Otherwise, it is possible that peripherals with the same address but of lower priority can never perform data exchange. The address is hard wired by jumpers to the device address decoder of the interface board. If necessary, this address can be changed by resoldering.

6.3.5 Bus Amplifier

Data transfer via a bus is limited by the length of the bus, the data transfer rate, the number of peripherals, and the technology of the integrated circuits used for the transmission system. For greater bus lengths and above a certain number of peripherals, bus amplifiers must be used (Fig. 6.14). Under certain conditions several bus drivers may be necessary.

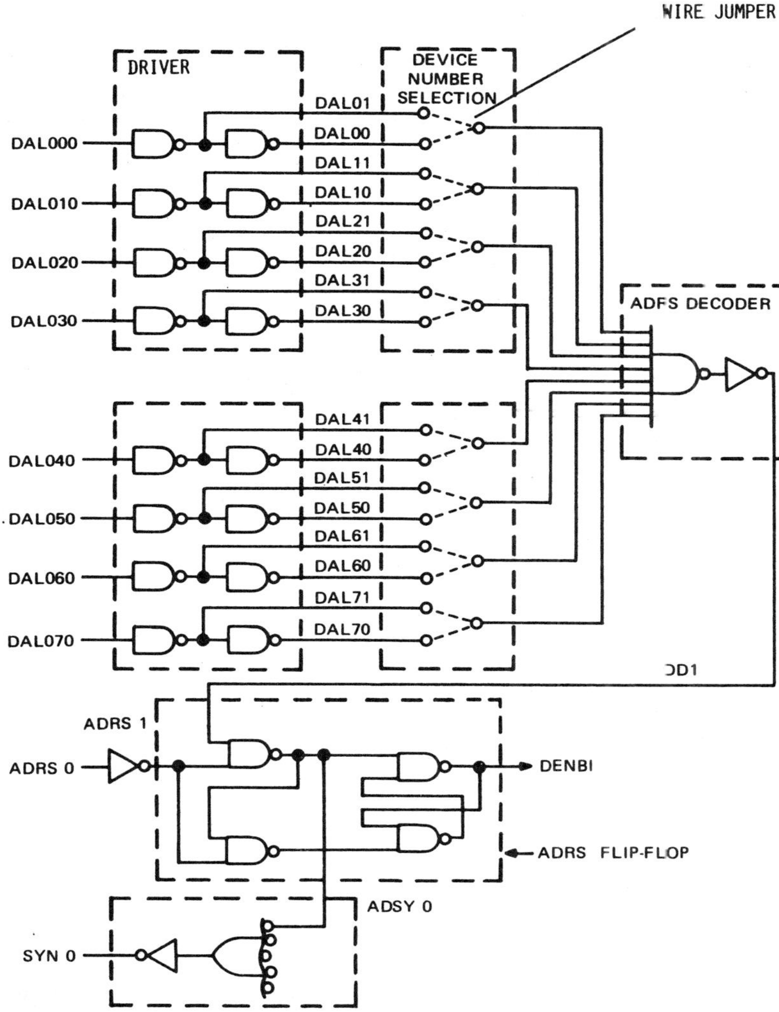

FIG. 6.13 Assignment of addresses to the address logic by means of wire jumpers. (Courtesy of Interdata, Perkin-Elmer, Oceanport, N.J.)

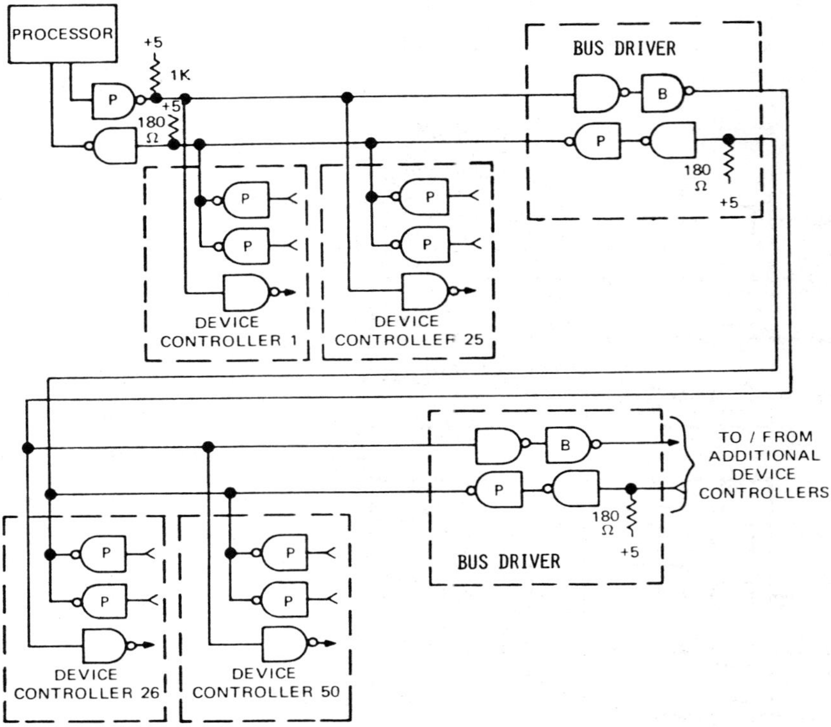

FIG. 6.14 Bus amplifier. (Courtesy of Interdata, Perkin-Elmer, Oceanport, N.J.)

6.4 DMA CHANNEL

6.4.1 Assignment of Bus Lines

A schematic diagram of the DMA channel was shown in Fig. 6.1. The number of DMA lines can vary, depending on the type of computer. In our model computer there are 41 lines (Fig. 6.15). Their functions are described in Table 6.4.

6.4.2 Description of Line Functions

DMAC, Device Selection Lines: These lines transfer control signals which determine whether a DMA channel or the CPU requests the next memory cycle.

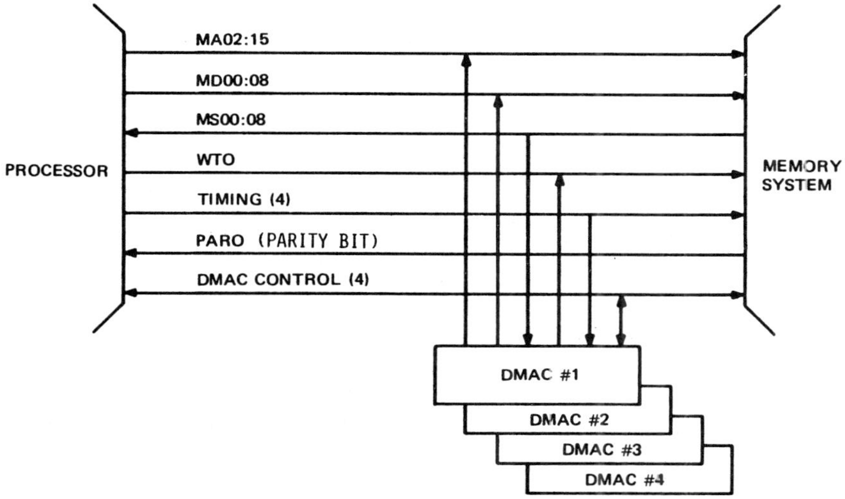

FIG. 6.15 Line assignment in a DMA channel. (Courtesy of Interdata, Perkin-Elmer, Oceanport, N.J.)

TIMING, Timing Lines: These four lines are ncessary to inform the CPU as to whether or not it exchanges data during the present memory cycle.

Early Read (ERO) and Late Read (LRO): Control the read current in the addressed memory module.

Inhibit (INHO) and Write (WTO): Control the write current in the addressed memory module.

WTO, Read/Write Control Line: The read/write line determines whether the current memory cycle is a read or write cycle.

MS00, ..., 08, Memory Strobed Readout Lines: Transfer the strobed data from the memory. During the write operation these lines are not used.

MD00, ..., 08, Memory Data Lines: Transfer data into the memory.

MA00, ..., 13, Memory Address Lines: With these 14-bit address lines 16,384 memory locations can be addressed.

PARO: This line enables the parity control logic when a parity memory module is addressed.

The memory can be accessed by abiding to the following priority rules. Four DMA channels are connected to the computer in a daisy-chain arrangement (Fig. 6.15). The CPU has the lowest priority. Only

TABLE 6.4 Tasks of Bus Lines for a DMA Channel

Number of lines	Function	Abbreviation	Direction of data transfer
4	Device selection	DMAC	
	CPU		CPU ⇄ Memory
	DMA channel		DMAC ⇄ Memory
4	Timing lines	Timing	
	Memory		CPU → Memory
	DMA channel		CPU → DMAC
1	Read/Write	WTO	
	DMA channel		CPU → Memory
			DMAC → Memory
9	Memory strobed readout (including one parity bit line)	MS00, ..., 08	
	CPU		Memory ← CPU
	DMA channel		Memory ← DMAC
9	Memory data (including one parity bit line)	MD00, ..., 08	
	CPU		Memory → CPU
	DMA channel		Memory → DMAC
14	Memory address lines	MA00, ..., 13	
	CPU		CPU → Memory
	DMA channel		DMAC → Memory
4	Parity enable line	PARO	Memory → CPU

one peripheral can use the memory bus at any given time. As soon as a DMA channel uses the bus, the CPU disconnects itself from the bus and at the same time issues an enable signal. Thus the DMA channel can connect itself to the bus. All timing signals are issued by the CPU. The DMA channel occupies the lines MA00, ..., 13; MD00, ..., 08; and WTO.

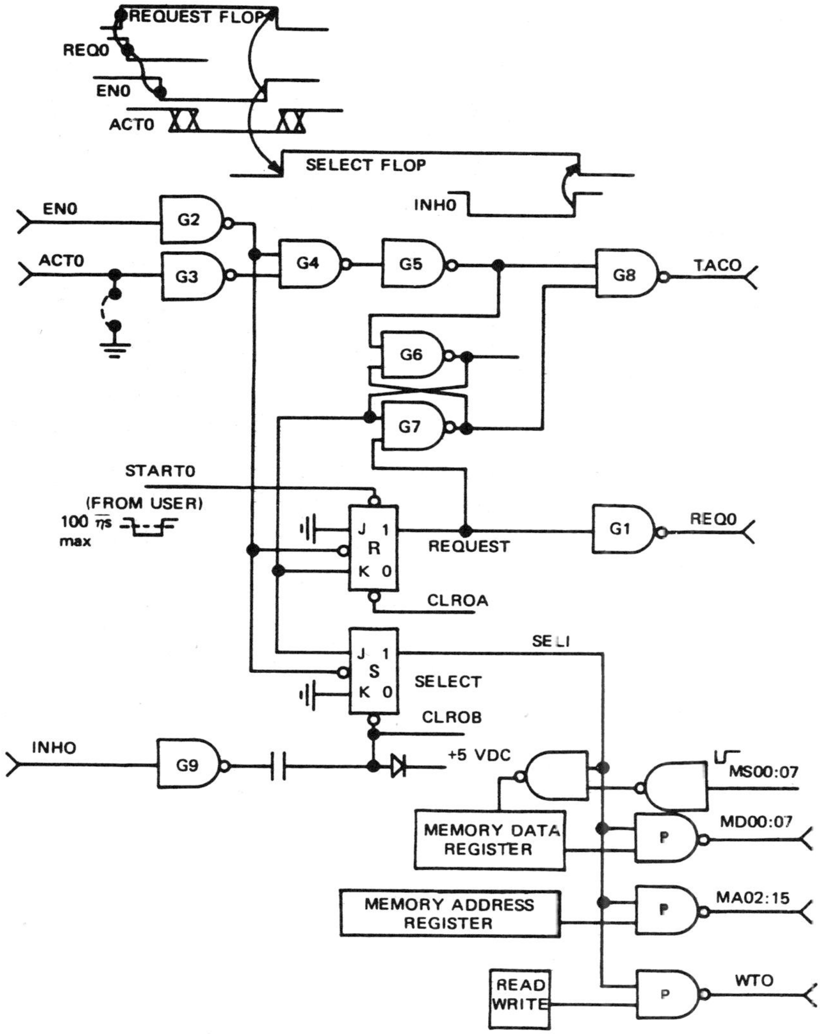

FIG. 6.16 Daisy-chain interrupt logic for DMA. (Courtesy of Interdata, Perkin-Elmer, Oceanport, N.J.)

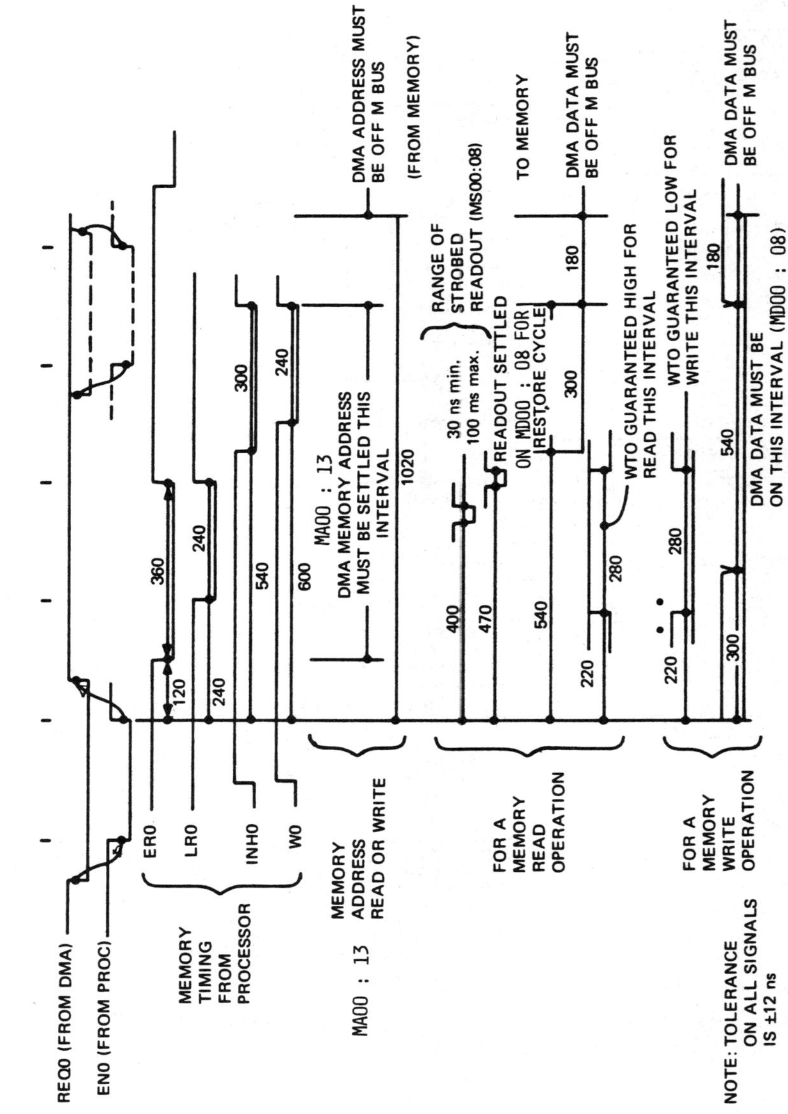

FIG. 6.17 DMA read and write cycle. (Courtesy of Interdata, Perkin-Elmer, Oceanport, N.J.)

6.4.3 The Function of the DMA Interface

When the signal STARTO sets a request flip-flop R, the request for data exchange from the DMA channel is initiated. The peripheral places its request LOW signal on the line via gate G1 (Figs. 6.16 and 6.17) and the CPU responds with ENO. Gates G2 through G7 are the interrupt circuit. Gates G6 and G7 determine whether or not the interrupt originates in this specific DMA channel. If yes, the DMA channels with lower priority on the daisy chain are turned off and data transfer is initiated. If not, the ACTO signal is passed on as a TACO signal and is investigated by the interrupt circuit of the next DMA channel in order to determine who has requested the interrupt. This process is completed when the DMA channel requesting the service has been found. The request flip-flop S is turned to HIGH. Thus the DMA channel has been connected to the bus. After completion of this transfer cycle the signal INHO turns the select flip-flop back to LOW and disconnects the channel.

For the DMA channel with the highest priority the ACTO input is grounded. For the other channels the signal TACO of the proceeding DMA channel determines the state of the signal ACTO of the following channel (Fig. 6.18).

Upon start-up of the computer system or after a power failure, the flip-flops R and S can be initiated by the signals CLROA and

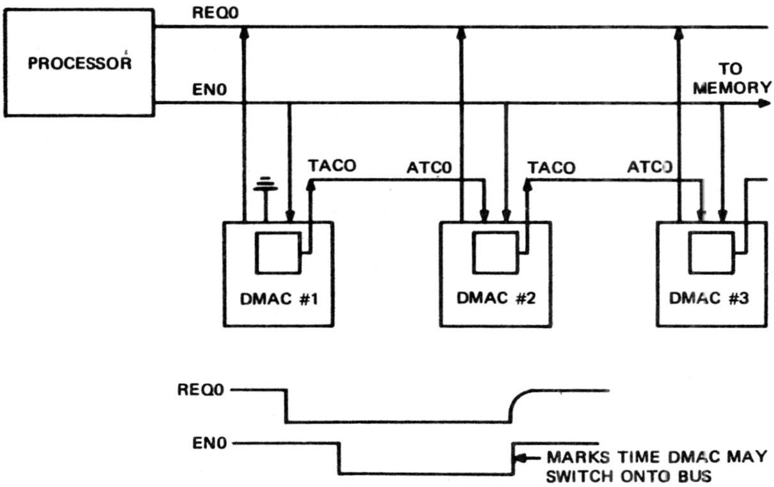

FIG. 6.18 Interrogation of interrupting units in a DMA channel. (Courtesy of Interdata, Perkin-Elmer, Oceanport, N.J.)

CLROB. The lines INHO and WTO control the circuits which activate or deactivate the write current to memory. If the WTO line is LOW, a read cycle takes place. If it is HIGH, a write cycle is initiated. The memory address lines are triggered by the rising edges of the ENO signals. The memory strobed readout lines MS00, ..., 08 transfer informaton from the memory to the DMA channel. The latter is also responsible for regenerating (over lines MD00, ..., 08) information erased during reading. Data is written into memory via lines MD00, ..., 08.

6.5 DESCRIPTION OF A UNIVERSAL BUS

In this section the function of the universal bus of the Dietz computer 621 will be explained. A description is only given for the signal transfer on the bus. The individual functions of the interface and the peripheral equipment will not be discussed. The interface for a universal bus is characterized by its design and is different from that used for a two-bus system. Many individual functions, however, are similar (for example, addressing of the peripheral, transfer and storage of control signals). A universal bus concept is also used in the PDP 11 and many modern microcomputers. However, microcomputers usually have a simpler interrupt architecture.

6.5.1 Address Space

The two-bus system described in Secs. 6.3 and 6.4 has access to the external peripherals via computer registers. With the universal bus it is possible for individual components of the computer system to exchange data directly among themselves (Chaps. 2 and 4). The external bus is an extension of the internal bus. All external peripherals have an address and can be addressed like memory addresses. Memory address instructions can also be used for addressing peripherals and special input/output instructions are not necessary. This method of addressing peripherals is called *memory mapping*.

To address the peripherals the address space of the computer is partitioned accordingly (Fig. 6.19). In the Dietz computer special address space is provided for data as well as for process peripherals. For example, if a new process peripheral is connected to the bus, its address has to be between 2000_{16} and 4000_{16}.

The interrupt circuit of the model computer with the two-bus system of Chap. 5 was designed as a daisy chain. The priority of each connected interface is determined by its location on the bus. Thus, it is not possible to change under program control the priority of peripherals. With the design of the Dietz bus system, however, different priorities can be assigned to peripherals under program control.

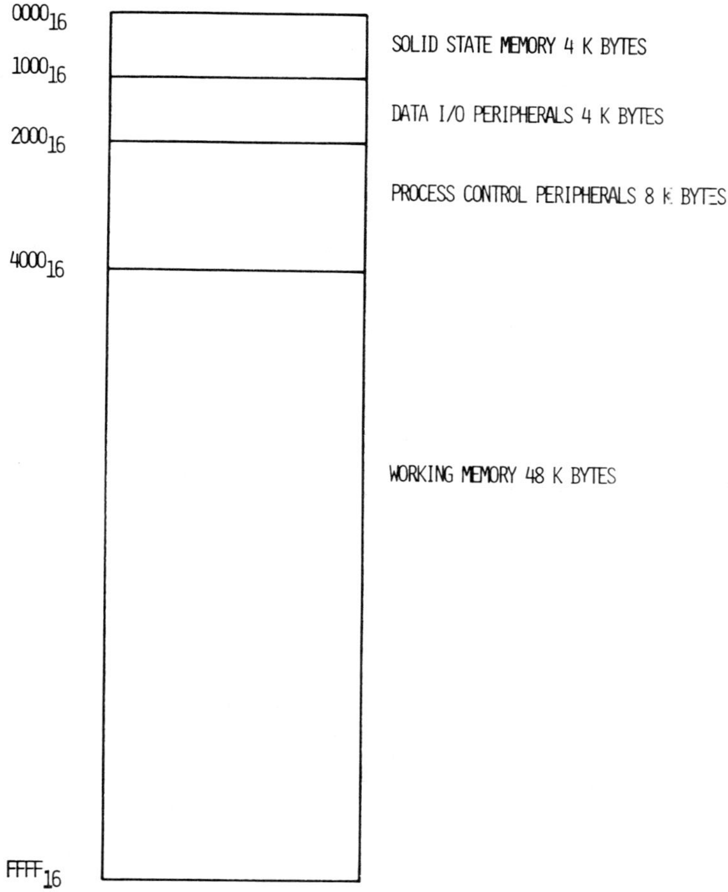

FIG. 6.19 Assignment of address space in a universal bus computer. (From Ref. 2, courtesy of Dietz Computer Systems, Mülheim a.d. Ruhr, Federal Republic of Germany.)

6.5.2 Bus Management

The fundamental concept of the bidirectional bus system of the Dietz computer can be seen in Fig. 6.2. The active peripherals have DMA capabilities. They are directly connected to the bus, whereas the passive peripherals only perform data exchange via a device selector under program control.

Communication over the bus is performed in the master-slave mode. The bus system has been conceived as follows:

1. Active elements request the bus on their own and perform data exchange. The CPU is responsible for the bus management and works only as an active element.
2. The active elements are connected directly to the universal bus. Program-controlled data input/output and direct memory access can be performed via this data channel.
3. At any given time only one active element can control the bus. A busy flag signal prevents further access.
4. If several active elements desire to use the bus, the CPU controls the priority of data transfer. Each peripheral must have a predetermined priority.
5. Passive elements cannot request the bus. They are called upon by active elements for data exchange and announce completion of the exchange by an "END" signal. The transfer time is independent of the bus length.
6. All passive elements can be connected to the minibus by a device selector.
7. Active elements can be declared passive elements and behave as such, when requested by the CPU to do so.
8. Memory modules of the working memory are always passive.
9. The bus can be used in bidirectional mode so that a direction signal must be issued.

6.5.3 Assignment of Bus Lines

The universal bus of this computer has 63 signal lines. Their assignment is shown in Table 6.5. The computer is byte oriented and therefore has only eight data lines. The address field of an instruction consists of an 8-bit main address by which the peripherals connected to the universal bus can be addressed (including the device selectors). With the help of the remaining 4 bits it is possible to call each of the 16 subaddresses of the peripherals connected to the individual minibusses. A total of 768 peripherals can be connected to the computer.

The level lines LOO, ..., L15 indicate at which of the 16 possible levels the computer is operating. This information is needed for selecting level-bound interfaces. Interrupt lines SOO, ..., S15 allow performance of an interrupt request by a peripheral at the corresponding desired level. The other lines shown in Table 6.5 need no further explanation.

6.5.4 Program-Controlled Data Transfer

Data transfer can only be initiated (and controlled) by an active element. If the CPU operates as an active element, program-controlled data transfer takes place. If another element parallel to the program which is being executed exchanges data with a memory, it is done under direct memory access. The bus management is the same in both cases.

TABLE 6.5 Assignment of Functions to the Lines of the Universal Bus

Number of lines	Function	Initialized by	Abbreviation	Direction of data transfer CPU — peripherals
8	Data		DOO, ..., O7	→ / ←
16	Address	Master	AOO, ..., A15	→ / —
1	Direction of data transfer	Master	RK	→ / ←
1	Bus busy	Master	BE	→ / ←
1	Data transfer completed	Slave	FE	→ / ←
1	Bus wanted	Master	GE	→ / —
1	System clear	CPU	N	↔
16	Level lines, shows level at which computer operates	CPU	LOO, ..., L15	→
16	Interrupt lines	Master Slave	SOO, ..., S15	←
1	Parity error	Master Slave		←

Data Output Cycle

The CPU requests the bus with the help of the BE signal and puts the address of the desired peripheral on the address lines AOO, ..., A15 (Fig. 6.20). At the same time, the direction indicator RK is issued. The level information can be issued independently of the bus request on lines LOO, ..., L15. When the data signals have reached steady state, after a predetermined time the addressed peripheral strobes the data in and confirms their receipt by the signal FE. After receipt of the FE signal the active element deactivates the address, data, and direction lines by removing the BE signal. The

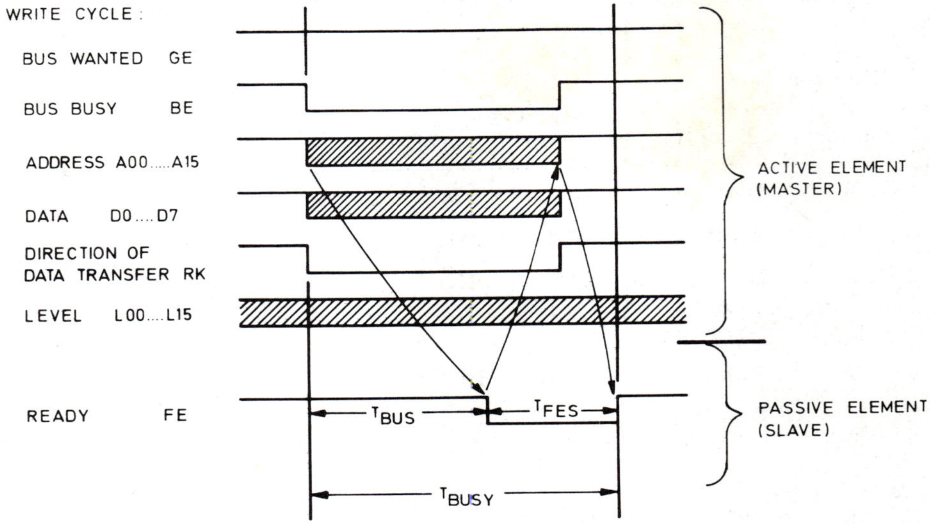

FIG. 6.20 Data output cycle of a universal bus. (From Ref. 2, courtesy of Dietz Computer Systems, Mülheim a.d. Ruhr, Federal Republic of Germany.)

passive element removes the signal FE and data transfer is completed. The CPU itself does not have to issue a bus request signal.

Data Input Cycle

The active element requests the bus and places a direction signal and an address on the bus. The addressed passive element initiates an internal read cycle and makes the data available after the time period T_{BUS} (Fig. 6.21). Thereafter, a delayed confirmation signal FE is issued in order to signal the availability of data. After recognizing the completion signal FE, the active element strobes the data in and turns the direction indicator RK, the bus-busy signal, and the address off. The passive element removes the FE signal and makes the bus available for further use.

In order to avoid blocking of the bus a watchdog timer is initiated during activation of the bus-busy signal BE. If no FE signal is issued after 1 sec, data transfer is interrupted and an error message is given.

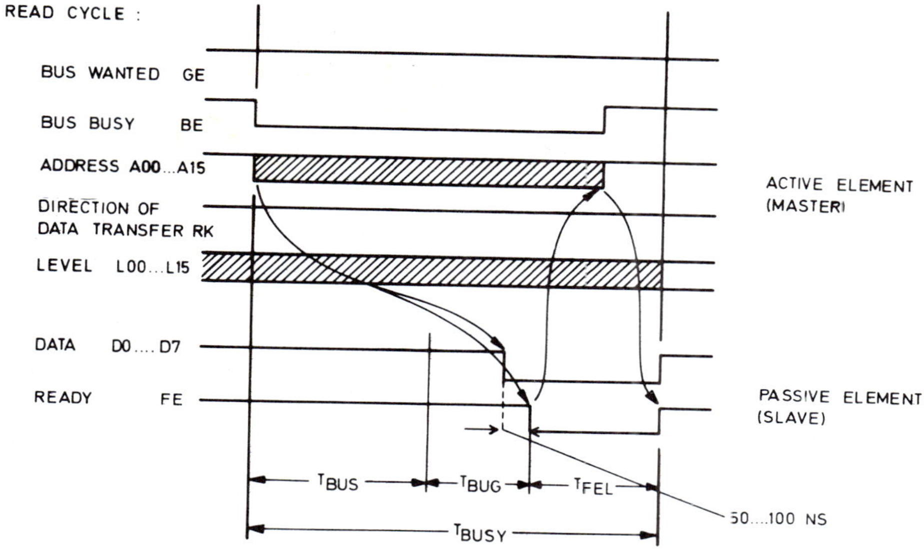

FIG. 6.21 Data input cycle of a universal bus. (From Ref. 2, courtesy of Dietz Computer Systems, Mülheim a.d. Ruhr, Federal Republic of Germany.)

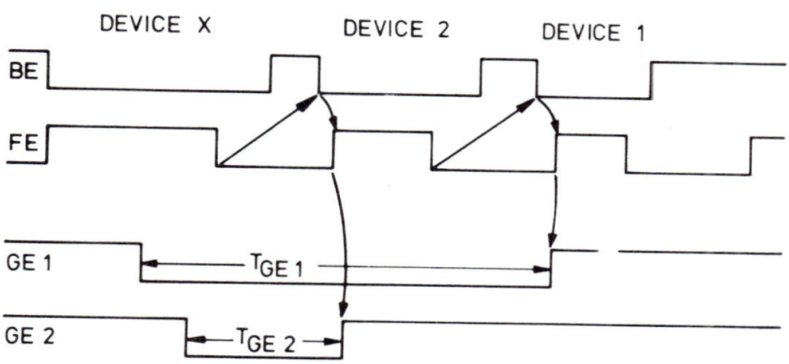

FIG. 6.22 Priority control of a universal bus. (From Ref. 2, courtesy of Dietz Computer Systems, Mülheim a.d. Ruhr, Federal Republic of Germany.)

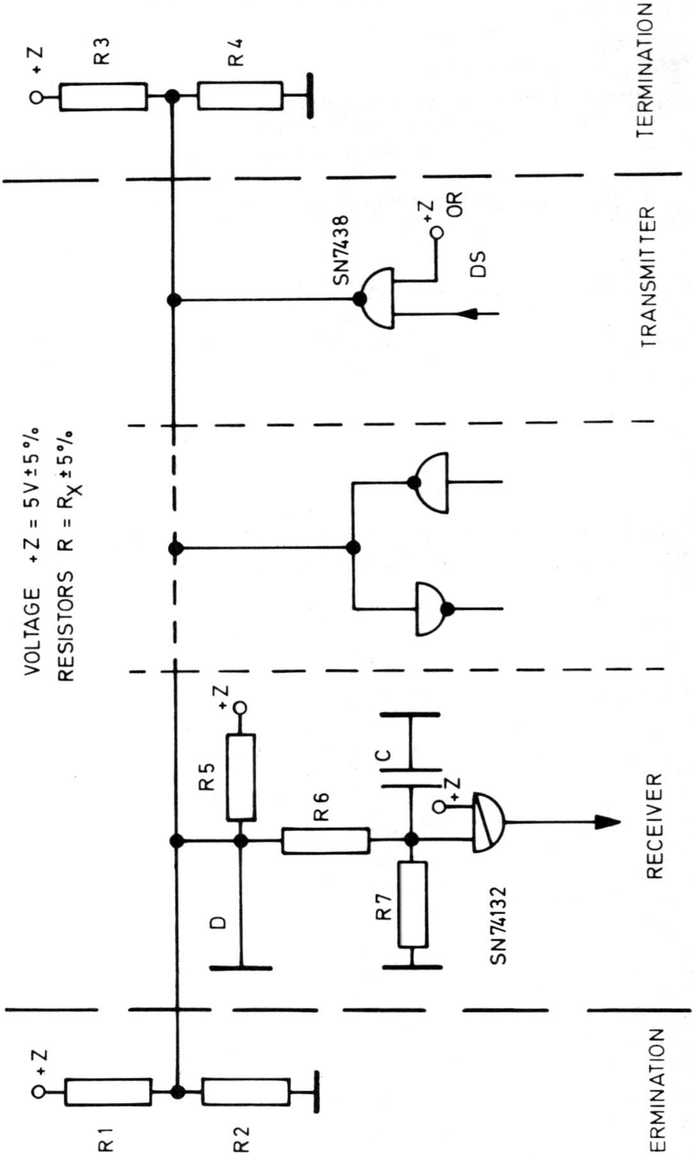

FIG. 6.23 Interfacing devices to a universal bus. (From Ref. 2, courtesy of Dietz Computer Systems, Mülheim a.d. Ruhr, Federal Republic of Germany.)

Priority Control

If several peripherals attempt to occupy the bus, the priority is controlled by the signal GE (Fig. 6.22). The bus is requested by the active element with the BE signal and is released by the passive element with the FE signal after completion of data transfer.

In Fig. 6.22 both peripherals 1 and 2 attempt to request the bus at the same time; in other words, the signals GE1 and GE2 are turned on. Since peripheral 2 has a higher priority, peripheral 1 must wait until 2 makes the bus available again. Peripheral 1 can only request the bus when the signals BE and FE have not been issued.

6.5.5 Hardware Characteristics of the Bus

Since the entire data communication with the CPU and memory is done over the bus, fast transfer speeds are a prerequisite. A transfer rate of 1×10^6 to 3×10^6 words per second is obtainable. The bus wires must be carefully terminated at their ends to prevent reflections (Fig. 6.23). The impedance of the bus is approximately 50 to 100 Ω. To guarantee noise-free data transfer for each group of signals, careful attention has been given to the design of the individual groups of signal lines. If more than six transmitter and receiver stations are connected to the bus, every 4 or 5 m the bus signals must be reshaped by the bus amplifier.

REFERENCES

1. INTERDATA, Model 1 User Manual, Interdata, Inc., Palo Alto, Calif., 1970.
2. Dietz 921 Handbuch, Dietz Computer Systeme, Heinrich Dietz, 433 Mühlheim Ruhr, 1976.
3. PDP11, *Processor Handbook*, Digital Equipment Corporation, Maynard, Mass., 1979-1980.

7

SOFTWARE FOR INTERFACING

In this chapter the fundamental structure of software for input and output will be discussed. These program modules make it possible to exchange data between the computer and the interface. A program which initiates data exchange, transfers the data, and completes the exchange will be called the *driver*. Since computer configurations differ considerably from one manufacturer to another, drivers must be designed specifically for each computer. Quite often even drivers of computers of the same make are not interchangeable. In general, the design of interface software is not a task for the beginner. To be able to program a driver, the computer hardware, the instruction set, and the operating system must be well understood. Software for interfacing can be fairly simple or very complex, this depending on the application, the operating mode of the peripheral, the architecture of the computer system, and the configuration of the peripherals connected to it.

7.1 SOFTWARE/HARDWARE TRADEOFF

In Sec. 1.3 it was mentioned that it is often difficult to determine which functions should be performed by hardware and which by software. When building interfaces, there often exists a combination where a well-conceived tradeoff between hardware and software leads to the most cost-effective design (Fig. 7.1). Unfortunately, this solution is very difficult to determine. One needs excellent software and hardware experience in order to find a good design. In practical applications one often finds that the electrical engineer trends toward a hardware and the computer programmer toward a software solution. Pure hardware or pure software designs in general have their specific advantages or disadvantages (Table 7.1). The hardware solution offers fast speed, and the software solution offers simplicity in programming.

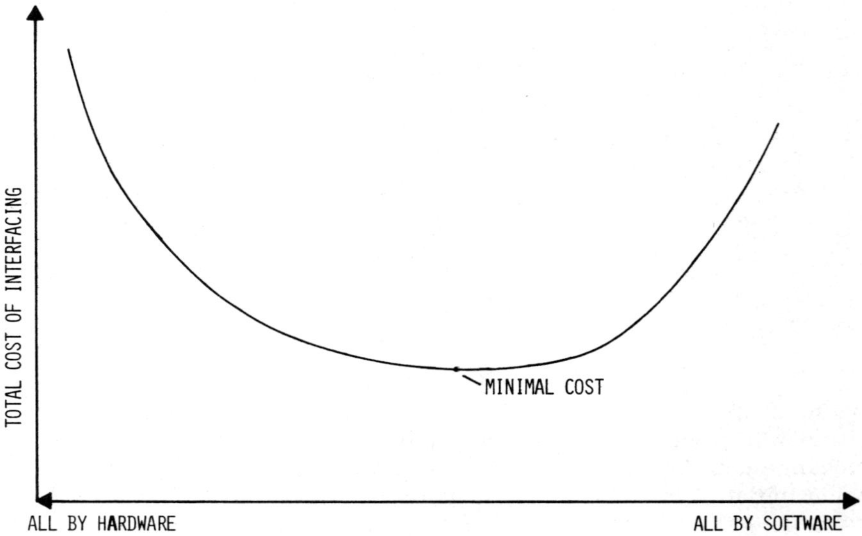

FIG. 7.1 Minimizing interfacing cost by proper hardware/software tradeoff.

Hardware has its advantages where complete functions of an interface can be designed into an integrated circuit. For example, nowadays universal asynchronous receiver transmitter (UART) circuits are commonly used to perform data transfer. This hardware does a multitude of different functions such as the synchronization of transmitter and receiver, parity check, and parallel-to-digital conversion. In old systems data communication procedure and control had to be done by software, which, in turn, could occupy the computer for a long period of time. Likewise, today characters are seldom generated by software on a display screen. For this we can employ a hardware circuit, called a *character generator*. It is addressed by software and creates the desired character on the screen.

Figure 7.2 shows an application where a robot is directly controlled by a computer [1]. A dividing line indicates which tasks are performed by hardware and which by software. In this system a large part of the control function is actually done by software. When software is required to perform sample and hold functions at high speed (Fig. 7.2), the response of the computer may be too slow to control the process. This is often the case with machine tools. In complex

TABLE 7.1 Software/Hardware Tradeoff

Advantages	Disadvantages
Hardware	
Very fast	Expensive
Parallel operation of several units possible	Difficult to change
	Susceptible to noise
Frees computer from routine operations, thereby increases its computational capacity	Long development times
	Ages quickly
Software	
Easy to alter	Under certain conditions slow, since there is serial execution
Inexpensive	
After initial start-up period very reliable	Ties computer down to routine tasks, reduces capacity

applications such tasks are often performed by analog circuits to obtain fast control circuits.

Table 7.2 and Fig. 7.3 display different possibilities for the conversion of BCD code into binary code. Such code conversion, for example, is necessary when digital voltmeters are connected to the computer. The measuring parameters will be presented in BCD code to the computer by the digital voltmeter. For further processing the computer has to convert this code into binary information. The software routine shown in Table 7.2 can easily be written. If the voltmeter has to perform many analog measurements, code conversion has to be done for each parameter measured, and this can be very time consuming. An alternative solution is code conversion with the help of the hardware circuit, shown in Fig. 7.3b [2]. This code converter is designed for six BCD decades. When employed with a 16-bit computer, however, there are only five decades necessary. In this case the circuit is simpler. It is very easy to see that the development effort for this hardware, in comparison to writing software, is rather expensive. For this converter a circuit board must be manufactured, parts have to be soldered to it, and it must be tested. The hardware solution, however, would offer a considerable reduction in computing time.

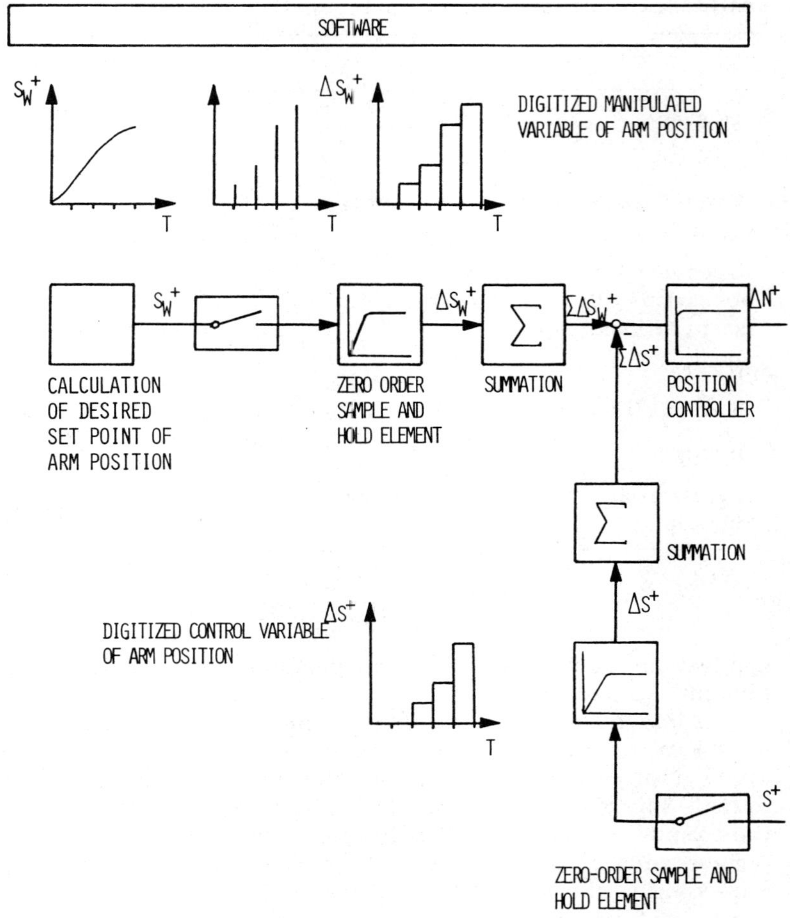

FIG. 7.2a Software/hardware tradeoff for a digital control loop of a robot: S_W^+, set point of arm position; ΔS_W^+, digitized set point of arm position; S, analog control variable of arm position; S^+, control variable to be digitized; ΔS^+, digitized control variable of arm position; ΔN^+, digitized manipulated variable of arm position; N_W, analog manipulated variable of arm position. (From Ref. 1.)

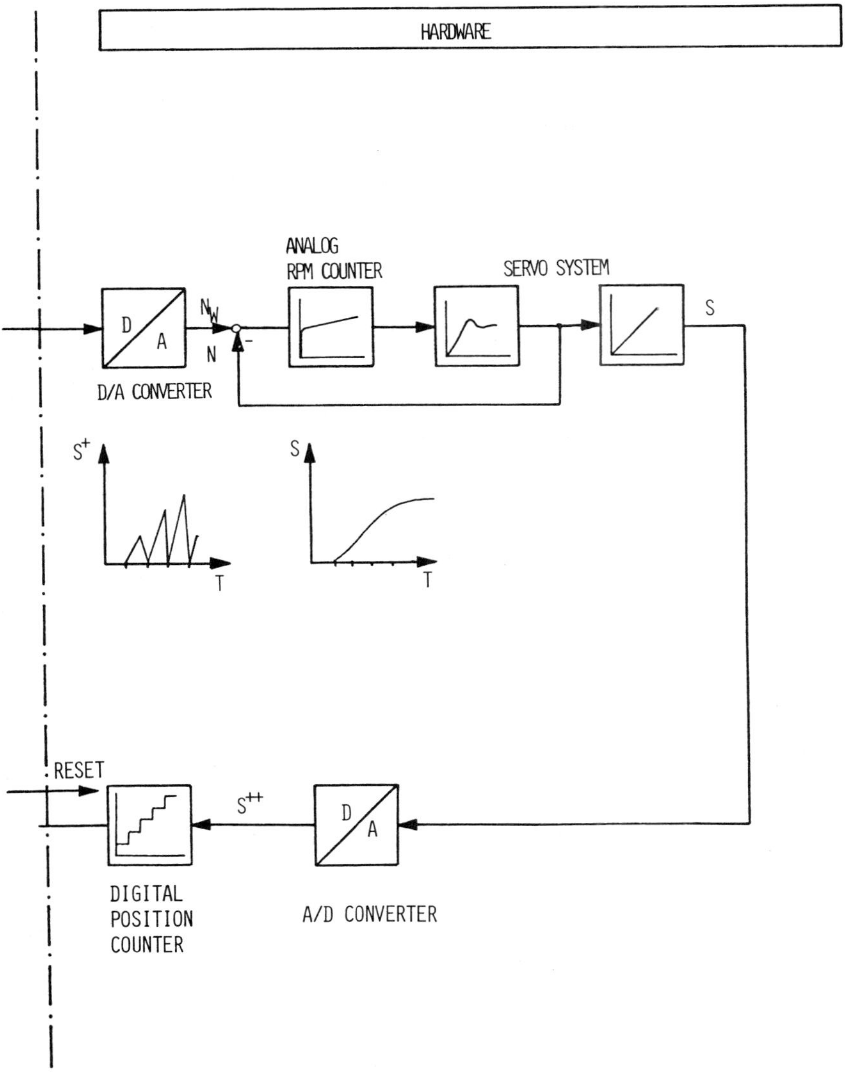

FIG. 7.2b (Continued.)

TABLE 7.2 Routine to Convert BCD Code to Binary Code

```
 BCD ROUTINE   LOAD BCD VALUE IN ACCUMULATOR
               STORE IN MEASURE
               LOAD COUNTER IN ACCUMULATOR
               STORE IN BIT COUNTER
               LOAD FILE ADDRESS IN ACCUMULATOR
               STORE IN POINTER
               LOAD ZERO IN ACCUMULATOR
               STORE IN RESULT
       BEGIN   LOAD MEASURE IN ACCUMULATOR
               CHECK MOST SIGNIFICANT BIT
               JUMP TO CONTINUE IF = 0
               LOAD POINTER IN ACCUMULATOR INDIRECT
               ADD RESULT
               STORE IN RESULT
               LOAD POINTER IN ACCUMULATOR
               ADD 1
               STORE IN POINTER
    CONTINUE   LOAD MEASURE IN ACCUMULATOR
               ROTATE ACCUMULATOR ONE BIT LEFT
               STORE IN MEASURE
               LOAD BIT COUNTER IN ACCUMULATOR
               ADD 1
               STORE IN BIT COUNTER
               CHECK ACCUMULATOR
               JUMP TO END IF = 0
               JUMP TO BEGIN IF NOT = 0
         END   JUMP BCD ROUTINE INDIRECT

     MEASURE   DATA  ....
     COUNTER   DATA  -16
 BIT COUNTER   DATA  ....
FILE ADDRESS   DATA  1000
     POINTER   DATA  ....
        ZERO   DATA     0
      RESULT   DATA  ....
        1000   DATA  8000
               DATA  4000
               DATA  2000
               DATA  1000
               DATA   800
               DATA   400
               DATA   200
               DATA   100
               DATA    80
               DATA    40
               DATA    20
               DATA    10
               DATA     8
               DATA     4
               DATA     2
               DATA     1
```

6-BIT CONVERTER

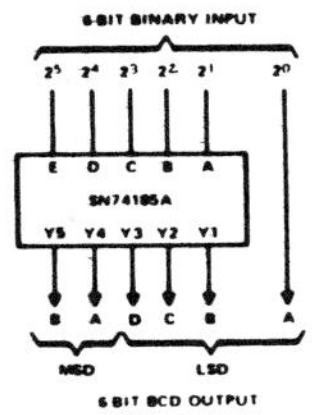

FUNCTION TABLE

BINARY WORDS	INPUTS						OUTPUTS							
	BINARY SELECT					ENABLE								
	E	D	C	B	A	G	Y8	Y7	Y6	Y5	Y4	Y3	Y2	Y1
0 · 1	L	L	L	L	L	L	H	H	L	L	L	L	L	L
2 · 3	L	L	L	L	H	L	H	H	L	L	L	L	L	H
4 · 5	L	L	L	H	L	L	H	H	L	L	L	L	H	L
6 · 7	L	L	L	H	H	L	H	H	L	L	L	L	H	H
8 · 9	L	L	H	L	L	L	H	H	L	L	L	H	L	L
10 · 11	L	L	H	L	H	L	H	H	L	L	H	L	L	L
12 · 13	L	L	H	H	L	L	H	H	L	L	H	L	L	H
14 · 15	L	L	H	H	H	L	H	H	L	L	H	L	H	L
16 · 17	L	H	L	L	L	L	H	H	L	L	H	L	H	H
18 · 19	L	H	L	L	H	L	H	H	L	L	H	H	L	L
20 · 21	L	H	L	H	L	L	H	H	L	H	L	L	L	L
22 · 23	L	H	L	H	H	L	H	H	L	H	L	L	L	H
24 · 25	L	H	H	L	L	L	H	H	L	H	L	L	H	L
26 · 27	L	H	H	L	H	L	H	H	L	H	L	L	H	H
28 · 29	L	H	H	H	L	L	H	H	L	H	L	H	L	L
30 · 31	L	H	H	H	H	L	H	H	L	H	H	L	L	L
32 · 33	H	L	L	L	L	L	H	H	L	H	H	L	L	H
34 · 35	H	L	L	L	H	L	H	H	L	H	H	L	H	L
36 · 37	H	L	L	H	L	L	H	H	L	H	H	L	H	H
38 · 39	H	L	L	H	H	L	H	H	L	H	H	H	L	L
40 · 41	H	L	H	L	L	L	H	H	H	L	L	L	L	L
42 · 43	H	L	H	L	H	L	H	H	H	L	L	L	L	H
44 · 45	H	L	H	H	L	L	H	H	H	L	L	L	H	L
46 · 47	H	L	H	H	H	L	H	H	H	L	L	L	H	H
48 · 49	H	H	L	L	L	L	H	H	H	L	L	H	L	L
50 · 51	H	H	L	L	H	L	H	H	H	L	H	L	L	L
52 · 53	H	H	L	H	L	L	H	H	H	L	H	L	L	H
54 · 55	H	H	L	H	H	L	H	H	H	L	H	L	H	L
56 · 57	H	H	H	L	L	L	H	H	H	L	H	L	H	H
58 · 59	H	H	H	L	H	L	H	H	H	L	H	H	L	L
60 · 61	H	H	H	H	L	L	H	H	H	H	L	L	L	L
62 · 63	H	H	H	H	H	L	H	H	H	H	L	L	L	H
ALL	X	X	X	X	X	H	H	H	H	H	H	H	H	H

H = high level, L = low level, X = irrelevant

(a)

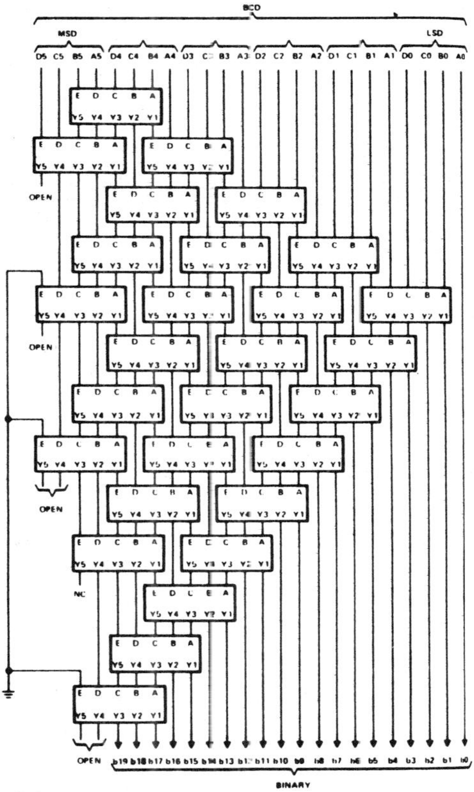

(b)

FIG. 7.3 Conversion of BCD code to binary code by hardware. (From Ref. 2, courtesy of Texas Instruments Deutschland GmbH, Freising, Federal Republic of Germany.)

7.2 INPUT/OUTPUT PROGRAMMING

7.2.1 Program-Controlled Input/Output

The input/output architecture of a process control computer is considerably more complex than that of a commercial computer. Fast execution of a process signal is required. Furthermore, the computer must perform useful control functions and should not wait for slow peripheral devices. These requirements are met with the help of elaborate interrupt circuits. The execution of a interrupt request is done by the operating system (software) and the hardware circuit of the interrupt system. In more advanced computers this task is often performed by hardware since execution of software is relatively slow. The number of available interrupts in a computer is not necessarily a measure for the quality of the interrupt system. The time elapsing from the announcement of an interrupt until the jump into the subroutine for processing this interrupt is approximately 2 to 70 μsec in contemporary computers. For time-critical processes this response time must be taken into consideration.

In the following paragraphs the various possibilities of data input/output will be described. It is assumed that the interface will have a temporary buffer. During data output the computer fetches information from its accumulator and transfers it into the temporary buffer. The hardware logic of the interface provides further processing of the data. In the reverse direction during data input the interface first places its data into the temporary buffer and the CPU fetches the information from here to transfer it into its accumulator. With this arrangement the CPU does not have to wait for the slow peripheral. During configuration of the process control system, each peripheral obtains its individual address. Thus it becomes possible to address the peripheral with an instruction and to transfer data via the temporary buffer.

Unconditional Data Transfer Using No Interrupt

The simplest input/output system can be designed with an unconditional data transfer. During input the CPU assumes that the interface has data available for transfer. Likewise, during output, it is assumed that the temporary buffer is able to accept data. The following example (Table 7.3) shows data input from the temperature measuring instrument (Sec. 6.3.1).

First the temperature measuring instrument holding the address 10 is turned on. Then the address of the first thermocouple is brought from the memory location TC1 to the accumulator. The next instruction transfers the contents of the accumulator into the instruction register of the interface. This in turn initiates the measurement of the thermocouple TC1. The measured parameter is stored in the

TABLE 7.3 Unconditional Program-Controlled Data Transfer Using No Interrupt

Memory address	Contents	Comments
START	ADDRESS 10	The temperature measuring instrument 10 is activated
	LOAD TC1 IN ACCUMULATOR	Loads the subaddress of the thermocouple TC1 into the accumulator
	TURN ON 10, TC1	Turns on thermocouple TC1
	INPUT 10, TC1	Fetches the measured parameter from the temporary buffer and puts it into the accumulator
	STORE ACCUMULATOR IN MEASURE 1	Stores the measured parameter in storage location MEASURE 1
	LOAD TC2 IN ACCUMULATOR	Loads the subaddress of thermocouple TC2 into the accumulator
	TURN ON 10, TC2	Turns on thermocouple TC2
	INPUT 10, TC2	Fetches the measured parameter from the temporary buffer and puts it into the accumulator
	STORE ACCUMULATOR IN MEASURE 2	Stores the measured parameter in storage location MEASURE 2
	HALT	
TC1	DATA 001	Instruction codes for selecting thermocouple
TC2	DATA 010	
MEASURE 1	DATA...	Storage places for measured parameters
MEASURE 2	DATA...	

temporary buffer. Thereafter the CPU issues an input instruction and transfers the measured parameter from the temporary buffer to the accumulator. From here the measured data is stored in storage location MEASURE 1. Thereafter the measurement of thermocouple TC2 is initiated. It is easy to see that this input/output operation can lead to difficulties. There is the possibility that during initiation of the data transfer the interface is already busy and can neither strobe data in nor out. Without a protected buffer it would be possible to read incomplete data or to overwrite unread data. For this reason an unconditional data input/output is hardly ever used.

Conditional Data Transfer Using No Interrupt

In order to eliminate the disadvantage of unconditional data transfer the interface can be provided with a flag flip-flop. During data output, the CPU checks the flag flip-flop and initiates transfer only when the peripheral has announced that it is ready to do so by raising the flag bit. During data input, the peripheral first stores information in the temporary buffer and then signals to the CPU with the help of a flag flip-flop that data are ready for transfer. The program discussed above can therefore be improved as follows (Table 7.4).

After the thermocouple has been turned on, the status of the measuring instruments is checked. If there is a problem, the program branches to an error search routine. If no problem exists, the program checks whether a flag flip-flop has been set. If yes, the next instruction is skipped, and the data transfer to the accumulator is initiated. Otherwise, the program enters a loop and waits until the bit has been set. This type of input/output has the disadvantage that the computer has to wait in the loop until a flag flip-flop has been set. Under certain conditions it is possible that it never leaves the loop and that the computer is blocked.

In conditional program-controlled input/output one has to make certain that the sequence of processing instructions is done correctly. Figure 7.4a shows a case in which 256 measured parameters are read from a data buffer to a printer. The program fetches one measured parameter and attempts to send it to the printer. Thereby it waits until the printer is ready. An improvement can be obtained by the program shown in Fig. 7.4b. Here the measured parameter is sent to the printer only when it is not busy [3].

Program-Controlled Input/Output Using Interrupt

Waiting in a loop can be eliminated by adding an interrupt circuit to the computer. For example, if a program being executed advises the computer to perform a measurement, it first turns on the instrument and then initiates the reading of the parameter. During this period, the execution of the main program continues since the instru-

TABLE 7.4 Conditional Program-Controlled Data Transfer Using No Interrupt

Memory address	Contents	Comments
START	ADDRESS 10	
	LOAD TC1 IN ACCUMULATOR	
	TURN ON 10, TC1	
	CLEAR FLAG 10	The flag flip-flop which was set during the preceeding input is reset.
	CHECK STATUS 10	Checks status; in case of an error, execute the next instruction; otherwise the next instruction is to be skipped.
	JUMP HELP	Jump to error search routine.
	CHECK FLAG 10	Checks flag flip-flop; if not set, go to the next instruction; if set, the next instruction is to be skipped.
	JUMP - 1	
	INPUT 10	
	STORE ACCUMULATOR IN MEASURE 1	
	LOAD TC2 IN ACCUMULATOR	
	TURN ON 10, TC2	
	CLEAR FLAG 10	
	CHECK STATUS 10	
	JUMP HELP	
	CHECK FLAG 10	
	JUMP - 1	
	INPUT 10	
	STORE ACCUMULATOR IN MEASURE 2	

TABLE 7.4 (Continued)

Memory address	Contents	Comments
HELP	...	Error search routine.
	HALT	
TC1	DATA 001	
TC2	DATA 010	
MEASURE 1	DATA...	Storage for measured data.
MEASURE 2	DATA...	

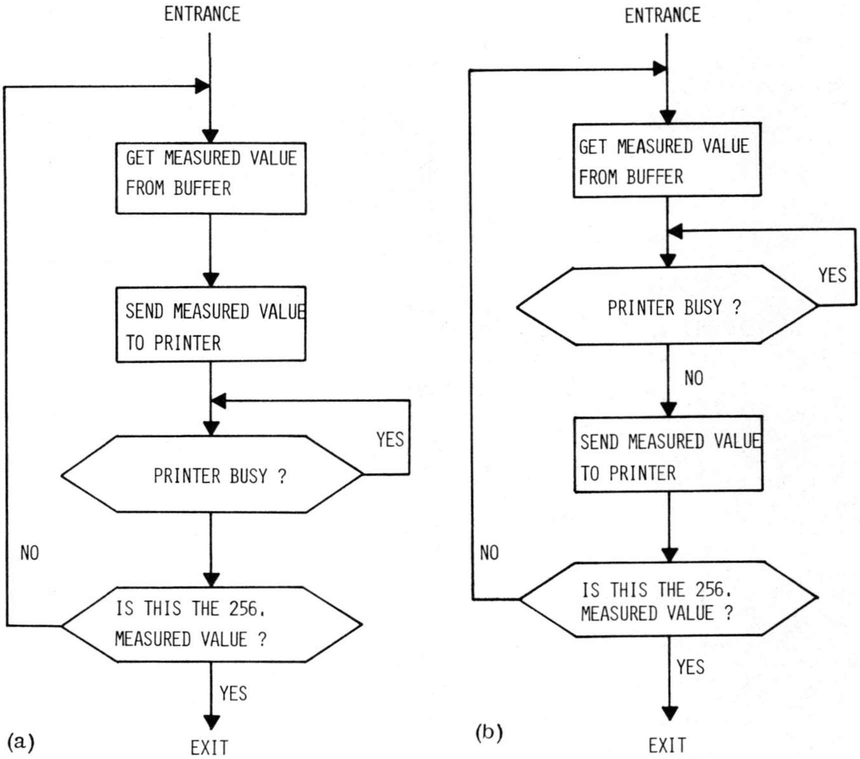

FIG. 7.4 (a) Synchronization of output of measured data to printer. (b) Overlap of output of measured data with printer operation.

ment can perform the reading by itself. As soon as the measured parameter is available in the temporary buffer, the main program is interrupted, and the computer is then able to transfer the data. The following steps are necessary for this procedure:

Recognition of the interrupt request.
Verification of the interrupt.
Check whether interrupt is permitted.
Completion of the current instruction of the interrupted program.
Interruption of the program.
Storage of the contents of all registers and counters used by the current program.
Disable interrupt system.
Determine priority of the interrupting peripheral.
Jump to the subroutine to perform data transfer.
Enable interrupt system.
Restore the former contents of the registers and counters.
Continue the main program.

Data exchange with interrupt is also conditional data transfer; however, it brings the essential advantage that the computer must not wait for the slow peripheral. Today almost all process control computers are provided with interrupt logic. In the following sections the structure of programs for typical interrupt architectures of different complexity will be described.

Interrupt Structure with One Hardware Priority.

One peripheral. In the simplest interrupt system all peripherals are connected via the same interrupt line to the computer. There is only one hardware interrupt level for all external peripherals. Now, assume that there is only one peripheral and that it requests service; the following operation then takes place:

1. The peripheral sends an ATTENTION signal to the computer (Sec. 6.3.3).
2. The computer answers with an ACKNOWLEDGE signal.
3. The CPU completes the execution of the current instruction. The contents of the program counter, which contains the address of the next instruction to be executed, is stored in memory cell with the address 0 by hardware (Table 7.5).
4. The instruction located in memory cell 1 is the next one to be executed. It causes a jump to the subroutine SERVICE.
5. The interrupt circuit is enabled so that the initiated data transfer cannot be interrupted by another request.
6. The program saves the contents of the accumulator in memory cell SAVE.

TABLE 7.5 Interrupt-Controlled Data Transfer for One Peripheral

Memory address	Contents	Comments
0000	2378	The address of the next instruction to be executed by the main program.
0001	JUMP SERVICE	Jump to the subroutine.
1000	(Main program)	Starting address of the main program.
2377	(Current instruction)	Interrupt occurs here.
SERVICE	TURN OFF INTERRUPT	Disables interrupt.
	STORE ACCUMULATOR IN SAVE	Saves contents of accumulator.
	INPUT 10	Service peripheral.
	STORE ACCUMULATOR IN MEASURE 1	Stores measured parameter.
	LOAD SAVE IN ACCUMULATOR	Restores original contents of accumulator.
	TURN ON INTERRUPT	Enables interrupt.
	JUMP INDIRECT TO 0000	Jumps to the next instruction in the main program.
SAVE	DATA...	Temporary data storage.
MEASURE	DATA...	Storage for measured parameters.

7. The data of the peripheral to be transferred is loaded into the accumulator and then deposited into memory cell MEASURE.
8. The contents of memory cell SAVE are transferred back into the accumulator.
9. The interrupt circuit is enabled.
10. The program jumps indirectly into memory cell 0 and continues processing of the main program.

Several peripherals. Now we assume that we have several peripherals connected to the same interrupt system. During an interrupt request, we must first determine which peripheral has set its flag

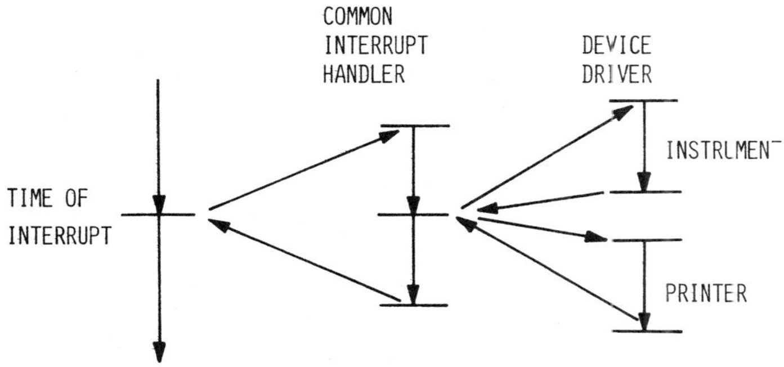

FIG. 7.5 Processing of an interrupt via common interrupt handler and device driver.

(Sec. 6.3.3). In this case the processor must interrogate each peripheral in sequence in order to detect which unit requested service. This task is assigned to the common interrupt handler. As soon as the peripheral with the highest priority is located, data transfer can be performed with the help of a jump-to-a-driver instruction (Fig. 7.5).

The interrupt unit can only process one interrupt at any given time. All other interrupts with lower priority must wait until the currently running service routine has been completed. Afterward, the peripheral with the next highest priority is searched. The system is structured as follows:

Each peripheral has a priority assigned to it.
Different service routines are provided for the various peripherals.
Peripherals with lower priorities are serviced after those with higher priority.

The common interrupt handler for the service of several peripherals is structured as follows (Table 7.6).

The peripheral with the highest priority is serviced first. All other peripherals must wait until this peripheral has completed its data transfer. For this reason it is important that the fastest peripheral is the first one connected in the priority chain. Otherwise, it is possible, if a peripheral is not immediately serviced, that the same peripherals data are lost when it places an interrupt request. The successive search for the requesting peripheral is slow.

An improvement of the interrupt system can be achieved by adding an interrupt register. The requesting peripheral sets 1 bit in the interrupt request register (Fig. 7.6) [4]. This bit points to the be-

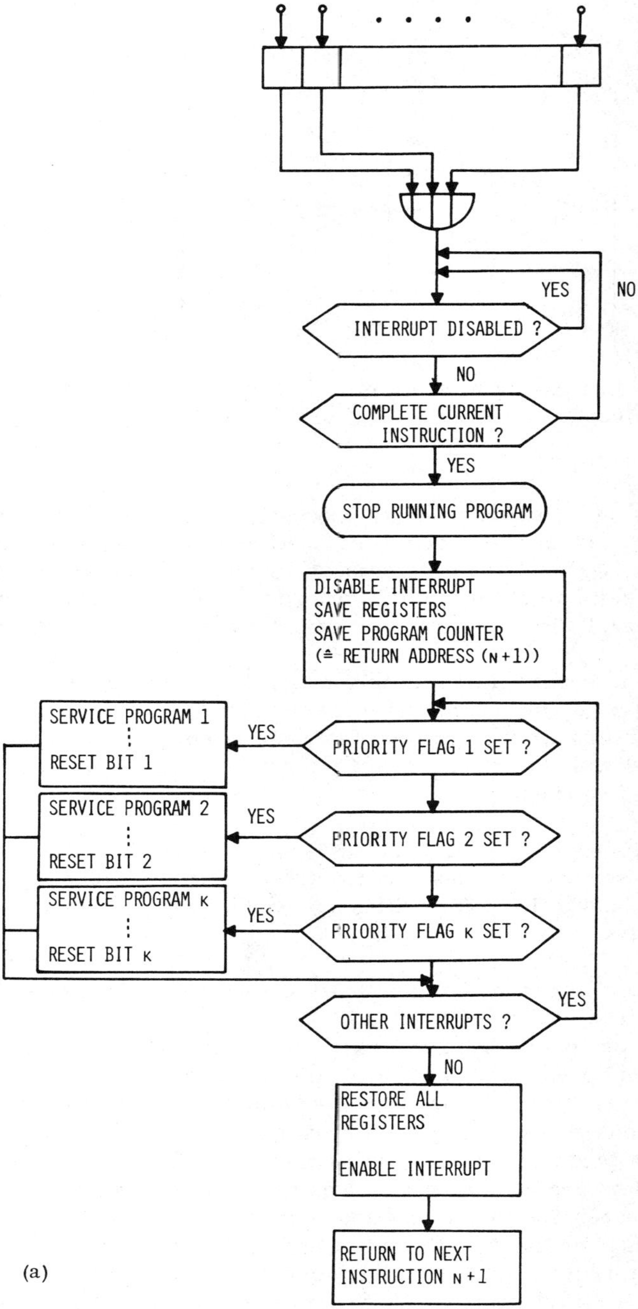

(a)

FIG. 7.6 Software-controlled interrupt service.

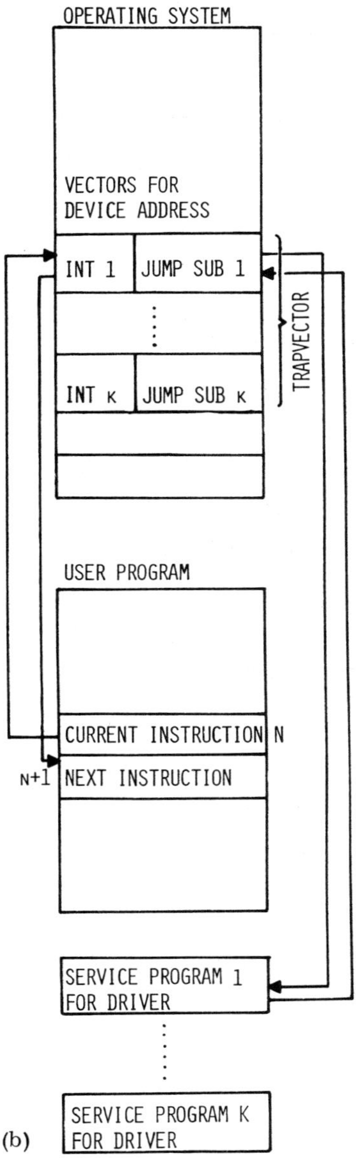

(b)

FIG. 7.6 (Continued.)

TABLE 7.6 Common Interrupt Handler to Initiate Data Transfer for a Computer with Several Peripherals

Memory address	Contents
0000	2378
0001	JUMP SEARCH
2000	(Main program)
2377	(Current instruction)
SEARCH	STORE ACCUMULATOR IN SAVE SKIP IF DEVICE 1 FLAG LOW JUMP SERVICE 1 SKIP IF DEVICE 2 FLAG LOW JUMP SERVICE 2 SKIP IF DEVICE 3 FLAG LOW JUMP SERVICE 3 JUMP ERROR (ERROR MESSAGE)
SERVICE 1	TURN OFF INTERRUPT FLAG 1 INPUT DEVICE 1 STORE ACCUMULATOR IN MEASURE 1 JUMP EXIT
SERVICE 2	TURN OFF INTERRUPT FLAG 2 INPUT DEVICE 2 STORE ACCUMULATOR IN MEASURE 2 JUMP EXIT
SERVICE 3	TURN OFF INTERRUPT FLAG 3 INPUT DEVICE 3 STORE ACCUMULATOR IN MEASURE 3 JUMP EXIT
ERROR	ERROR SEARCH ROUTINE
EXIT	LOAD SAVE IN ACCUMULATOR TURN ON INTERRUPT JUMP INDIRECT TO 0000
SAVE	DATA...
MEASURE 1	DATA...
MEASURE 2	DATA...
MEASURE 3	DATA...

ginning address of the service program for this peripheral. The beginning addresses is located in the so-called trap vector of the operating system. This trap vector in turn points to the service routine. At the end of the service routine an instruction causes a return jump to the vector location in the operating system. Here again the priority of the interrupt to service must be determined.

Masking of Interrupt Levels. In many applications it is necessary to remove or reconnect the peripheral from, or to, the priority chain. This in particular is important when certain parts of the main program have to be protected from specific interrupts or if during interrupt processing peripherals with a lower priority have to be prevented from accessing the processor. For this purpose the interrupt unit of the computer is equipped with a mask register (Fig. 7.7) [4]. This mask can be enabled and disabled through program control. The mask register can be addressed and its contents can be changed with the help of the accumulator during program-controlled input/output. To enable or disable an interrupt, the masking word with the corresponding bit pattern is loaded into the mask register. An AND operation is performed with the contents of the interrupt register and mask register, thereby blocking out all peripherals which may not interrupt the ongoing input/output operation.

For fast processing of an interrupt request of high priority one does not wish to wait until the service of a peripheral with low priority has been completed. Peripherals with high priority must therefore be able to interrupt those of lower priority. Thus several contents of program counters and registers of interrupted service routines must be stored. The following example shows how a peripheral with higher priority interrupts the service of a peripheral with lower priority with the help of a mask, in order to be serviced. By using a mask register, the SERVICE 3 routine of the example of Table 7.6 can be changed to the routine shown in Table 7.7 [5]. It is assumed that all interrupt signals are originally disabled.

If the contents of a mask register have to be altered during the service of the peripheral with a higher priority, its old contents must be saved. Likewise, the contents of the program counter must be saved. After processing the higher priority request the contents of the old register must be restored. In addition, the program must make certain that the interrupt system is enabled and disabled in the correct sequence.

Interrupt System with Multiple Hardware Priorities. With this interrupt system all peripherals have a common address bus (Sec. 6.4.3) on which the requesting peripheral puts the address of its service routine during an interrupt. First, the contents of the accumulator are stored in memory cell TEMP 1. The address of the service pro-

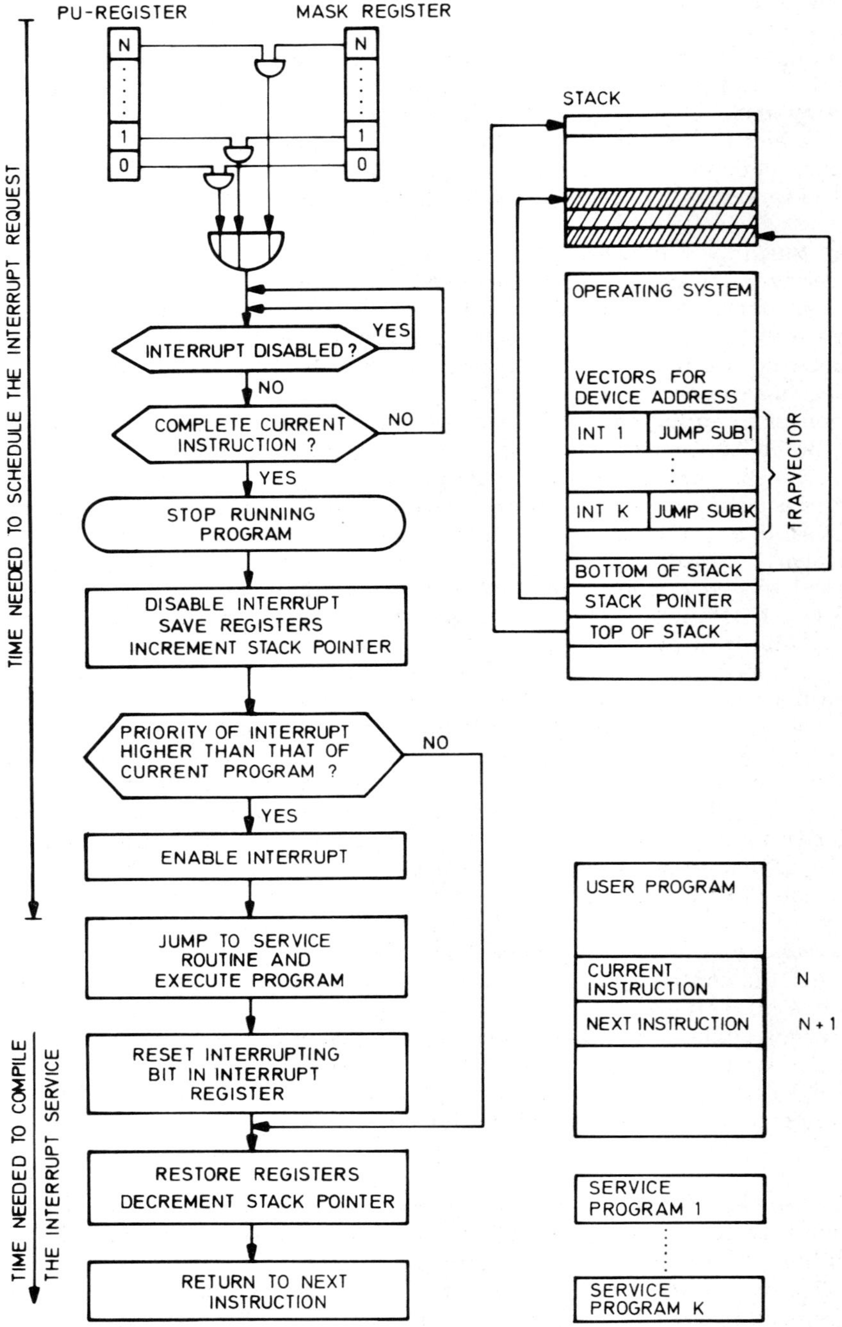

FIG. 7.7 Hardware-controlled interrupt service.

TABLE 7.7 Processing Interrupts with the Help of a Mask Register

Memory address	Contents	Comments
SERVICE	STORE ACCUMULATOR IN TEMP1	Save the contents of the accumulator
	LOAD 0000 IN ACCUMULATOR	Save the program counter contents from memory cells 0000
	STORE ACCUMULATOR IN PC1	
	LOAD MASK REGISTER IN ACCUMULATOR	Save the present mask
	STORE ACCUMULATOR IN TMASK1	
	LOAD MASK 3 IN ACCUMULATOR	Bring the new mask into the masking register
	STORE ACCUMULATOR IN MASK REGISTER	
	TURN ON INTERRUPT	
	INPUT DEVICE 3	
	STORE ACCUMULATOR IN MEASURE 3	Transfer the measured parameters from peripheral 3
	TURN OFF INTERRUPT	
	LOAD TMASK1 IN ACCUMULATOR	
	STORE ACCUMULATOR IN MASK REGISTER	
	LOAD TEMP1 IN ACCUMULATOR	
	TURN ON INTERRUPT	
	JUMP INDIRECT PC1	

Source: Adapted from Ref. 5.

TABLE 7.8 Interrupt Processing in a Computer with Several Hardware Priorities

Memory address	Contents
LOCATE	STORE ACCUMULATOR IN TEMP 1
	READ INTERRUPT ADDRESS
	STORE ACCUMULATOR IN POINTER
	JUMP INDIRECT POINTER
SERVICE 3	LOAD 0000 IN ACCUMULATOR
	STORE ACCUMULATOR IN PC1
	LOAD TEMP 1 IN ACCUMULATOR
	STORE ACCUMULATOR IN TEMP 2
	LOAD MASK REGISTER IN ACCUMULATOR
	STORE ACCUMULATOR IN TMASK1
	LOAD MASK 3 IN ACCUMULATOR
	STORE ACCUMULATOR IN MASK REGISTER
	TURN ON INTERRUPT
	INPUT DEVICE 3
	STORE ACCUMULATOR IN MEASURE 3
	TURN OFF INTERRUPT
	LOAD MASK 1 IN ACCUMULATOR
	STORE ACCUMULATOR IN MASK REGISTER
	LOAD TEMP 3 IN ACCUMULATOR
	TURN ON INTERRUPT
	JUMP INDIRECT PC1

Source: Adapted from Ref. 5.

gram SERVICE 3 is brought into the accumulator and from there stored in a memory location POINTER by an input/output instruction READ INTERRUPT ADDRESS (Table 7.8) [5]. By an indirect jump to the memory location POINTER the service program SERVICE 3 is enabled. This service routine first loads the contents of memory cell 0000 con-

TABLE 7.9 Automatic Hardware-Controlled Interrupt

Memory address	Contents
SERVICE 3	(Return address)
	STORE ACCUMULATOR IN TEMP 2
	LOAD MASK REGISTER IN ACCUMULATOR
	STORE ACCUMULATOR IN TMASK1
	LOAD MASK 2 IN ACCUMULATOR
	(For continuation, see the program in Table 7.8)

Source: Adapted from Ref. 5.

taining the address of the next instruction to be executed of the main program and deposits it in memory location PC1. This must be performed in order to be able to interrupt the service routine itself when necessary. Also the original contents of the accumulator are brought from memory location TEMP 1 to location TEMP 2.

Now the contents of the mask register belonging to the interrupt service routine are temporarily saved for later use in location TMASK 1. The mask of the interrupting peripheral is placed into the mask register, and the program then proceeds as shown in Table 7.7. At the end of the service routine, the original states of the registers are restored. In some computers it is possible to change the interrupt system from one interrupt level to several interrupt levels by adding a new circuit board to the interrupt circuit.

Automatic Hardware-Controlled Interrupt Systems. In more advanced computers the hardware determines the priority of the interrupt request. The computer, in this case, points directly to the cell in the operating system which initiates processing of the interrupt. The program does not have to determine the identity of the interrupt. For hardware-controlled interrupt processing the input/output bus needs additional lines. This interrupt system does not require saving the contents of the program counters and the registers. In our example, the hardware points directly to the address of the service routine SERVICE 3. At the same time the return address of the interrupted program is stored here. The service routine is simplified considerably (Table 7.9) [5].

Since saving and restoring register and memory contents is still very cumbersome, in high-speed computers this task is done completely automatically.

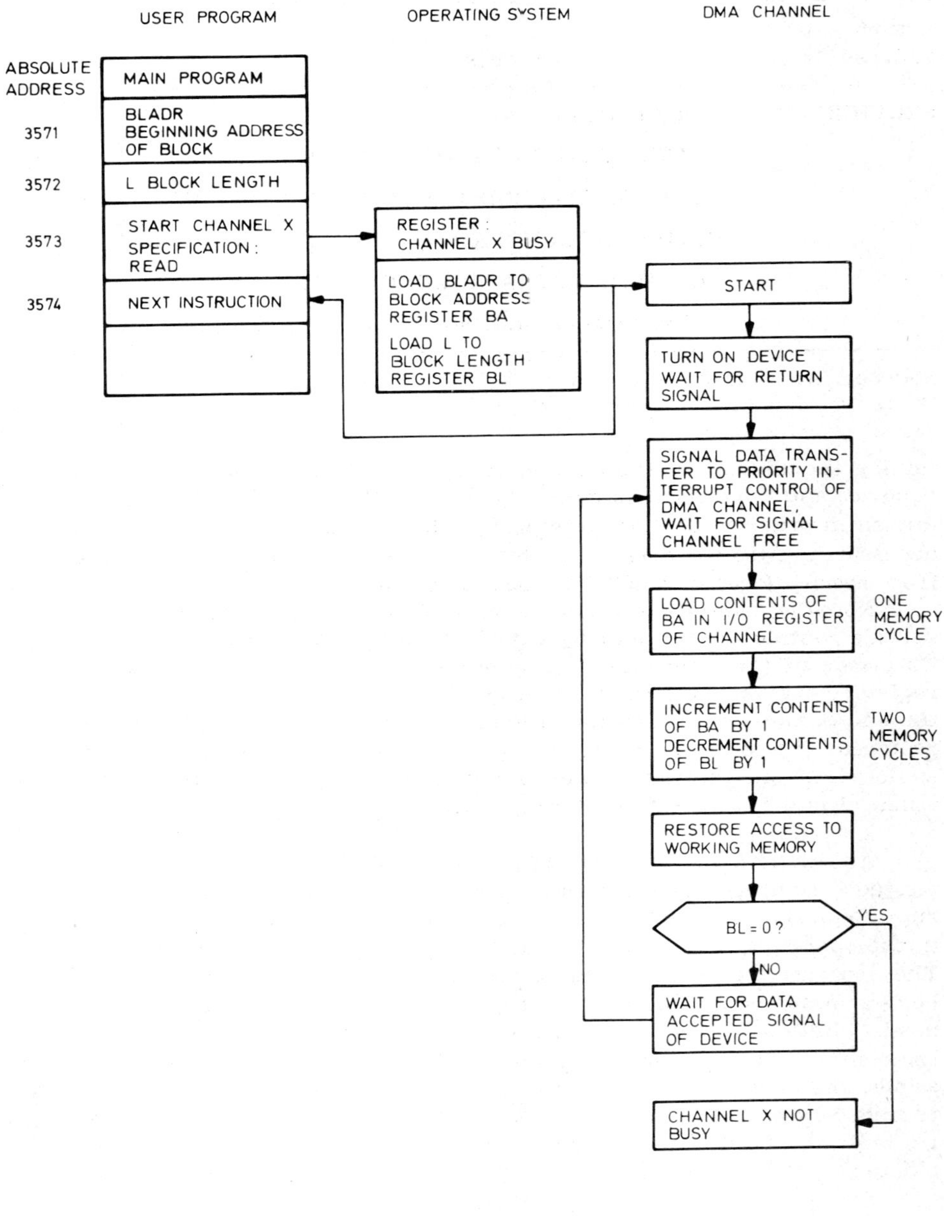

FIG. 7.8 Operation of a DMA channel.

7.2.2 Direct Memory Access

The function of the DMA channel was discussed in Sec. 6.4. When a DMA transfer is initiated, the DMA channel has to obtain information concerning the direction of the data transfer, the number of information units to be transferred, and the location in memory where the information is to be stored or from where it has to be taken. These facts must be made known to the DMA channel by a small subroutine which is program controlled. The following instructions are to be given:

START CHANNEL
LOAD STARTING ADDRESS
LOAD LENGTH OF BLOCK

The typical function of a DMA transfer is shown in Fig. 7.8 [4]. In this example information is read from the working memory and sent to a peripheral for output. After the channel has been started, a jump into the main program is performed in order to continue the program. The channel starts the peripheral and waits for the ready signal. As soon as this signal arrives, the first unit of information will be called from storage with the help of a cycle steal process and will be transferred to the peripheral.

The memory address register of the DMA channel is incremented by one, and the number of information units to be transferred is decremented by one. The use of the memory is made available to the CPU again. The course of these events is repeated until the whole data block has been transferred. Thereafter the channel is made available for other data transfers. During DMA data transfer the computer works parallel with the channel. The cycle steal does not noticeably affect processing of the program. The program just skips one cycle at the end of the current memory cycle and is then continued. It is not even necessary to complete an instruction. There will be no register or memory contents saved and restored. Since the DMA channels have a higher priority than the CPU, the CPU is disconnected during cycle steal from the memory.

7.3 DRIVER

7.3.1 General Task of Drivers

Data transfer between a peripheral and the computer should be performed as quickly as possible and without error. The reaction time of the computer to serve an interrupt depends on various factors:

The design of the interrupt system.
The time needed for processing interrupts.

The time required for execution of individual instructions.
The time needed for execution of the driver program.

In the preceding chapter the influence of the first two factors was discussed. It was shown that processing time for interrupts depends, to a great extent, on the architecture of the interrupt system itself. For example, computers with several hardware interrupt levels are much faster than those having only one hardware level where the priority of a request is determined by software. The execution time of the individual instructions in general cannot be changed by the user, except in the case where microprogramming is provided. The effect of execution time must be considered when selecting a computer. The driver program, however, can be optimized. Typical functions which the driver has to perform during an input/output operation are as follows:

Initializing data transfer.
Performing data transfer.
Transmitting control instruction to the peripheral.
Checking whether instructions to the peripherals are valid.
Supervising the peripheral functions.
Reporting problems with the peripheral to the operating system.
Reporting the driver status to the operating system.
Storage of the information to be transferred in temporary buffers.
Administrating the temporary buffer.
Packing and unpacking of information in memory cells.
Counting of information units to be transferred.
Formating information.
Attaching or removal of redundant information.
Parity checking.
Filtering of measuring variables.
Linearizing of measuring variables.
Transformation of code.

Some of the above mentioned assignments may be performed by the operating system, it is also customary to assign some of the driver tasks to the application program. This is possible, for example, with the last five functions listed. This division of assignments has the advantage that the driver has only a minimum of functions to perform and thus is much faster. The less important functions are done by the application program when it has additional time available.

If there are more of the same type of data peripherals, common drivers can be employed. If one peripheral is just performing an input/output operation and is interrupted by a similar peripheral with higher priority, the common drivers must be reentrant. Because the peripheral with higher priority transfers its own parameters to the driver, the old parameters of the peripheral which formerly used the driver are destroyed. Therefore the old parameters must be saved before the driver can be used by the new peripheral. Generally, saving the parameters occurs with the help of a stack, whose administration is time consuming. For time-critical processes it is much easier to use an individual driver for each peripheral. When data transfer time is not critical, it is very common to block the access to a driver by a mask until the peripheral using the driver has completed its job.

7.3.2 Driver for Interrupt-Controlled Data Transfer

The principle of data input/output is shown in Fig. 7.9. During execution of the main program, an instruction is given requesting input of the value of a measured parameter. The program jumps into the common interrupt handler, which first determines the priority of the in-

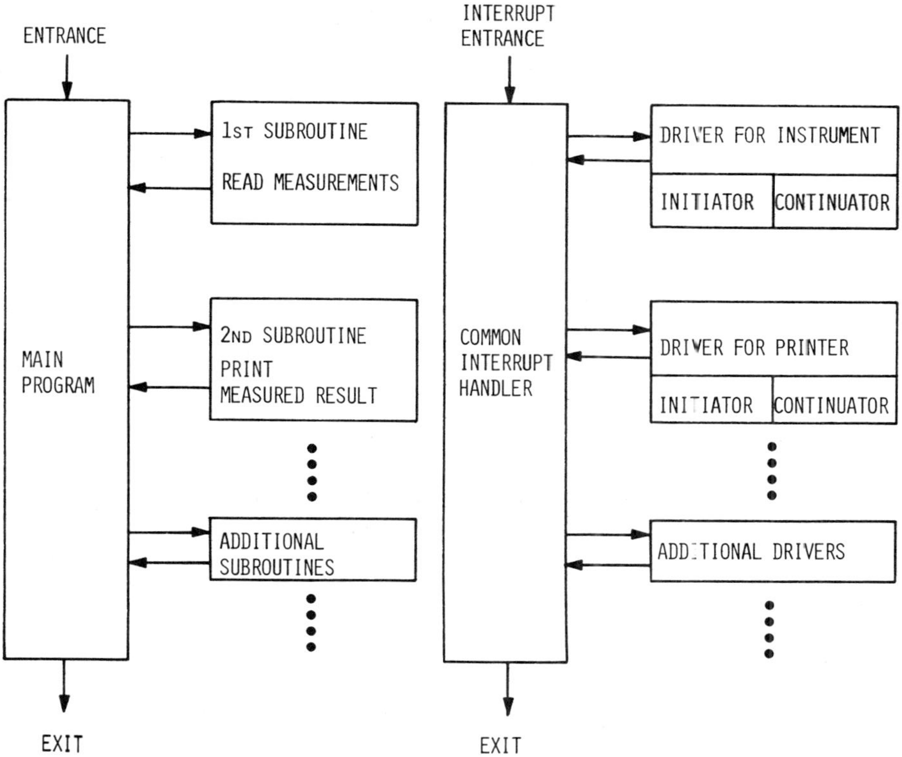

FIG. 7.9 Principle of interrupt-driven data input/output.

terrupt (Sec. 7.2). A check is made to determine whether the peripheral can be serviced immediately or if its request should be put into a queue. As soon as the interrupt request can be serviced, the driver of the instrument is activated. This driver consists of an initiator and continuator to perform data transfer. The data to be brought to the computer must be placed in a data buffer which is administered by the driver. During further processing of the above main program, an instruction is given to output the value of the measured parameters, for example, on a typewriter. Again, the common interrupt handler is entered. The priority of the request is determined and the driver of the printer is entered as soon as it is available. It, in turn, outputs data to the printer head. The data buffer must also be administered accordingly.

Initiator

The following functions are assigned to the initiator:

1. Transfer of all operating parameters to the driver, in order to be able to perform the input/output operations, for example, the type of input/output, address of the data buffer where the measured parameters are to be stored, and the number of measurements to be performed must be known.
2. To investigate whether or not the peripheral's driver is already busy.
3. If the driver is not busy, a signal is given that the driver will now be occupied.
4. Transfer the call parameters.
5. Set a link in the interrupt system for the continuator.
6. Activate the peripheral and initiate the data transfer.
7. Return control via a jump instruction to the main program.

Many of the tasks of the initiator are identical for all devices. Therefore, it is advisable to divide the initiator into a common part and a part specifically designed for the peripheral. The common part, for example, can check whether the function to be performed is a status-controlled input/output or an initialization instruction. When the common interrupt handler has completed its task, the driver for the specific peripheral is called upon.

Continuator

The continuator will be called on when the peripheral is ready to input or output data. The following tasks are performed:

Save the contents of the working registers and program counter.
Input and output data.
Administer the data buffer.

Enable the driver.
Restore the original contents of the registers.
Perform the return jump to the location at which the main program was interrupted.

Parameter Transfer Table

Upon initiation of an input/output call, the user program transfers the contents of a parameter transfer table to the initiator of the driver (Fig. 7.10). Typical parameters transferred are as follows:

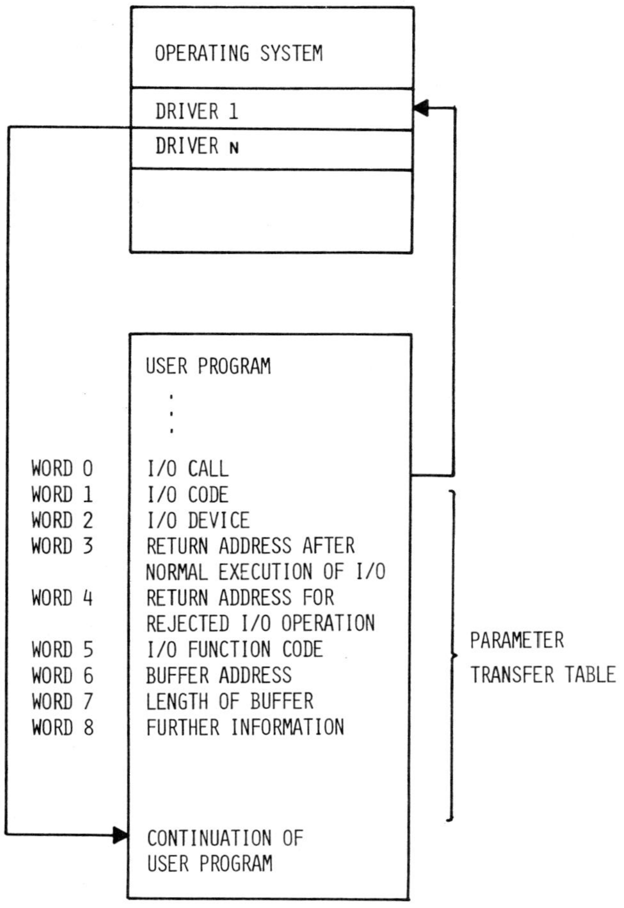

FIG. 7.10 Parameter transfer table to transfer operating parameters from the program to the driver.

1. I/O code.
2. Identification of the I/O operation to be performed, such as
   a. Status request
   b. Function code
   c. Input/output operation
   d. Initialization of the peripheral
3. Input/output peripheral: description of the input/output peripheral.
4. Return address for a normal execution of an I/O.
5. Return address if an error occurs: in this case an address is given to the program where it finds an error handling routine, which then attempts to evaluate the type of error.
6. Input/output function: function instruction to the peripheral, for example, rewind a magnetic tape.
7. Address of data to be read or written.
8. Number of data to be transferred.
9. Further information.

The length of the parameter transfer table differs for different computers. It also depends on the operation to be performed. There are different parameter transfer blocks for data transfer, for control instructions, for function tests, and for the initialization of peripherals.

Execution of a Data Input/Output

The initiator turns the peripheral on so that the device can prepare itself parallel to the main program for data transfer. As soon as a peripheral is ready for data transfer, the main program is interrupted by the peripheral, and the continuator supervises and completes the data transfer. The exact operation of data transfer, for example, for a measurement is as follows (Fig. 7.11):

1. The main program requests the input/output for a peripheral by means of an instruction (Sec. 6.3.3).
2. The control instruction initiates the measurement by setting a control flip-flop (Sec. 6.3.3).
3. The flag flip-flop is set on LOW to be able to recognize the availability of data (Sec. 6.3.3).
4. The main program is continued until an interrupt announces the availability of data (Sec. 6.3.3).
5. As soon as data are available for the peripheral, the flag flip-flop is set on HIGH (Sec. 6.3.3).
6. The main program is interrupted and the continuator of the driver for the peripheral is activated with the help of the information provided by the interrupt register in order to input data (Sec. 6.3.3).
7. After completion of data transfer, the main program is continued.

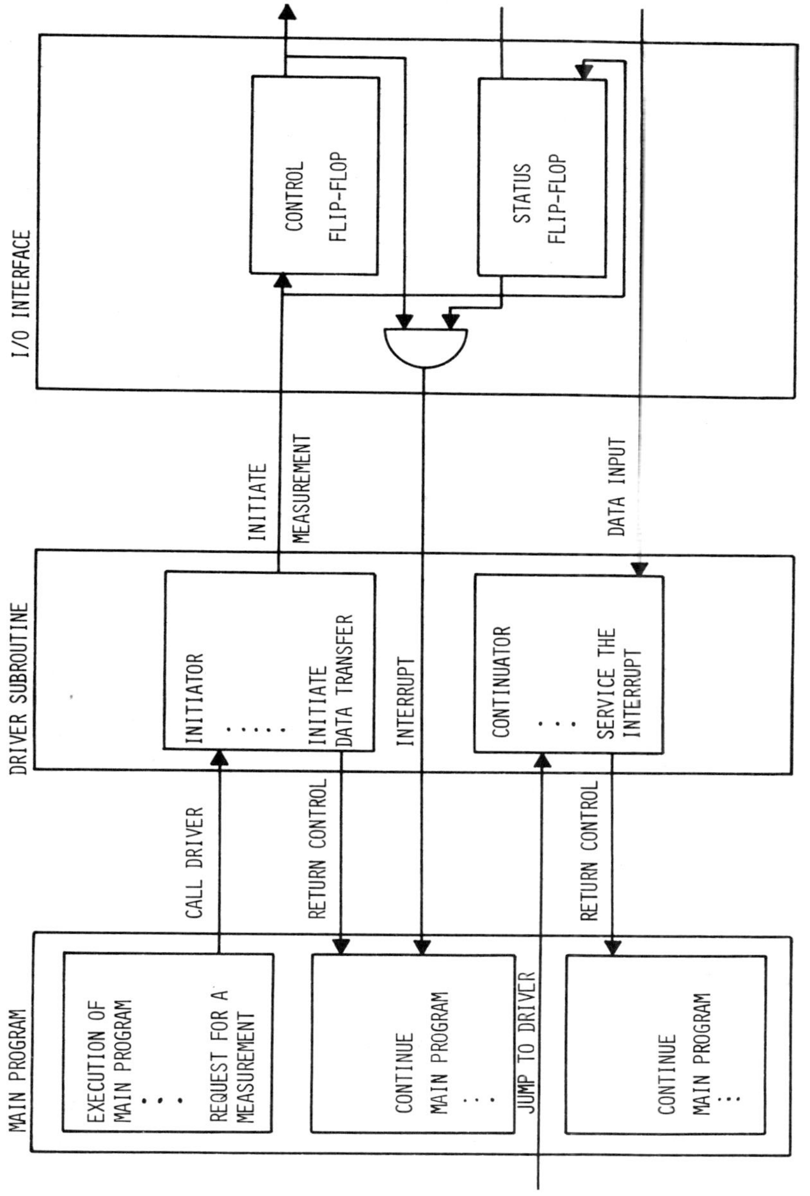

FIG. 7.11 Initiation and continuation of an interrupt-controlled data transfer.

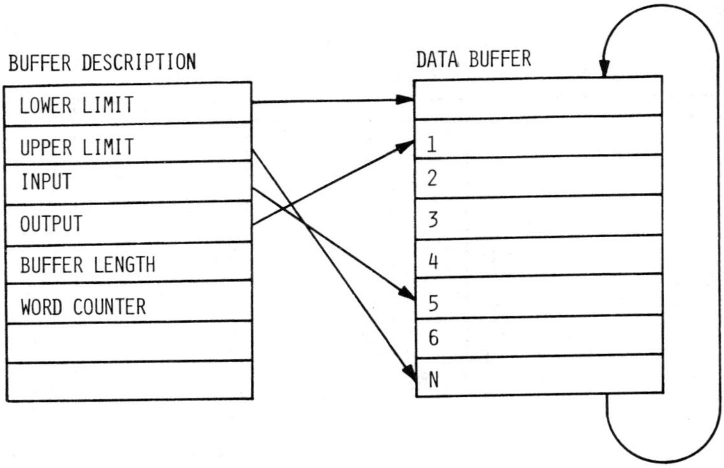

FIG. 7.12 Ring buffer for temporary buffering of data.

Data Buffer

To facilitate the input/output operation, a data buffer must be available in the working memory. Since storage space is limited in the computer, the buffer must be layed out as a ring buffer, Fig. 7.12. The administration of this buffer is done by the first-in first-out method. The upper limits of the buffer must be known. Likewise, a pointer which points to the input and output operations is necessary. The word counter is needed to count data during input or output. In order to effectively administer the buffer the pointer must return to the beginning of the buffer as soon as it reaches the end (Fig. 1.12 and 1.13). In general buffer administration should be the task of the operating system. However, with some very small computers the system design has to do it.

7.4 OPERATING SYSTEM

7.4.1 General Tasks of the Operating System

The operating system is a program package which coordinates and administers the individual tasks of the computer and the execution of the computing operations. For this reason it is necessary to determine the priority of each task to be performed and to provide the necessary resources for a task. The most important assignments of the operating system are as follows:

Task management of the computer system
- Scheduling of resources
- Interrupt management
- Tasking and scheduling
- Time allocation
- Coordination of internal task communication

Memory management
- Working memory
- External storage

Input/output scheduling
- Scheduling of peripherals
- Scheduling of communication equipment
- Supervision and control of data transfer

The efficiency of a computer depends to a great extent on the comfort the operating system offers. The design of the operating system is mainly the job of the computer manufacturer. For large, generally also expensive computers the operating system offers many functions and simplifies programming of the computer. For microcomputers there are often only primitive operating systems available. For these the user must write his or her own monitor.

Within the context of this book the administration of the interrupt requests and the input/output peripherals is of main interest. A driver particularly designed for the peripheral must be integrated into the operating system in order to control the peripheral. The operating system must know what the peripheral looks like, which priority it possesses, how it is to be operated, what type of peripheral parameters are to be transferred, and how its transferred data are to be treated and stored.

Software-controlled processing of interrupt signals is shown in Fig. 7.6. The right half of the figure explains the coordination between the main program and the operating system. For example, if a common interrupt handler determines which peripheral has to be serviced, the program jumps to the trap vector INT 1 in the operating system, and this vector points to the address of the driver program to be used. After execution of the driver program, the program jumps back via the address in the trap vector to the main program. A new interrupt can only take place after the previous peripheral has completed its data transfer.

Hardware-controlled interrupt processing is shown in Fig. 7.7. After an interrupt request, the program jumps via the trap vector to the service routine. If presently a peripheral with lower priority is being serviced, it will be interrupted, and the contents of the registers will be stored in a stack. After the peripheral with higher priority has been serviced, the old contents of the registers are fetched from the stack and brought back to the registers. The previously in-

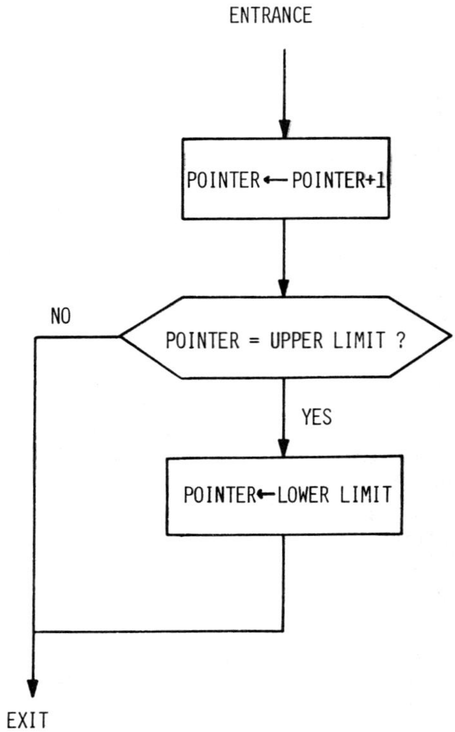

FIG. 7.13 Control logic for circular buffer.

terrupted work can continue. For this purpose it is necessary to know the upper and lower limit of the buffer; also a pointer has to indicate the next free locations in the buffer (Figs. 7.12 and 7.13) [2]. To avoid an overflow of the buffer the operating system must keep track of the buffering operations and report difficulties with it to the computer operator (Fig. 7.14).

7.4.2 Logical and Physical Unit

If several peripherals are connected to a computer installation, the flexibility in writing programs can be greatly improved by not addressing a peripheral by its physical addresses, but by a given logical address. This makes it possible to assign peripheral addresses dynamically, and drivers can be used for more than one input/output unit. In addition, it is much easier to have programs written by different programmers (Fig. 7.15).

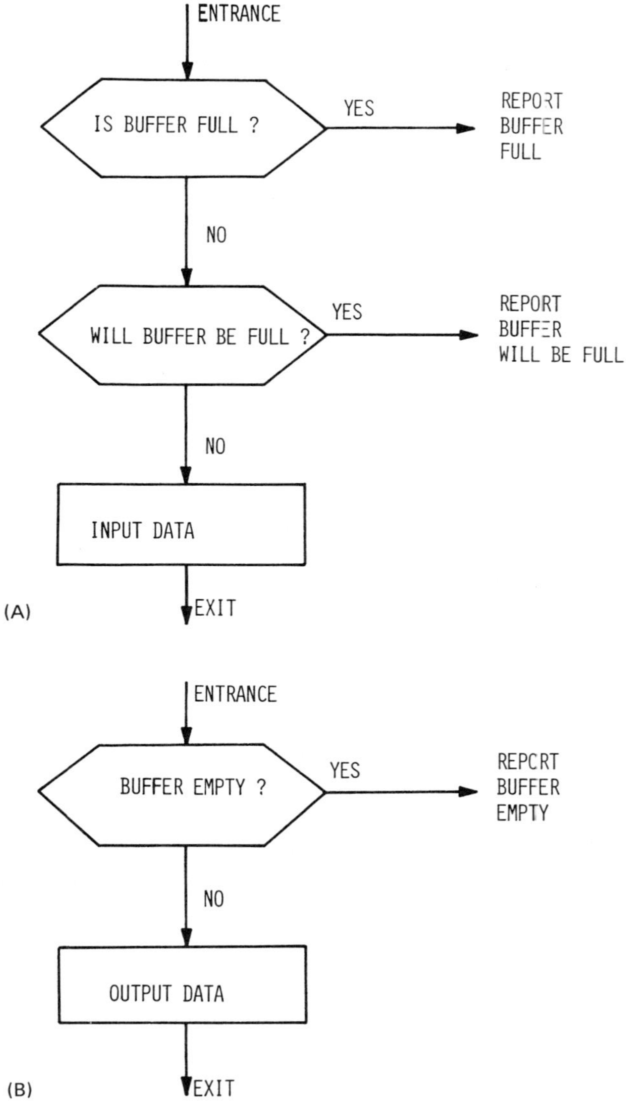

FIG. 7.14 Control of a circular buffer: (a) input to buffer and (b) output from buffer.

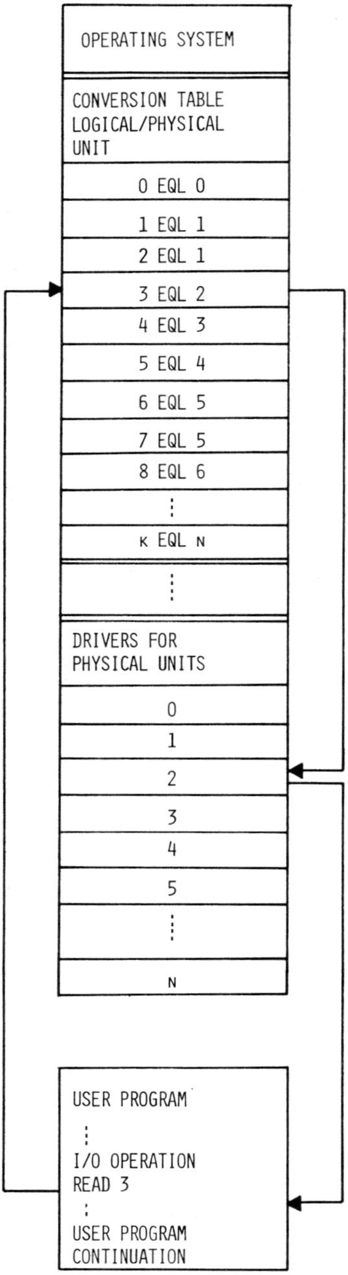

FIG. 7.15 Logical and physical unit.

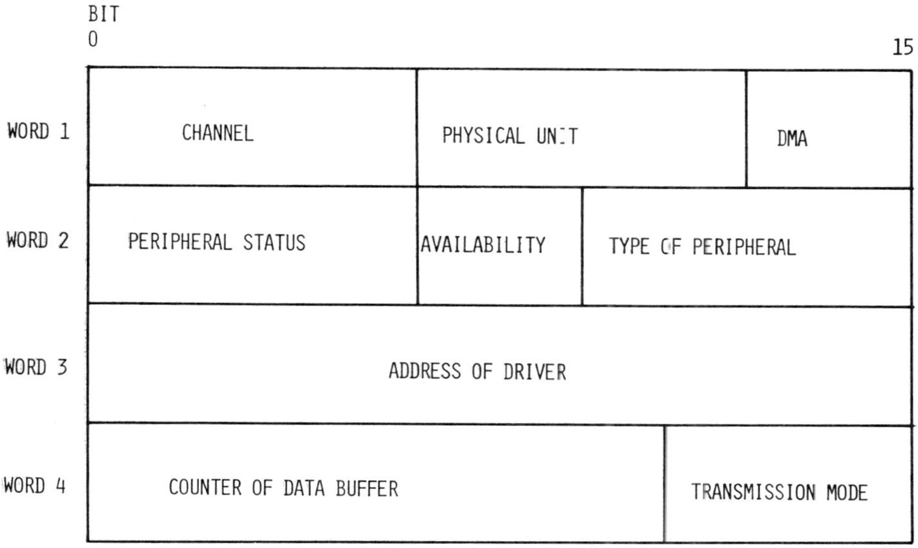

FIG. 7.16 Equipment table.

7.4.3 Equipment Table

For each peripheral connected to the computer an equipment table is provided. This equipment table is tied into the operating system during configuration. If a new peripheral is introduced to the computer system, a new equipment table must be generated and tied into the operating system. Figure 7.16 shows a typical equipment table. Again, it is very specific for each individual computer. In this example, the equipment table has been designed as follows:

First Word — Number of the channel to which a peripheral is connected. In this case, for example, for a computer having a radial I/O structure (Sec. 4.3.4).
Physical address of the peripheral, in case several peripherals are connected to the same channel.
DMA, yes or no.

Second Word — Status of the peripheral (see Table 6.3).
Identification of type of peripheral.
Availability of the peripheral (see Table 6.3).

Third Word — Address of peripheral driver.

Fourth Word — Buffer counter, counts the number of characters or words to be transferred.
Indicates whether a byte or word transfer has to be performed.

The status of the peripheral, DMA, availability, and buffer counter can be changed dynamically during program execution. The program has access to these storage cells, can analyze their contents, and thus is able to interrogate the status of the peripheral or the buffer counter. All other storage locations of the equipment table contain fixed information.

7.4.4 Example of a Simplified Driver

In this section the structure of simplified driver for data input from a measuring instrument will be exemplified. The driver consists of an initiator and continuator. A common initiator part will not be integrated into this example; this makes the program more understandable. It should also be mentioned that the described driver is incomplete. The main object of this discussion is to explain the fundamental structure of the driver and show the parameter interchange between the parameter transfer block of the user program, the working registers, and the equipment table of the operating system. The user program or the driver is able to access the parameters from the working registers and use this information for certain prescribed procedures. The computer conceived for the example has several working registers; five of these will be used by the driver.

Initiator (Fig. 7.17)

1. First the interrupt system is enabled.
2. Words 1 and 2 of the parameter transfer table (PTT) (Fig. 7.10) are transferred to registers 1 and 2, respectively. Thus the driver obtains the address of the peripheral and the instructions for the input/output operation to be performed.
3. A check is made concerning whether or not the input/output code and the peripheral address have been defined. If the result is negative, the computer issues an error report (4).
5. If the result is positive, a check is made to determine whether or not it was a status request. If yes, the status words for the peripheral and the data buffer are read from the equipment table (Fig. 7.16) into the register (6). At the same time, during normal program execution, a return address for branching back to the user program is read from the parameter transfer block (Fig. 7.10) and placed into a register (6). The application program is thus able to perform further processing with the status information.
7. Upon a negative result, a check is made whether the request was a "clear" request. If yes, the instruction code is read into a register (10).
8. If not, a check is made on whether or not the peripheral is busy.

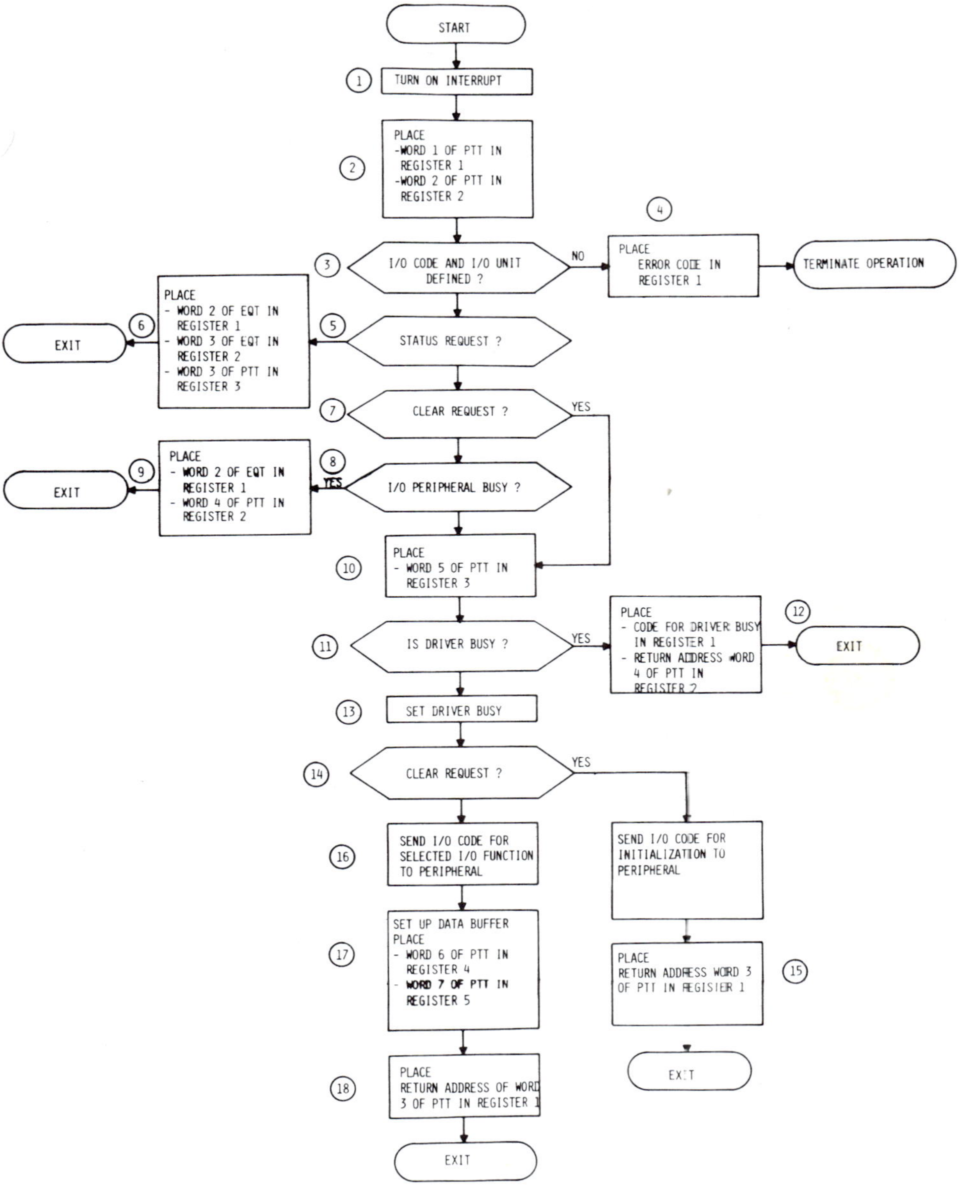

FIG. 7.17 Simplified initiator of a driver.

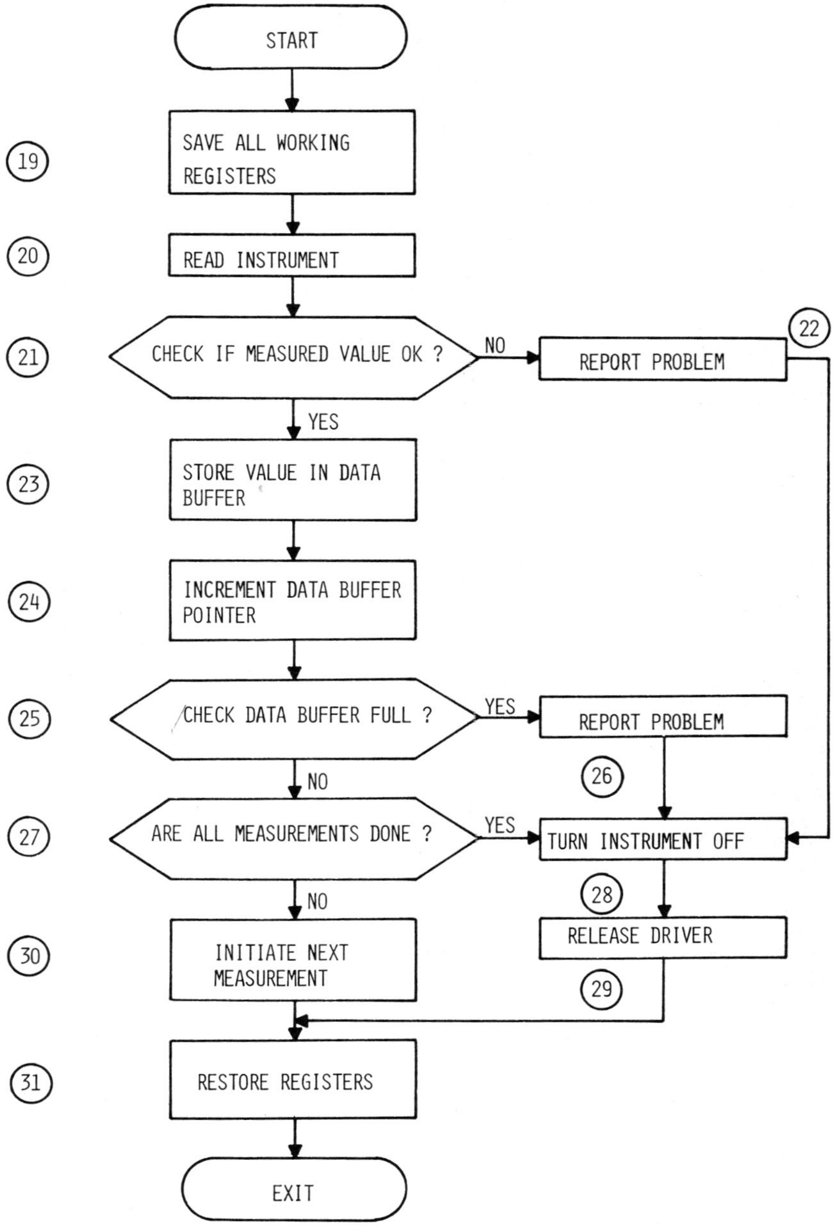

FIG. 7.18 Simplified continuator of a driver.

9. If yes, the status of the peripheral and a return address for an error search routine are deposited in a register. Thus a program is able to investigate the peripheral's status.
10. If the peripheral is not busy, the instruction code word 5 of the parameter transfer block is read into a register.
11. A check is maae to determine whether or not the driver is busy. If the result is positive, an investigation is made of why the driver is busy.
13. If the driver is not busy, it will be disabled.
14. If a "clear" request exists, the instruction code for a clearing operation is issued and read into the main program via a return address (15) .
16. In case the required operation is not a clear request the function of the peripheral to be performed is executed.
18. Now the main program is continued until the peripheral is ready for data transfer.

Continuator (Fig. 7.18)

The continuator is activated as soon as a peripheral is ready for data transfer. A very simplified continuator is shown in Fig. 7.18.

19. As soon as data are available all registers are saved.
20. The measured parameters are transferred.
21. The measured parameters are investigated, for example, to determine on whether or not a plausible signal has been transferred. If not, a thermocouple might be broken or grounded. In this case, the source of the problem is printed out (22) .
23. A plausible measured parameter is stored in the data buffer.
24. The pointer of the data buffer is incremented to prepare the peripheral for the next data input.
25. If a data buffer is full, an error message is issued (26) .
27. When all measurements are completed, the peripheral is turned off (28) , and the driver is released (29) .
30. In case not all of the measurements are completed, the next measurement is initiated.
31. The contents of the registers are restored and the main program is continued. When a new measured parameter is to be transferred, this part of the driver must be enacted again.

7.5 HIGHER PROGRAMMING LANGUAGES

The selection of a suitable programming language for controlling the process is of utmost importance. In the context of the goals set in this book it is of interest how the peripheral can be programmed easily and effectively. There are a large number of programming languages available, which can be divided as follows:

Assembly language
Compiler language
Interpreter language
User-oriented language

Most process control computers today are programmed in assembly language. With this it is possible to write effective programs with short execution times. In particular, digital control systems can be coded with little effort. The disadvantage is the large programming effort. For this reason the compiler languages are gaining more and more importance. With the help of these languages statistical calculations and optimizing tasks can be performed very easily. Some process control computers allow insertion of assembly code between the instructions of a higher order language. Thus it is possible to combine the advantages of both programming languages. In recent years a trend has been seen toward user-oriented languages. In this case the user only has to learn a code tailored to his or her problem. The computer is responsible for the interpretation of the programs. To adapt a higher order programming language for process control computer applications, language elements for real time operations must be provided. Some typical language elements, for example, for real time BASIC are as follows:

ON INT <i>	:	As soon as interrupt i has occurred.
AFTER <msec>	:	Wait for a given time in milliseconds.
INW (<D_i>)	:	Read data from peripheral device D_i. (Example: LET A = INW (<D_i>)
OUT (<C_j>)	:	Output to a counter C_j. (Example: LET OUT (<C_j>) = C * B)
ENAB i_1, i_2, i_3, ...	:	Enable interrupt for interrupts i_1, i_2, i_3, ...
DISAB i_1, i_2, i_3, ...	:	Disable interrupt for interrupts i_1, i_2, i_3, ...
PLOT (x,y,z)	:	Advise plotter to travel to the coordinate x,y. Activate coordinate z to engage writing pen.

Taking a look at the last instruction PLOT (x,y,z), we can easily see that three instruction parameters must be transferred to the driver of the plotter when the call is made with this statement in the higher order language. The parameters x and y indicate the coordinates of the writing pen. The z coordinate is needed to give instructions for the writing pen so that it should be engaged for writing or that it should be lifted. As a rule, the addresses of the storage location of the parameters to be transferred are given to the driver and not the parameter themselves. From this storage location the driver can access the parameters. It must be observed that the instruction parameters are entered in a prescribed sequence and that both the main program and the driver are programmed to recognize the correct meaning of each parameter. Similarly, the parameter taken in from a measurement by the driver must be made available at a known storage location to the calling statement of the higher order language.

REFERENCES

1. H. Becker and G. Leusch, Freiprogrammierbare Steuerung für Industrieroboter, Siemens Zeitschrift Nr. 51, Vol. 6, 1977.
2. *The TTL Data Book for Engineers*, Texas Instruments, Inc., Dallas, 1973.
3. H. F. Jordan and B. J. Smith, *The Hardware-Software Interface*, Arbeitsberichte des Instituts für Mathematische Maschinen und Datenverarbeitung, Universität Erlangen-Nürnberg, Erlangen, 1976.
4. M. Paul and E. V. Puttkamer, *Struktureller Aufbau von Prozessrechnern*, VDI-Verlag GmbH, Düsseldorf, 1974.
5. Granino A. Korn, *Minicomputer for Engineers and Scientists*, McGraw-Hill, New York, 1973.

8

APPLICATION PROBLEMS IN INTERFACING

8.1 LEVEL ADAPTATION

Since it is important that each functional unit of a computer system be designed with the most suitable technological components, problems of signal level adaptation occur quite often in interface circuits. In large and complex circuits it may be necessary to perform several signal level adaptations. Similar problems may arise in matching integrated circuits of different manufacturers, since they are often designed to operate at different voltage levels. Most modern digital circuits are based on TTL technology; therefore in the following description of signal level adaptation the TTL level will be taken as the base reference.

8.1.1 Interfacing of TTL Components with Relays

The simplest level adaptation is accomplished by a relay (Fig. 8.1). The application of this method, however, is limited to very slow changing signals, and therefore it is only employed for special circuits. For reed relays the upper switching frequency is 1 kHz. In most cases, a driver circuit is necessary to operate a relay with the TTL signal. It assures that enough current is available to actuate the coil of the relay. Figure 8.2 shows such a driver circuit. Here the current for the relay coil is taken from the TTL power supply. The value of the resistor R_1 depends on the driving current to switch the relay coil as well as on the current amplification factor β of the Darlington circuit:

$$I_c = \beta I_B \approx \frac{(V_E - 1.4\ V)\beta}{R_1} \qquad (1)$$

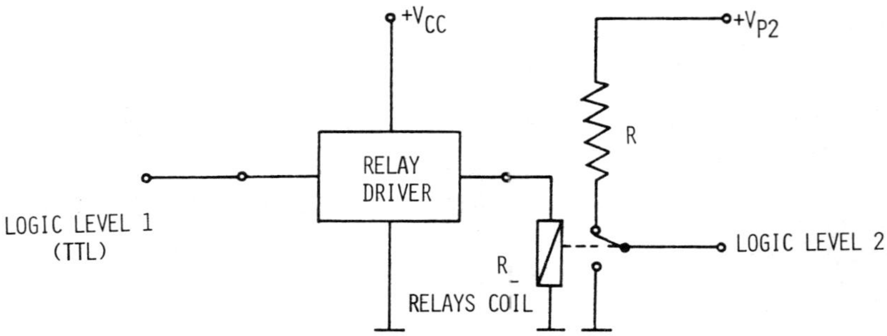

FIG. 8.1 Voltage level converter with relay: V_{cc} = TTL supply voltage for driver, and V_{P2} = any supply voltage.

V_E is the emitter voltage of the output stage of the Darlington transistor circuit. Its value is determined by the highest voltage level of the driving TTL gate. For example, if this level is 3.5 V and the forward voltage differential across the two transistors of the Darlington stage is 1.4 V, then V_E is approximately 2 V.

In the circuit of Fig. 8.2 the logic level is inverted. It is often of advantage to drive the relay coil with a voltage higher than the

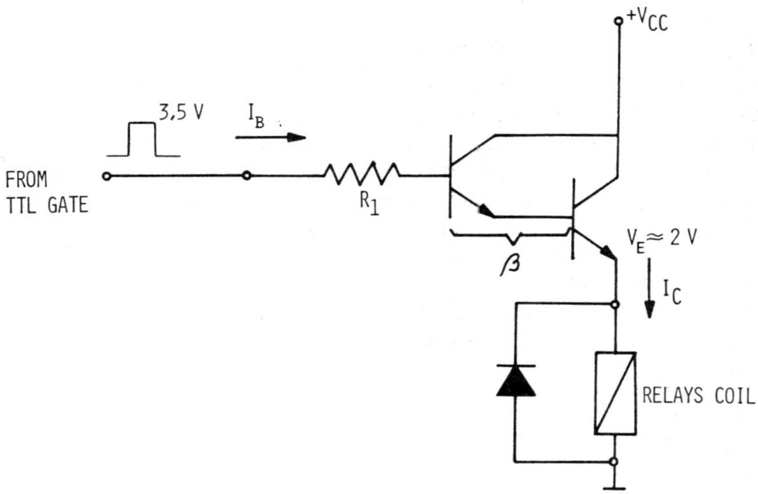

FIG. 8.2 Relay driver: β = current amplification of a Darlington transistor circuit; $I_c \approx \beta I_B$

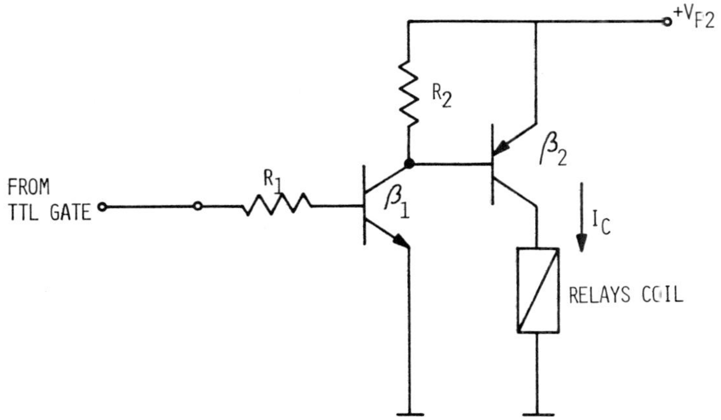

FIG. 8.3 Relay driver with separate supply voltage.

TTL voltage; thus lower coil currents can be obtained. Figure 8.3 shows an example of such a driver. The value for the resistors can be calculated as follows:

$$R_1 \sim \frac{2.5\beta_2}{I_c} \qquad R_2 = \beta_2 \frac{V_{p2}}{I_c} \quad \text{with } \beta_1 = 2\beta_2 \tag{2}$$

It is very important when applying drivers that the current load in the circuit does not exceed its maximum permissible value. With TTL circuits this value lies at approximately 400 μA (H level) or 16 mA (L level), respectively.

8.1.2 Interfacing of Electrical Power Devices with Solid-State Circuits

There are presently many integrated power-switching components available, which can be used for interfacing with electrical power devices. For example, the status of a 230-V switching component can be sensed with the circuit shown in Fig. 8.4. The 230 V are stepped down to approximately 5 V with the help of a voltage devider. The ac voltage thereafter is rectified, and the resulting dc drives a light-emitting diode which is part of a photoelectric isolator. On the receiver side a TTL signal is supplied to the computer. A filter in the circuit protects the device from noise interference.

Similarly, a TTL level signal originating from the computer can be transformed to a 230-V signal to drive electrical power equipment (Fig. 8.5). The signal originating from the computer drives via an

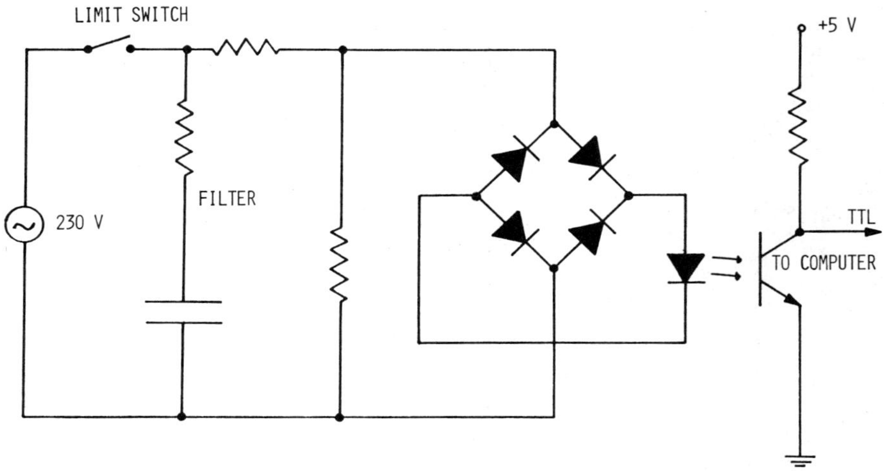

FIG. 8.4 Voltage level converter, 230 V → 5 V.

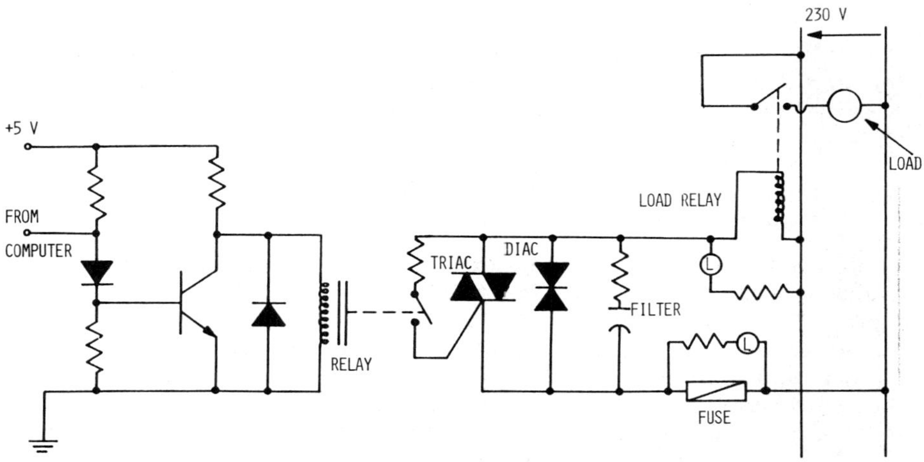

FIG. 8.5 Voltage level converter, 5 V → 230 V (solid-state relay).

amplification stage a small relay. Upon closure of the relay, a bidirectional triode thyristor (TRIAC) is energized which supplies the power needed to operate the 230-V equipment. The circuit is provided with a bidirectional diode (DIAC) to protect the TRIAC from voltage surges and a filter to reduce line interference. In addition, one indicator light shows the status of the circuit fuse, and an other light shows the presence or absence of power at the solenoid.

8.1.3 Interfacing of TTL with HTL Components

When interface circuits have to be designed for a noisy environment or an extremely fail-safe system, the use of TTL technology is not advisable. In these cases often HTL (high threshold logic) is required. HTL is more immune to noise, although it is slower.

Level adaptation by this method is from 5 to 12 V or, if necessary, to 15 V. For this task TTL gates with open collectors (for example, SN49701) or special driver modules (for example, SN75361A) are available. To avoid overdriving by the output signal a serial attenuation resistor in the output line may be necessary.

8.1.4 Interfacing of TTL with MOS Components

Converting from TTL to MOS levels is necessary when driving integrated circuits mainly of older design. Modern integrated MOS elements are frequently provided with TTL-compatible input/output circuits. In the following section some typical circuits for driving negative MOS levels will be discussed.

Several memory devices built from MOS technology and other circuits built from older technologies do need special input voltage levels for their operation (Sec. 5.5). For this reason they can not be directly connected to the outputs of TTL circuits. Modern CMOS circuits, however, are provided with buffers and thus can be operated with 5 V to be TTL compatible. Higher voltages such as 15 V are of advantage for the operation of MOS components. They increase switching speeds, are more immune to noise, and have improved driving capabilities. In this case interfacing to a TTL output is done with the help of an open collector which is connected in series with an collector resistor to the higher CMOS supply voltage.

In the opposite case when CMOS gates have to drive TTL circuits their low output current is not capable of switching the TTL device. For this level adaptation special CMOS chips were designed, for example, the chip 4050, which has six noninverting buffers or drivers.

In order to reduce the power consumption of dynamic MOS memories the high capacitance of the gate is utilized. In principle it is sufficient to charge the gate with the help of a short pulse in order to turn the transistor on or off. As long as the charge is maintained, the logic state of the device is also maintained. A driver to generate

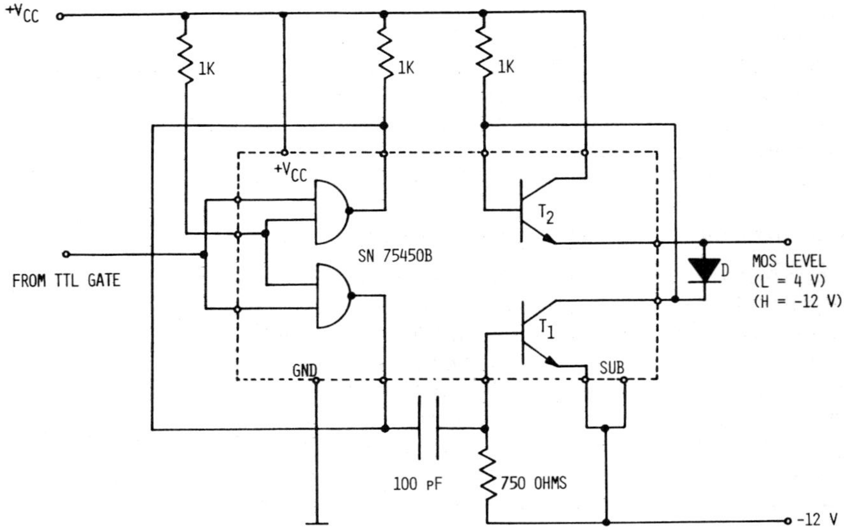

FIG. 8.6 Level converter TTL-MOS. (From Ref. 2, courtesy of Texas Instruments Gmbh, Freising, Federal Republic of Germany.)

control pulses with very steep edges is shown in Fig. 8.6. A typical representative of this device is the chip SN75450B. The circuit of Fig. 8.6 allows a voltage level shift from -10 to +4 V. This shift is generated by the capacitive coupling between the parallel connected outputs of the TTL gates and the output transistor T1. The coupling capacity in connection with the base resistor generates a time constant of 75 nsec, which determines the length of the pulse at the output. With the RC values of the circuit of Fig. 8.6 a clock signal of 4 to 8 MHz can be obtained. The function of this interface circuit is essentially determined by the gate capacity of the MOS register to be controlled. When the transistor T1 of the NAND gate is turned on and goes into saturation, it connects the output of the circuit with the negative level of the supply voltage. At this instance the load capacitor is charged. When transistor T1 is turned off after the reversal of the coupling charge, the negative charge of capacitance of the load circuit will cause transistor T2 to turn on, and the output will be connected to the positive level of the supply voltage. The charge reversal occurs very rapidly since T2 has a very low resistance when it is in saturation. When transistor T1 is turned on again, the forward voltage drop across diode D1 will cause the base-emitter voltage at T2 to assume a negative level. Thus this transistor will be turned off.

8.2 POWER SWITCHING

In most control computer installations electrical power devices have to be operated. They may be servo- or stepping motors, solenoids of valves, etc. (see Fig. 8.4 and 8.5). In the following section different methods of connecting such power devices with the help of interface circuits to the computer will be introduced.

8.2.1 Control of a Servomotor

The task of a servomotor is to adjust control devices, for example, a valve. The motor is part of the control loop. In the servomotor the momentary deviation of the manipulated variable from the control variable is constantly sensed. If necessary, the motor immediately adjusts the control device so that both variables become equal.

Figure 8.7 shows a control circuit for a servomotor. The comparison of the manipulated variable with the control variable is done by the computer. The interface sends the result back to the control circuit. A LOW signal at the input results in a clockwise rotation, and a HIGH signal, in a counterclockwise rotation. The entire system represents a two-position controller. The HIGH/LOW ratio at the input of the circuit determines the actual position of the motor. If the input gate is provided with a tristate output, a three-position controller can be realized.

The integrated operational amplifier is actuated by signals coming from the input gates. When the input control gate is in the off state,

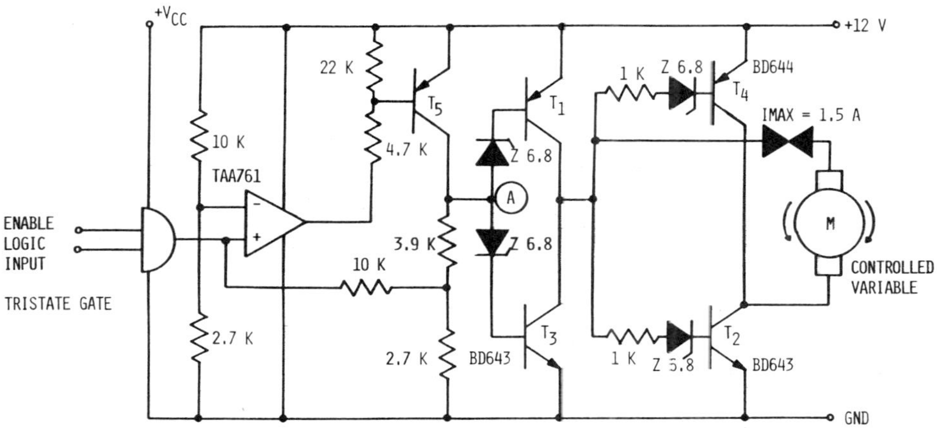

FIG. 8.7 Control circuit for a Servomotor. (From Ref. 3, courtesy of Siemens AG, Munich.)

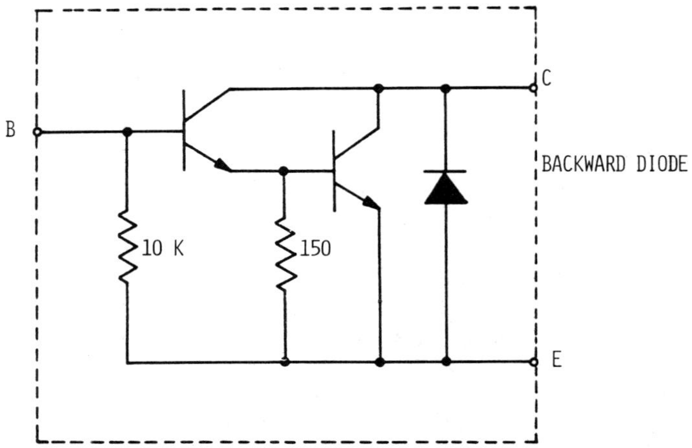

FIG. 8.8 Darlington transistor circuit for motor control.

the resistor combination of the input part of the circuit will hold the voltage level at point A at 6.8 V. In this state T_5 is partially turned on. At the bases of transistors T_1 and T_3 a voltage level of approximately 0 V will prevail and the motor will be in the idle state. When the input of the control gate is pulled down to LOW, transistor T_5 will turn on completely, and the voltage level at A will be greater than 6.8 V. The transistors T_3 and T_4 will be turned on and the motor rotates in one direction. When a HIGH signal is placed at the input of the control gate, transistor T_5 turns off, and the voltage level at A will drop below 6.8 V. Now transistors T_1 and T_2 will open and the motor turns in the other direction.

In this design the motor is controlled with a power stage consisting of a bridge circuit. Its design contains complementary Darlington transistors, and therefore it has a simple circuit. In addition, the transistors contain (as an integrated element) the so-called backward diodes, which provide a return-flow path for the inductive energy stored by the motor. Figure 8.8 shows the internal circuit of the Darlington stages.

8.2.2 Control of a Stepping Motor

Stepping motors are used where a shaft has to be rotated at exactly defined increments or where an incremental motion must be converted to a precisely defined linear motion. For this task polyphase stepping motors are commonly employed. The following example describes the operation of a three-phase stepping motor with an angular resolution of 30°. Twelve pulses are needed for one revolution. The polyphase motor has a star-connected stator. It can be energized by a driver circuit to rotate the armature either clockwise or counterclockwise. To control the motor, the interface sends six control signals (Fig. 8.9). When the motor is connected to a computer, a parallel output of at least 7 bits must be provided. Thus the software must supply 6 control bits and 1 start-and-stop bit to move the armature. A change in rotational direction is accomplished by software simply by changing the sequence of control signals supplied to the six control lines. Figure 8.10 shows the circuit of the driver. Via the digital inputs Q_1 and Q_2 and the switching transistors, actuation signals are provided to the motor to step it through the desired increments. In addition, the polarity of the windings can be reversed. Thus it is possible to control individual steps and the direction of rotation.

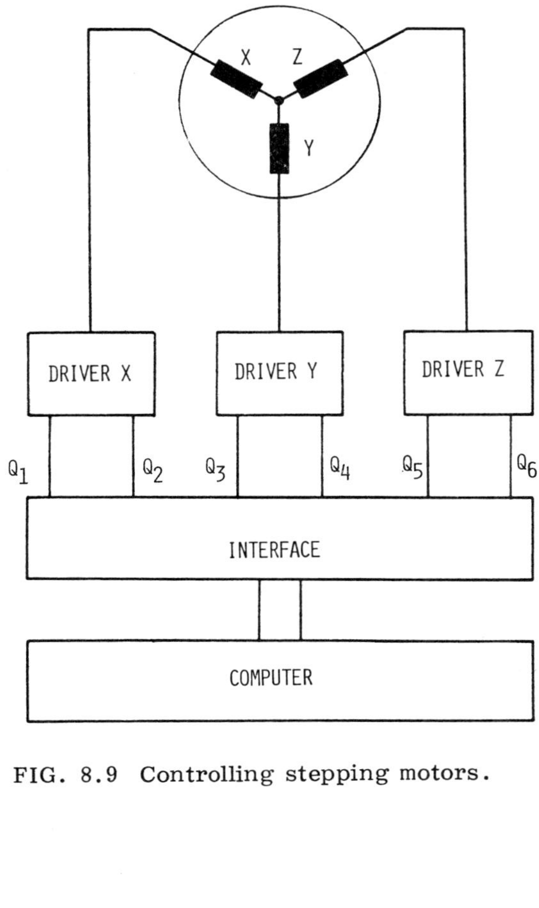

FIG. 8.9 Controlling stepping motors.

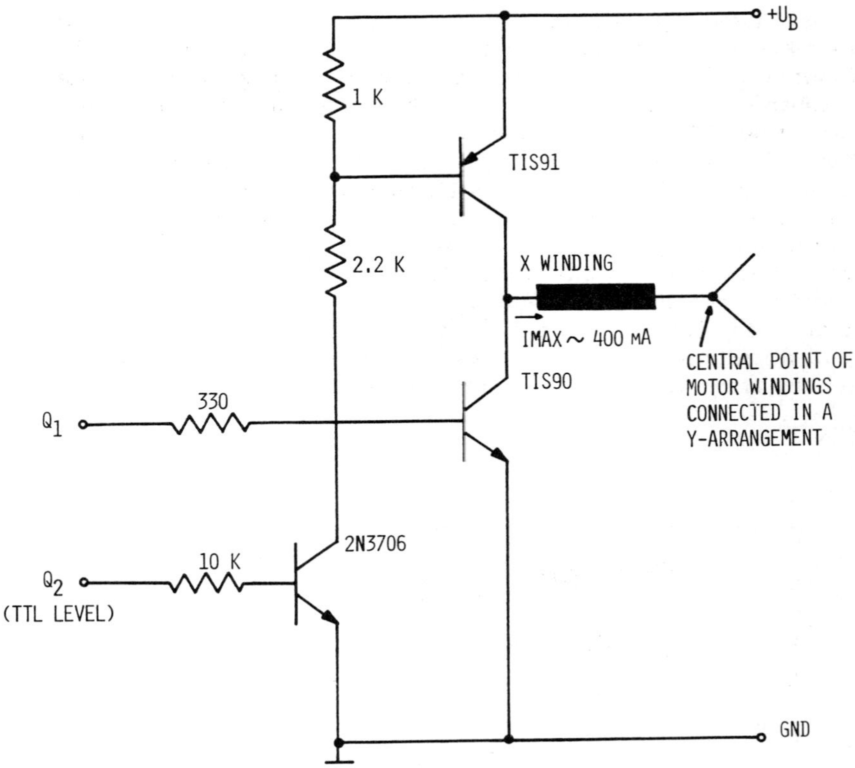

FIG. 8.10 Driver channel for a stepping motor. (From Ref. 2, courtesy of Texas Instruments GmbH, Freising, Federal Republic of Germany.)

8.2.3 Interfacing Pneumatic Controllers

When pneumatic controllers are used in a control system electric/pneumatic (E/P) converters are needed. In general such a converter contains a stepping motor which regulates the air flow, for example, for controlling a valve. The stepping motor can be connected directly to the computer. The valve controls the amount of supply air which is needed to actuate the pneumatic control element. To the stepping motor, a potentiometer is connected which feeds back the position of the valve to the computer. Figure 8.11 shows a schematic diagram of a E/P converter. The feedback control loop for direct digital control (DDC) using a pneumatic controller is shown in Fig. 8.12. The operating point of the process is reported back to the computer with the help of a P/E converter.

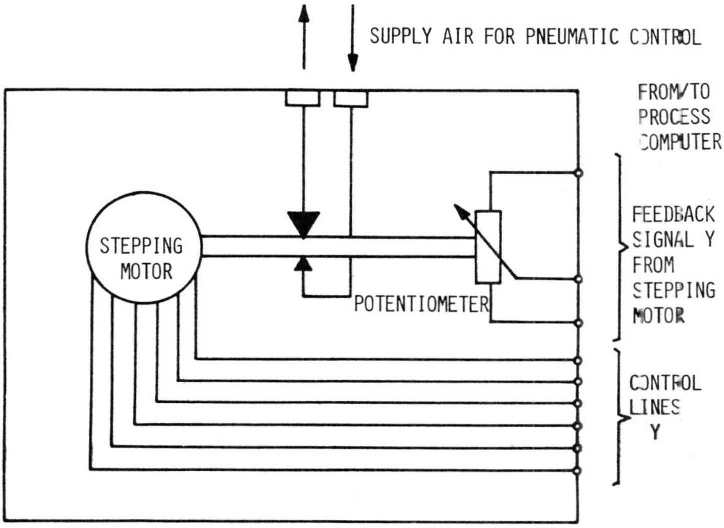

FIG. 8.11 Electronic to pneumatic converter: Y = manipulated variable.

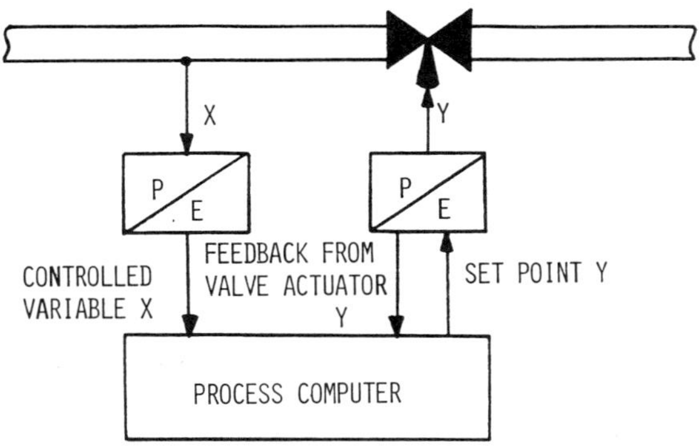

FIG. 8.12 DDC circuit for pneumatic valve.

8.3 PREVENTION OF ELECTRICAL INTERFERENCE AND NOISE

When developing automation systems for a rugged industrial environment, questions concerning correct wiring and protection against interferences are of utmost importance. The integrated-circuit components of complex interface circuits are very susceptible to interference caused by short voltage spikes; these are typically generated when heavy loads are switched. Even so, if one selects logic circuits such as HTL logic with high noise margins, the danger of noise interference cannot be completely eliminated. Interference spikes may reach the magnitude of several kilovolts. A safe system can only be guaranteed if preventative measures are taken during the design of the automating system to reduce noise interference.

8.3.1 Sources of Electrical Interference and Noise

Noise can be introduced into a computer system by the following physical phenomena:

1. AC frequency and its upper harmonics, caused by
   a. Inductive coupling
   b. Capacitive coupling
   c. Resistive coupling
2. Radio frequencies (high frequencies), caused by
   a. Sparks from brushes of rotating machines
   b. Arcs from switches and relays
   c. Arcs on the cathode of fluorescent lights
   d. Corona on high-power lines
3. Electromagnetic radiation, caused by
   a. Transformers
   b. Welders
   c. High current-carrying conductors
   d. Arcs of large contactors
4. Electrostatic fields, generated by capacitive coupling with
   a. Current conductors
   b. Switches
   c. Relays
5. Galvanic currents, originating from
   a. The thermoelectric effects when conductors of different materials are used
   b. Ground potentials

A common cause of noise is conductors which carry dc or ac currents for operating high-power switching gear and machine drives. The noise will be transmitted to the interface by inductive or capacitive coupling.

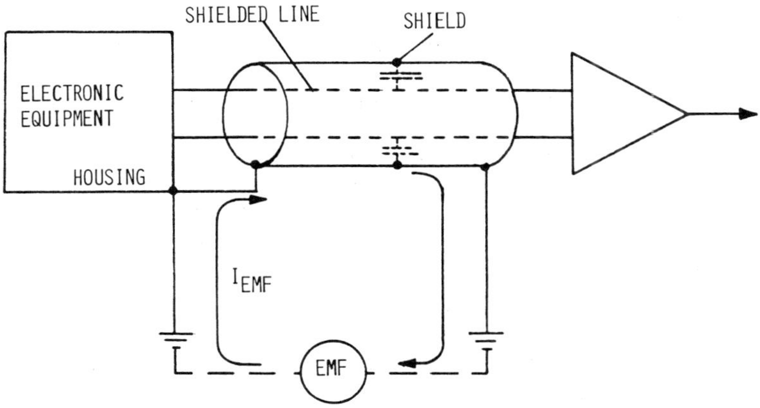

FIG. 8.13 Noise caused by a ground loop.

In addition, modern thyristor power elements commonly used in low-volt, high-current power supplies generally have a high noise band. Removal of this type of noise is possible by installing suitable filters in the connection leads to the interface. These filters are turned to eliminate the frequency of the main noise source. In severe cases the transmission of clean signals across a noisy area can be aided by the differential transmission techniques, which eliminate noise substantially (Chap. 4).

The second cause of noise is ground currents. The improper use of grounds may create severe problems. For example, if the shield

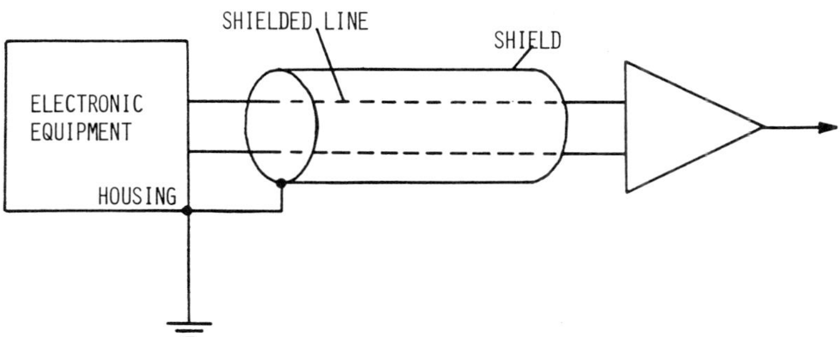

FIG. 8.14 Elimination of a ground loop by using a single ground.

of a device is grounded at different locations, very high electromotive forces (EMF) can be generated which cause interference or even destroy circuit components (Figs. 8.13 and 8.14). If the ground wire itself is connected to two different locations, a stray current may flow in the wire to the computer and alter the signal to be transferred. If the shield is connected to two different grounds, it may act as an antenna and induce a stray signal in the transducer leads (Fig. 8.13). The problem can be eliminated by using one ground only (Fig. 8.14).

8.3.2 Communication Lines

For the transmission of analog signals in the factory environment mostly a shielded twisted pair of cables are used. In general, low voltage signals are amplified. Often it is advisable to digitize analog signals at the source of origination and to transfer these as digital signals which are less sensitive to noise interference.

Signal lines should never be strung together with power lines in the same conduit. In case high-power lines have to be intersected, this should be done at a right angle.

For the transfer of digital signals potentials of 24, 48, 115, and 230 V or higher are preferred. Shielding is not necessary in all cases. In order to avoid high cable expenses, it is advisable to experiment with the communication wires to determine the correct signal lines, the signal potential and necessary shielding procedures.

A ground lead in a signal communication device serves as a supply for a fixed reference potential for measurement transducers or the distributed data communication systems. For this reason ground loops shoudl be avoided in all communication systems. It is also important to provide a fixed reference potential for a communication system which has very long communication lines. Otherwise, it is possible that between the termination points of the long leads a high potential can be generated by capacitive or inductive interferences. To avoid ground loops a galvanically decoupling device should be installed. There are several different types of decoupling devices available:

Transformers
Relays
Flying capacitors
Magnetic cores
Photoelectric isolators

Figure 8.15 shows the application of a photoelectric isolator which is most commonly used. In this case, two of these devices were installed, one on the transmitting and the other on the receiving end of the line.

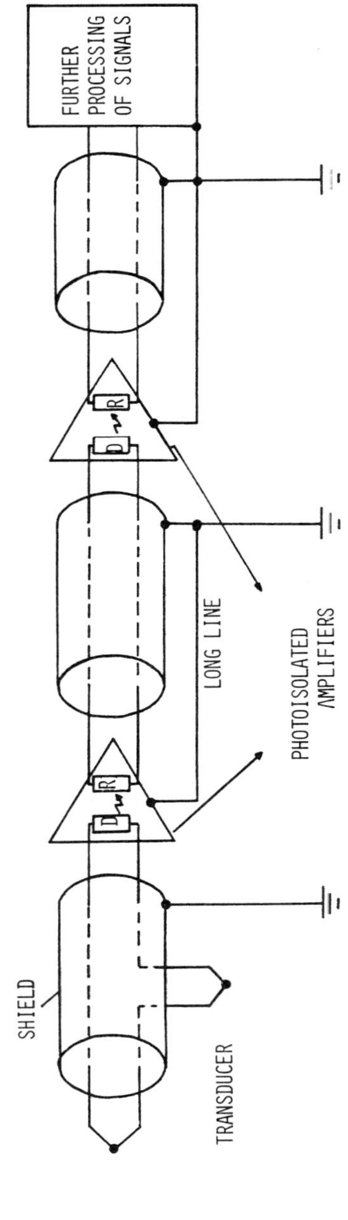

FIG. 8.15 The use of photoisolators in a transmission line.

BIBLIOGRAPHY

Bloomfield, N. J., Drive circuits for reed relays, *Machine Design*, Mar. 3 (1975).

Grounding and Noise Reduction Practices, Scientific Data Systems, El Segemdo, Calif., 1973.

Rembold, U., Prozessrechneranlagen, ihre Planung und Implementierung, Vorlesungsskript, Universsität Karlsruhe, Karlsruhe, 1978.

The Interface Circuits Data Book, Texas Instruments Learning-Center, Texas Instruments, Inc., Dallas, Tex., 1976.

Siemens Schaltbeispiele 1975/76, Siemens Bauelemente, Munich.

TTL Cook Book, Texas Instruments Learning-Center, Texas Instruments, Inc., Dallas, Tex., 1976.

9

SPECIAL PERIPHERAL BUS SYSTEMS

9.1 INTRODUCTION

Different designs of distributed computer systems for process control were discussed in Chap. 4. For the connection of peripheral equipment to a computer structures similar to that shown in Fig. 2.1 are usually chosen. The master computer is set up at a central location, and the peripheral subsystems of the distributed process are connected to the computer by a bus system. A computer control system which uses such a peripheral bus can easily be adapted to changing requirements of the process. Thus a necessary expansion of the process does not noticeably increase the cost of the computer control equipment, since only peripherals must be added.

The design of an external bus for a computer-based measuring and control system is similar to that of an internal bus system of a microprocessor system (compare Chap. 2).

The use of standardized interfaces and transmission systems further increase this flexibility. Thus a possible addition of more complex peripherals becomes more economical. Their compatibility allows easy integration into various systems, which provide the same standard data lines. A diversified market already exists for standardized peripheral components for some of the proven high-performance bus systems. If properly used, they may considerably reduce the implementation cost of a system.

The basic concept of the standardized bus system is usually based on the requirements of the peripheral equipment. There are two methods to conceive a bus system:

1. Only the transmission system for data exchange between various peripherals and the computer is defined. The interfaces of the transmission system are standardized, but not the design of the function of the peripherals.

2. All devices in the system are standardized. The dimensions of device modules, circuit boards, plugs, etc., are defined. In addition, the characteristics of the electronic circuits (voltage, timing) as well as the logic functions are defined exactly.

These basically different criteria in many cases lead to the design of a specific bus system for the solution of a particular task. Often, contradictory requirements in automation do not allow realization of a universal or ideal bus. This results in the problem of selecting the appropriate standard bus system for a particular application.

Currently, four peripheral bus systems which provide standard interfaces and device systems are used. Their characteristics include:

1. The PDV bus system is designed for serial transmission over a maximal distance of 3 km. Data is transferred over two or four lines which form a dual loop. It also possesses standard interfaces for up to 252 devices and achieves a transmission speed of 120 kbit sec^{-1}.
2. The IEEE 488 bus system, in this chapter called the IEC bus system, has 25 lines and is designed for short distances. Data transmission occurs byte serially over a standard bidirectional data line. The maximum transmission speed is 250 kbyte sec^{-1}. Up to 25 measuring devices can communicate with each other.
3. A similar architecture can be found in the CAMAC system. Here, however, the entire device system, consisting of modular functional units, is precisely defined by adherence to strict standard rules. This bus has 86 lines.
4. For locally distributed computing stations and other digital communication devices the Ethernet provides a communication system, similar to that of the PDV bus.

   The maximum data rate is 10 Mbits sec^{-1}. A total of 1024 stations with a maximum separation of 2.5 km can be connected.

The following list summarizes the characteristic criteria to be taken into consideration when evaluating the performance of a bus system:

Spatial distribution of the processes
Length of the data paths (distances)
Performance of the transmission (data rate)
Response time for requests and alarms
Possibilities of adding further peripherals
Availability and reliability of the system
Compatibility of the individual functions and the logical structure

9.2 PDV BUS

Industrial process control requires the availability of a data transmission system which is reliable and tolerant in regard to system failures, and allows coupling and decoupling of subsystems without interrupting the total system. In 1974 a committee was formed in Germany which proposed standardization of an industrial serial bus with hierarchical structure. Several pilot systems were installed for different applications. Preliminary results were very encouraging and led to a more detailed proposal for a standardized Prozess-Daten-Verarbeitung (PDV) bus.

The proposed bus has to comply with the following requirements:

Transmission efficiency
Variable data block lengths
Handling alarms
Connection and disconnection of stations without disturbing the system
Suitable data structure and transmission procedures to permit efficient control of the process
Overall reliability of the total system when one or several stations fail

The design of the bus is based on a hierarchical control concept in which subsystems are linked to a master station. This concept assures data processing in real time with a high degree of reliability. Some typical applications for the PDV buses are as follows:

Direct numerical control (DNC)
Computer-aided manufacturing (CAM)
Direct digital control (DDC)

For the proposed data transmission system a dual-loop design has been suggested with a message and answer line (Fig. 9.1). Generally, stations are coupled to the bus line via a bus coupling unit and transfer their information bit-serial using a synchronous handshake procedure. In order to remain open to further technological advances, the transmission path itself was not subject to the proposed standardization. Only the following characteristics of the peripheral bus system are defined:

System configuration
Interfaces between the PDV bus and stations
Communication procedure

The transmission speed is as follows:

10 kbaud at 1300 m	1 Mbaud at 100 m
100 kbaud at 1000 m	10 Mbaud at 10 m

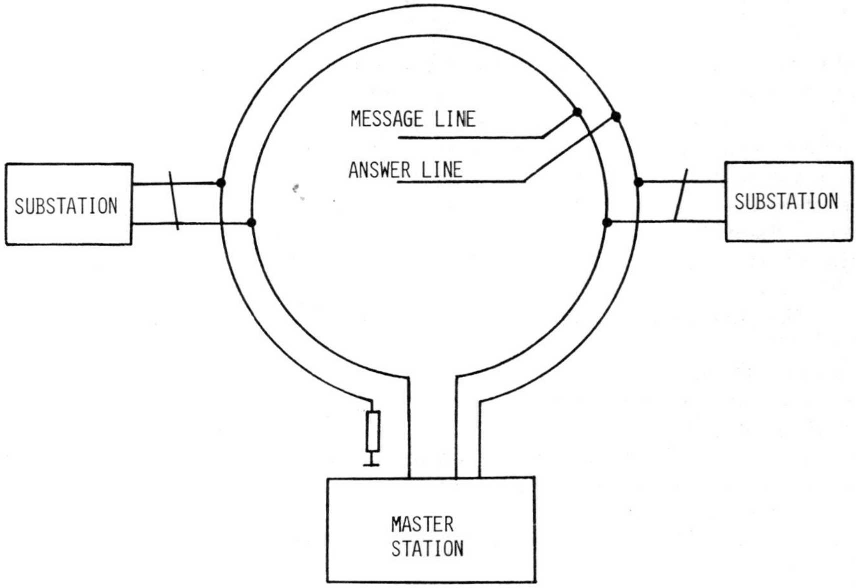

FIG. 9.1 Dual-ring structure of the PDV bus.

9.2.1 Configuration of the PDV Bus System

The basic idea is a hierarchically organized bus in which a master station controls up to 252 slave stations (substations). In principle the function of the privileged master station can be delegated to other stations. However, at any given time only one master station can be active. This usually is the central process control computer. Some privileged stations can generate alarms at any time and can receive information from the master station and answer accordingly. The function of the simpler slave stations is reduced to receiving and answering; they cannot generate alarms.

The transmission line shown in Fig. 9.1 is conceived in such a manner that it provides the possibility of recognizing interrupts on the line and checks transmitted information for errors. To achieve high data rates and short reaction times, the dual-loop ring is separated into a message line (master station → slave station → master station) and an answer line (slave station → master station). In addition to this hierarchical communication feature, direct communication between two slave stations (so-called cross-communication) is possible. However this must be initiated by the master station. Besides the dual-loop structure, a simple ring is suggested for longer distances,

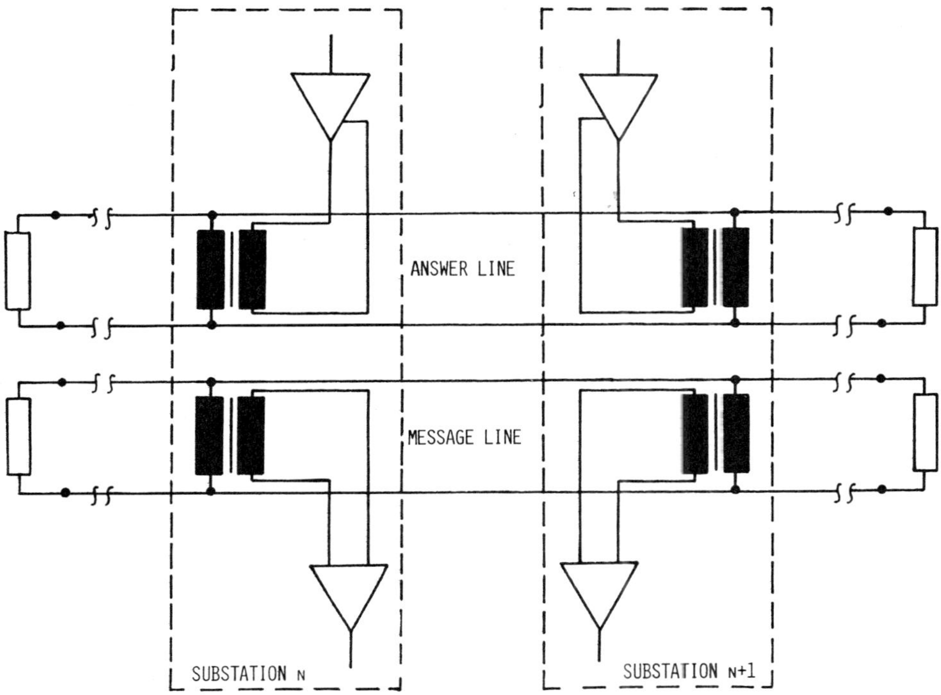

FIG. 9.2 Interfacing the PDV bus with isolation transformers.

in which messages and answers can be transmitted over the same line.

In both designs connecting to the line must occur without system interruption, which means that no active element may be implemented as a direct link into the line. Otherwise, a possible defect in a link will disturb the function of the entire bus. In most implementations the solution is found in the use of isolation transformers (Fig. 9.2). Individual stations are connected by the "plug-in principle" to the message and answer lines via a special bus-coupling unit.

9.2.2 Transmission Code

The code for the signal must be chosen so that the clock signal included in the information can easily be separated from the data. Appropriate codes are the so-called biphase code (Manchester II) or the pulse delay modulation (PDM) in which both logic states LOW and HIGH are represented by pulses of different length (Fig. 9.3).

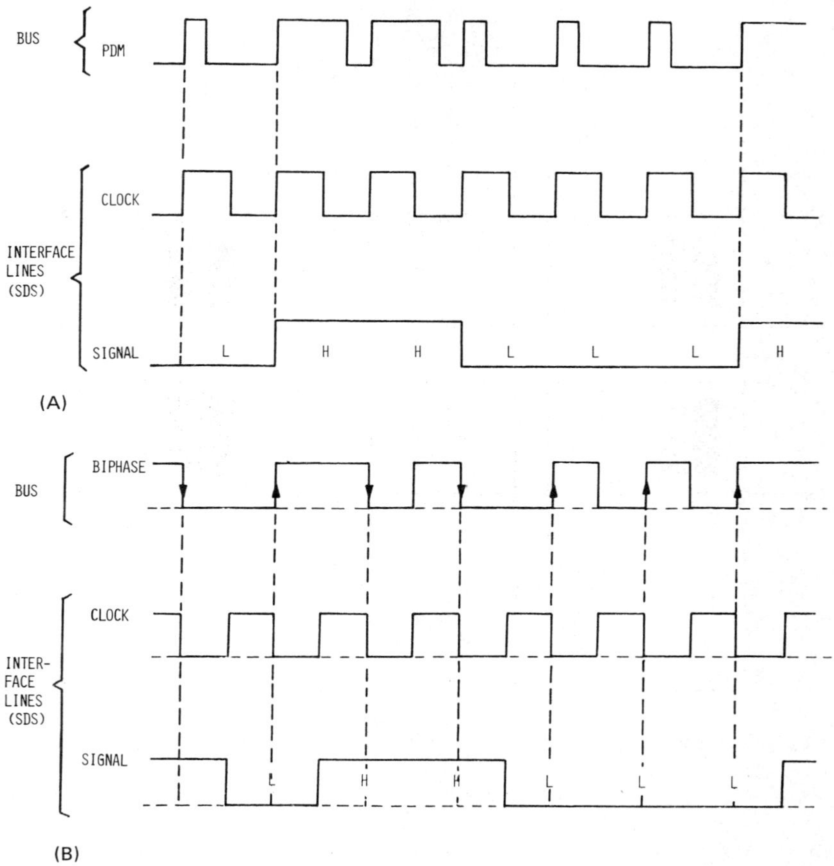

FIG. 9.3 Transmission codes for the PDV bus: (a) pulse delay modulation (PDM) and (b) biphase code (pseudoternary).

9.2.3 Interfaces

The functional components necessary for connecting the station to the bus are shown in Fig. 9.4. The bus-coupling unit provides the connection to the transmission channel and must be adapted to the technology used for signal transmission. In addition to the usual electrical transmission method over coaxial lines, there is also the optical method of transmitting information over fiber optics. The latter method offers the advantage that it is immune to noise.

The link between the bus-coupling unit and the process control or measuring units (measurement instruments, DDC processors, etc.) is

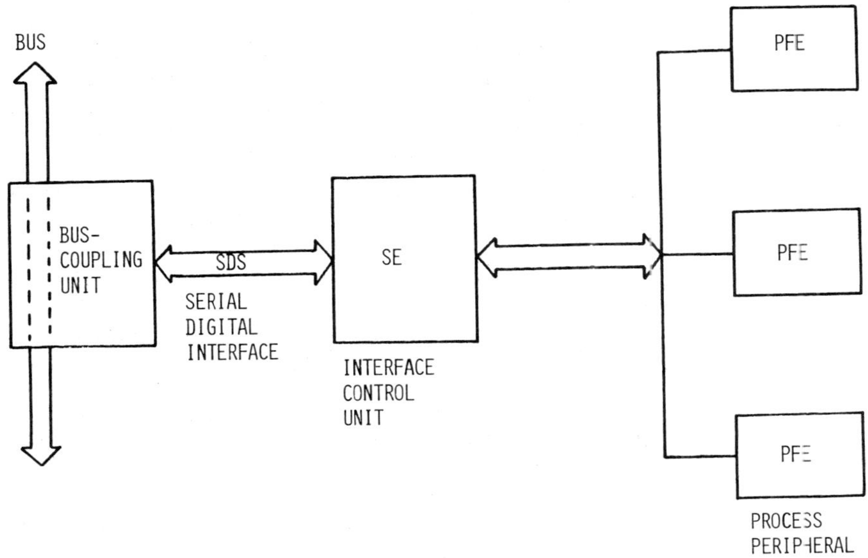

FIG. 9.4 Basic components of a substation.

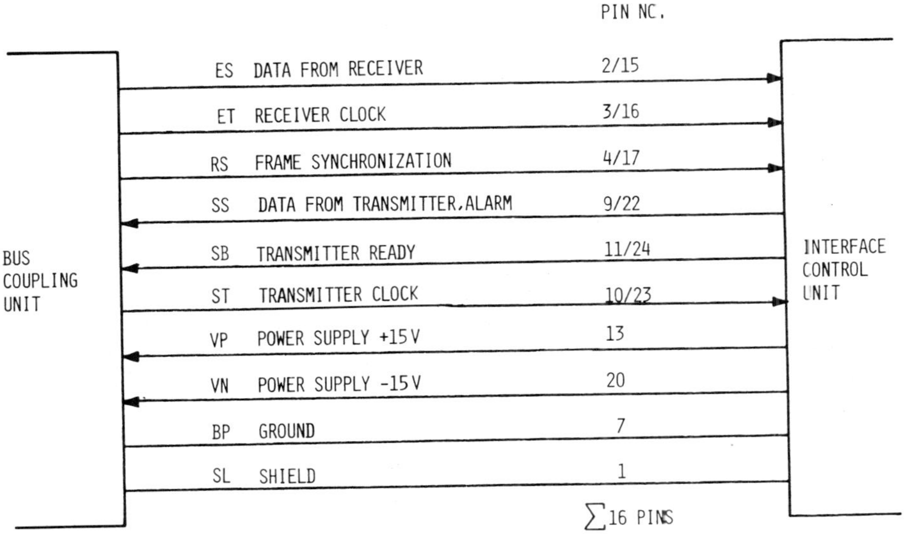

FIG. 9.5 Standardized interface between bus and control unit.

called a *serial digital interface* (SDS). The assignment of the individual lines which are tied together to a block with 16 pins as well as their functions are shown in Fig. 9.5. The received signal in the dual-loop system is always a message and is transmitted via line ET to the interface control unit. Receiver clock pulses are generated by the bus-coupling unit using the transmitted bus signal and and made available to the data interface via the line ET.

The bus-coupling unit detects the beginning and the end of a message and informs the data interface about this events with a signal for frame synchronization (RS). By using a signal "ready to send" (SB) the data interface indicates that a valid answer is available. The bus-coupling unit delivers via the line (ST) the clock signal for transmission over line (ST) and passes the information to be exchanged on to the bus via the transmitting line (SS). In addition, in the dual-loop ring system, alarm messages are transferred via the same line. Some specially equipped slave stations are capable of initiating alarms by themselves. In this case a slave station sets an alarm bit on the answer line within a specific time window. There are 24 possible bit positions to transmit different messages over the line. If an alarm service is needed, the interface control unit requests the bus-coupling unit to set its specific alarm bit on the message line. The station which has requested the alarm is recognized by the logic of the active master station. This logic detects the cause of the alarm by counting the bit string received from the message line and thereby determining the sender of the bit. To assure that transmission procedures can be performed independently of the actual transmission path, the interface control unit executes the following functions:

1. Checks for transmission errors and generates a check byte for the answers.
2. Decodes the address and the function.
3. Provides status information for alarm requests.
4. Handles data and alarms.

9.2.4 Organization of Data Transmission

The message transfer is done synchronously with variable block lengths by using a handshake principle. This means that each message coming from an active master station and sent to a slave station must be acknowledged by control information. Synchronization of the message and of the answer as part of the so-called transmission frame is done by a synchronization signal (SYN). It informs the bus-coupling unit about the beginning of a transfer. In Fig. 9.6 the structure of a transmission unit is shown. It consists of various transmission elements. The bytes may contain the following information:

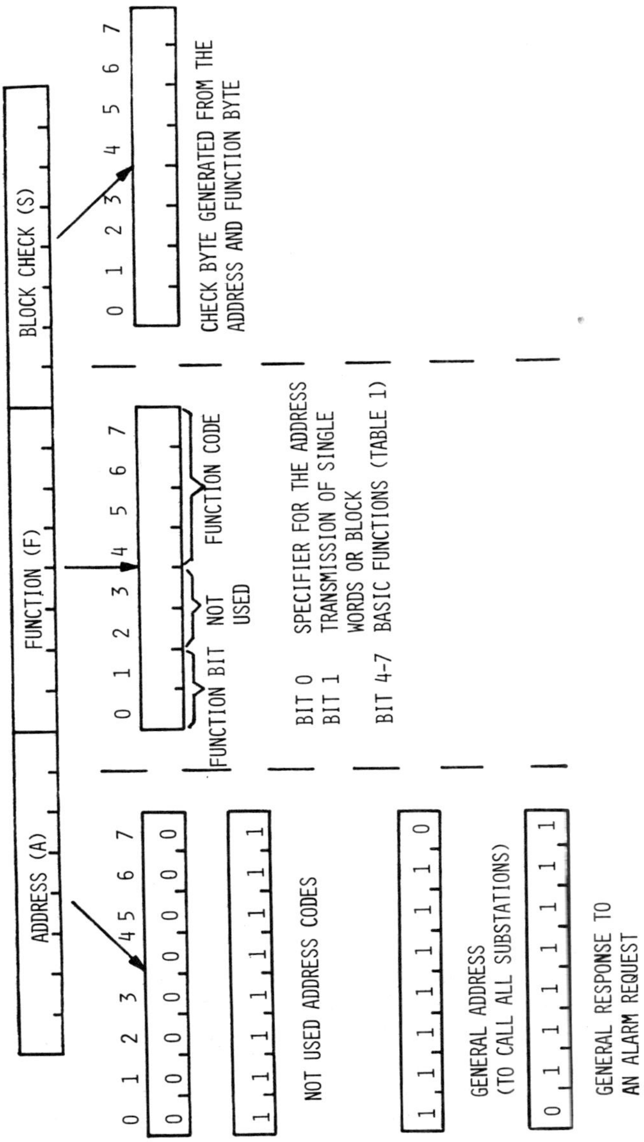

FIG. 9.6 Basic structure of a transmission unit.

TABLE 9.1 Basic Functions of the Transmission Procedures

Basic interface functions	Code Bit 5	6	7	8
Reset	L	L	L	L
Read operation	H	L	L	L
Write operation	L	H	L	L
Read and write operation	H	H	L	L
Repeat read operation	H	L	H	L
Repeat write operation	L	H	H	L
Repeat read and write operation	H	H	H	L
Remote enable	H	L	L	H
Remote disable	L	H	L	H
Request status	H	L	H	H
Start of sequential request	L	L	H	L
End of sequential request	H	H	L	H
Start of transmission between substations	L	L	H	H
End of transmission between substations	L	H	H	H
Not used	H	H	H	H
Not used	L	L	L	H

Up to 252 addresses can be coded by the address code (A).

The basic functions of various communication procedures are coded by the function code (F) (Table 9.1).

The data block contains 8 bits of code transparent information. One data word always consists of two data blocks (D,D).

The check byte(s) supervises transmission of two bytes by using a cyclic redundancy check (CRC).

Two examples of the logic interplay between the master station, which sends a message, and the addressed station, which receives the address, will be presented.

1. For a single address cycle the following rules apply:

   Message from the active master station: A F S

   Response from the addressed slave station: A F' S

   The address A placed by the master on the communication line indicates the desired address. The slave station verifies this message by using the address part as the origin address in its transmission unit. For verification a specifier for the address is placed into the returned function code F':

   H = destination address (message)

   L = origin address (answer)

2. The transmission protocol for writing a block is constructed as follows:

Message:	A	F	S	D	D	S	D	D	S
	word 0			word 1			word 2		
Response:	A	F	S						

The block transmission is indicated in the function code by a logic state of "bit 1 = H." The length, or in other words, the number of DDS groups in the block to be written, is indicated by the first byte of word 1. Subaddresses can be placed into the second byte.

9.2.5 Alarm Handling

It was already mentioned before that alarm message can be sent during a specified time frame. Slave stations having the alarm option request

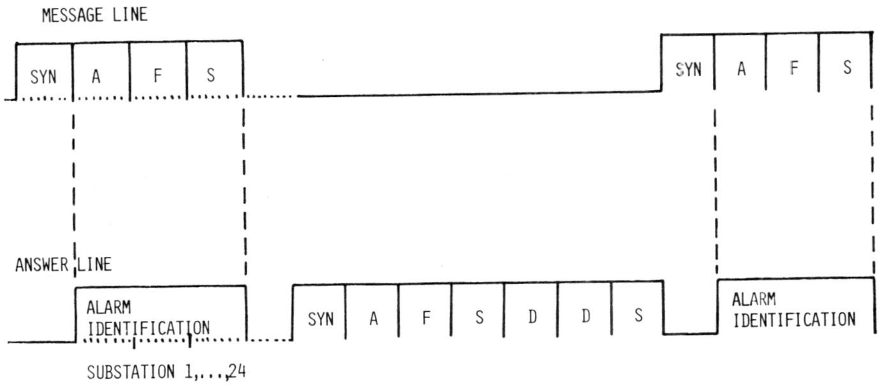

FIG. 9.7 Additional utilization of the answer line for alarm messages in the dual-ring system.

service from the master station via the answer line. When the head of a transmitted message passes by the station on the message line, this active station sets an alarm bit on the answer line. Each of the 24 preferred stations has a unique bit position assigned to it in the answer protocol (Fig. 9.7). The master station receives this alarm bit on the answer line and identifies the source of alarm. Thereafter the alarm request of the slave station will be served by the master station with a complete message-response operation.

In addition to this spontaneous alarm procedure, there is also a shortened cyclical interrogation mode provided. With the help of a special procedure all slave stations which are able to generate an alarm request will be addressed in sequence. This procedure actually is a very fast polling process. It is supplemented by a special procedure for direct communication between two slave stations and used for the treatment of system errors. These procedures, however, will not be explained here in more detail.

9.3 IEC BUS

In April 1975 the Institute of Electrical and Electronic Engineers published the IEE Standard 488/1975 "Digital Interface for Programmable Instrumentation." For this interface bus concept there are several names being used. In general it is called the *IEC bus*, the universal interface bus, general-purpose interface bus, or Hewlett Packard in-interface bus. An international standard defines the design (Fig. 9.8).

This bus will allow plug-compatible connection of a whole spectrum of instruments as well as controllers and a computer to a functional system, which can be tailored to the specific needs of an installation. There is already a standardization trend to be seen in industry. Approximately 200 manufacturers offer instruments and computers for various measurement applications, which can be connected to the standardized IEC interface. The instruments can be interexchanged and combined to a complex measuring system with the help of the IEC bus.

Functionally the interfaces are designed from different building blocks which are part of the standardized system. The development engineer can limit her- or himself to select the necessary functions needed for the design of the peripheral interface, without losing the compatibility of the complete system. For instruments to be used as controllers, the user is provided with higher programming languages, which will simplify the system design and enhance the range of application.

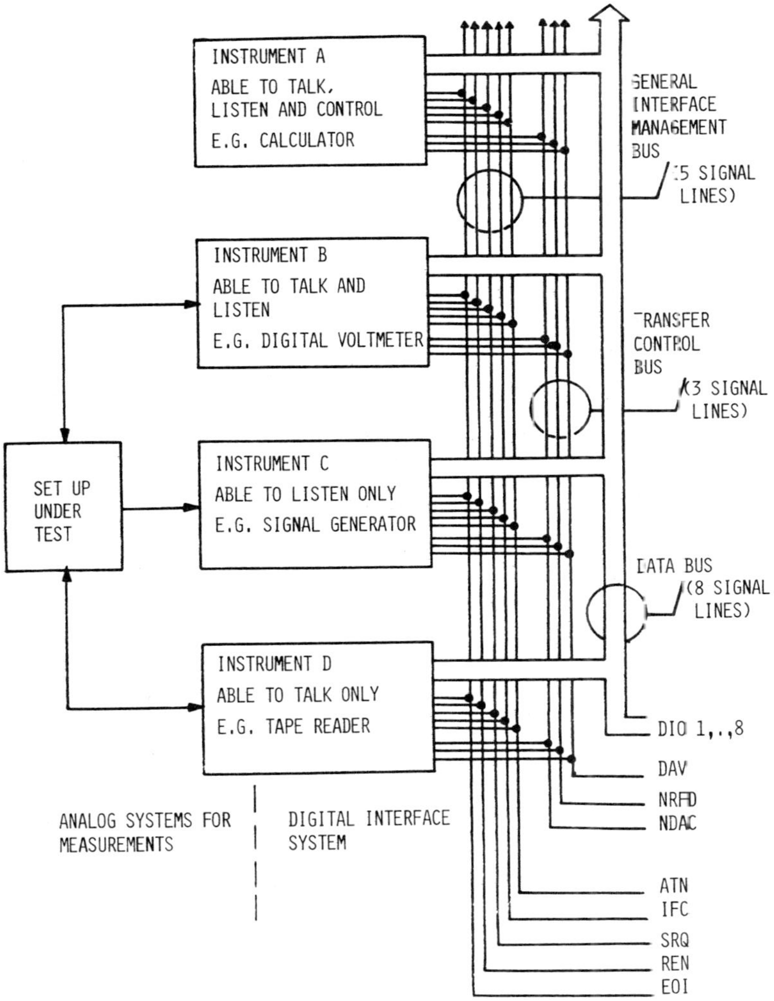

FIG. 9.8. Design of a measuring system using the IEC bus: DAV, data valid; NRFD, not ready for data+; NDAC, data not accepted+; ATN, attention; IFC, interface clear; SRQ, service request; REN, remote enable; EOI, end of identify. (+ indicates that negation is represented by a LOW state on these two lines only.)

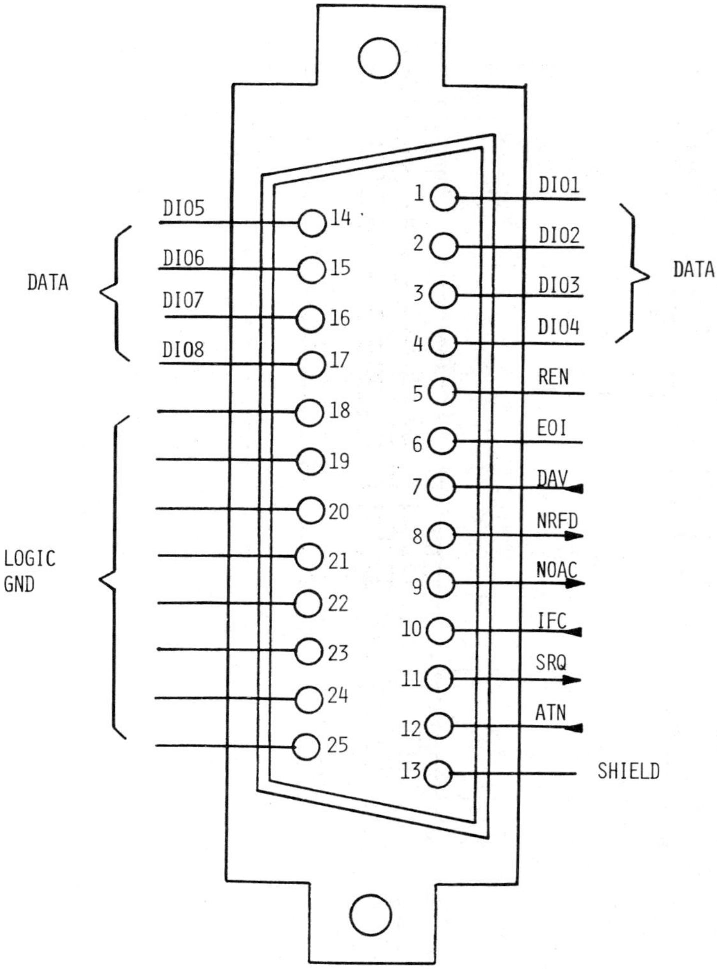

FIG. 9.9 Pin assignment in an IEC connector.

9.3.1 Design of the IEC Bus System

The standardized system is similar to the PDV bus. It is based on a unified transmission system, in which all participating instruments are connected to a standard interface. In this bus are 16 signal lines: eight lines for data transfer, five lines for control, and three lines for the handshake procedure. In order to prevent possible cross-talk all control lines are twisted together with a ground cable. Thus the bus has a total of 25 lines. This necessitates the use of a 25-prong plug, which is also standardized. The assignment of its various pins is shown in Fig. 9.9.

In this bus system all participating instruments are treated equally. Data communication takes place without a central control unit. All connected instruments can exchange information among themselves. In this case, communicating data sources and, if necessary, several receiving instruments must first be connected with each other through the systems control. The byte-serial data communication over the eight bidirectional lines is supervised with the help of control signals. The parallel lines allow transmission of 2^8 data combinations in both directions. Usually these are interpreted as ASCII code. A special request and handshake procedure assures the operation of even the slowest instrument without time loss. The transmission speed is 200 to 400 kbyte sec^{-1}. The bus is limited to a length of 20 m.

9.3.2 Interfaces

The object of standardization is the interface between the IEC bus and the connected measuring device. In addition the signal transmission over the cable and the logic function of the interface of the connected device have to comply to standard procedures. To illustrate this complex relationship on Fig. 9.10, an interface normally contained in the device is shown as a separate modular constructed IEC interface unit. It consists of a plug for the IEC cable, bidirectional drivers and receivers, a message coding logic, and the actual interface function blocks connected to the interface. Further processing of data in the instruments, which also implies processing of received data, is not standardized.

In order to be able to connect the bus to an instrument for a particular application, the development engineer may select from the spectrum of available interface functions. The control function is an exception, it gives an instrument the ability to send device addresses, universal instructions, and instructions to address other instruments over the bus. If this function exists in more than one instrument of the system, only one control function may be active at a given time. A so-called system controller determines the interpretation of the signals on the bus. This controller may only be implemented once in the bus system; usually it is the control computer which supervises all other control functions. If necessary it can deactivate a previously active instrument.

In the bus there are five lines reserved for bus management (Fig. 9.8):

1. Attention (ATN): It interprets signals on the bus as addresses, instructions, or data.

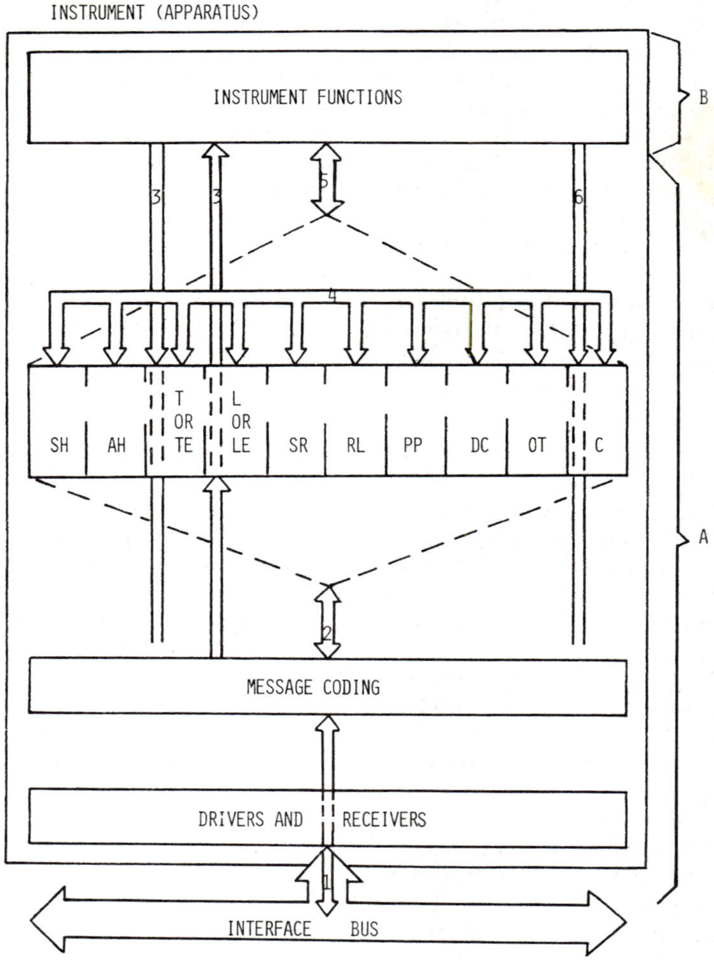

FIG. 9.10 Assignment of the logic functions and the data transmission lines in an IEC interface. A: Standard capability defined. B: Capability defined by designer. (1) Interface bus signal lines. (2) Remote interface messages to and from interface functions. (3) Instrument dependent messages to and from instrument functions. (4) State linkages between interface functions. (5) Local messages between instrument functions and interface functions. (Messages to interface functions are defined, messages from interface functions exist according to the designer's choice.) (6) Remote interface messages sent by instrument functions with a controller.

2. Interface Clear (IFC): It lets a system controller take the command from another controller and/or initialize the bus.
3. Remote Enable (REN): It allows the control of the instrument to be switched between its local (manual) control and remote control (programming codes received or addressed as a listener).
4. Service Request (SRQ): It lets an instrument indicate to the controller that an alarm has occurred and requests it to take some specific action asynchronously with respect to other bus operations. One SRQ function is necessary for each independent request for a service.
5. End or Identify (EOI): In connection with the ATN signal, EOI represents either the end of a data communication or the receipt of a status message as a result of a request.

Supervision and processing of alarm messages (SRQ) as well as the transmission of the signals (IFC) and (REN) is the task of the system controller. This is the reason why it has a principal supervising function to control all other instruments.

Three additional lines of the bus are reserved for the handshake process to initiate data communication:

1. Data Valid (DAV): It identifies information on the data bus as valid.
2. Not Ready for Data (NRFD)$^+$: It indicates the status of an interface that is not ready to accept data.
3. Data Not Accepted (NDAC)$^+$: It indicates that the interface has not yet received data.

The "+" sign indicates that negation is represented by a LOW state on these two lines only.

The time necessary to send data is determined by the characteristics of the transmitter. The time needed to accept data and to be ready for further data transfer depends on the receiver's characteristics. The characteristics of the entire bus system determine the settling time. This system consists of drivers, receivers, terminators in the instruments, and the bus cable.

Figure 9.11 shows the handshake principle with one talker and several listeners which uses the signal lines DAV, NRFD, and NDAC. Each of the signals NRFD and NDAC consists of superimposed signals derived from two or more listeners. Because of the varying lengths of the data communication paths and different data acceptance and processing times, they receive the same data, however, at slightly different times.

The repertoire of available interface functions characterizes the performance of an IEC instrument. It signifies to a great extent the flexibility and compatibility of the connected instruments. For the

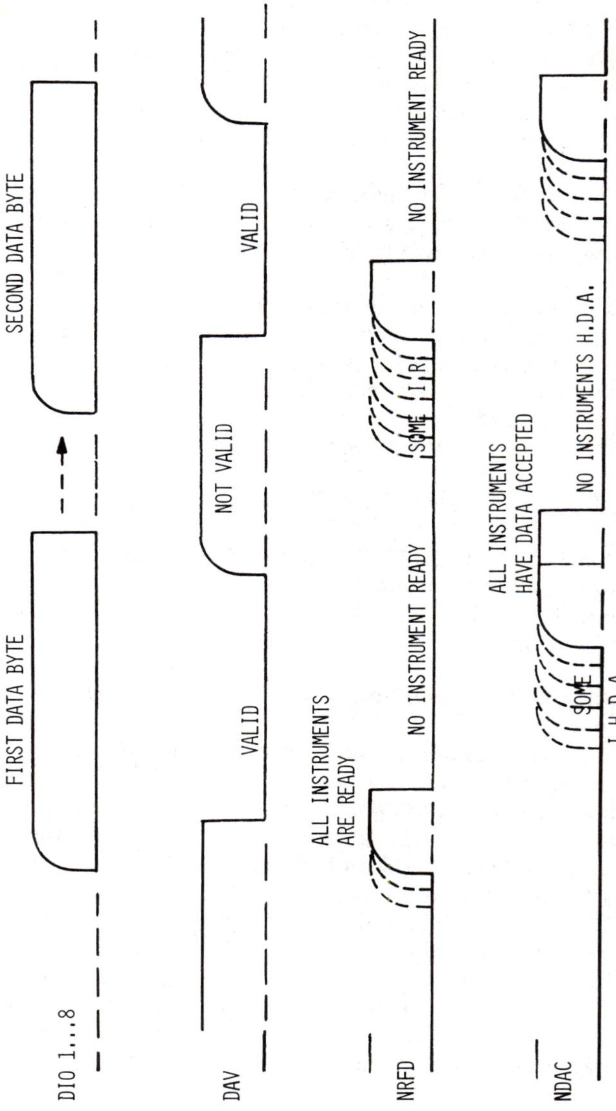

FIG. 9.11 Timing diagram for a handshake process with one talker and several listeners.

simplest configuration only two functions out of the entire list of functions shown in Tables 9.2a and 9.2b are necessary to perform a normal communication. For example, an instrument which is only able to talk needs, in addition to a talker function (T), the source handshake (SH). Similarly, an instrument which is only able to listen needs, in addition to a listener function (L), the acceptor handshake (AH). Both functions SH and AH drive the signals DAV, NRFD, and NDAC, which control timing of the communication process (Fig. 9.11).

The listener (L) and the talker (T) are activated by their addresses which are transferred together with the accompanying signal ATN over the data bus. This also holds true for the control function (C). The decoder in the interface of each instrument decides whether it receives its own address, and if yes, it activates the corresponding function. Five bits on the data bus are reserved for 31 primary addresses for talkers and listeners. The (C) function can be extended by several additional functions to comply with the requirements of varying configurations of the bus systems. For a minimum configuration the availability of the alarm signal (SRQ) and the clearance signal (IFC) to reset the interface functions is mandatory. Additional functions are described in detail in the standard published by IEEE [4].

Theoretically the configuration of a measuring system with various IEC bus instruments should create no problems. Nevertheless there may be difficulties. At present only the data transmission system is standardized, but not the functional code for actuating the

TABLE 9.2a Abbreviations of Interface Functions

Function of interface	Abbreviation
Source handshake	SH
Acceptor handshake	AH
Talker or extended talker	T or TE
Listener or extended listener	L or LE
Service request	SR
Remote/local	RL
Parallel poll	PP
Device clear	DC
Device trigger	DT
Controller	C

TABLE 9.2b Descriptions of Interface Functions

Function	Description
SH	Source handshake To synchronize the transmission of information on the data bus by the talker when sending instrument-generated data and by the controller when sending interface messages.
AH	Acceptor handshake To synchronize the receipt of information on the data bus for all interface functions when receiving interface messages and for the listener function when receiving instrument-generated data.
T	Basic talker To let an instrument send data to another instrument.
L	Basic listener To let an instrument receive data from another instrument.
SR	Service request To let an instrument indicate to the controller that some event has occurred and request it to take some specific action asynchronously with respect to other bus operations. (One SRQ function is required for each independent reason for requesting service.)
PP	Parallel poll To let instruments return 1-bit status to the controller. Up to eight instruments may respond simultaneously. More than one instrument may respond on the same status line so that logical operations (AND,OR) may be performed on a group of instruments.
RL	Remote local To let the control of the instrument be switched between its local (manual) controls and remote control (programming codes received while addressed as a listener).

TABLE 9.2b (Continued)

Function	Description
DC	Device clear To provide a means by which an instrument (device) may be initialized to a predefined state. All instruments are cleared concurrently.
DT	Device trigger To let instruments (devices), either singly or in a group, be triggered, or some action be started.

instruments. There is only a proposal for the code to be used, which is ASCII code. It is therefore necessary to agree upon a higher programming language which can be understood by all connected instruments. Possible candidates are process control programming languages, such as FORTRAN, PEARL, BASEX, etc. These may have to be extended for special-purposes applications in the IEC system.

9.4 CAMAC BUS

The CAMAC system is the result of several years of cooperative development of the European nuclear research centers who use computers to control measuring systems. The abbreviation CAMAC stands for "computer-aided measurement and control." The concept of this universal bidirectional interface system is based on the second criterion which was described in Sec. 9.1. In this bus concept all devices of the system are standardized. The CAMAC specifications define all device components, transmission paths, and signals. In addition to the compatibility of the hardware, communication instructions, data communication procedures, and handling of alarms are standardized. Also special CAMAC software was defined to simplify programming. Only the interfaces between the CAMAC controller and the input/output units of a connected computer are different.

9.4.1 Configuration of the Bidirectional Interface System

The basic configuration of the decentralized CAMAC peripheral system is shown in Fig. 9.12. The system controller is connected via a branch highway to several crate controllers. The system controller, also called *branch driver*, controls the data flow between the branch

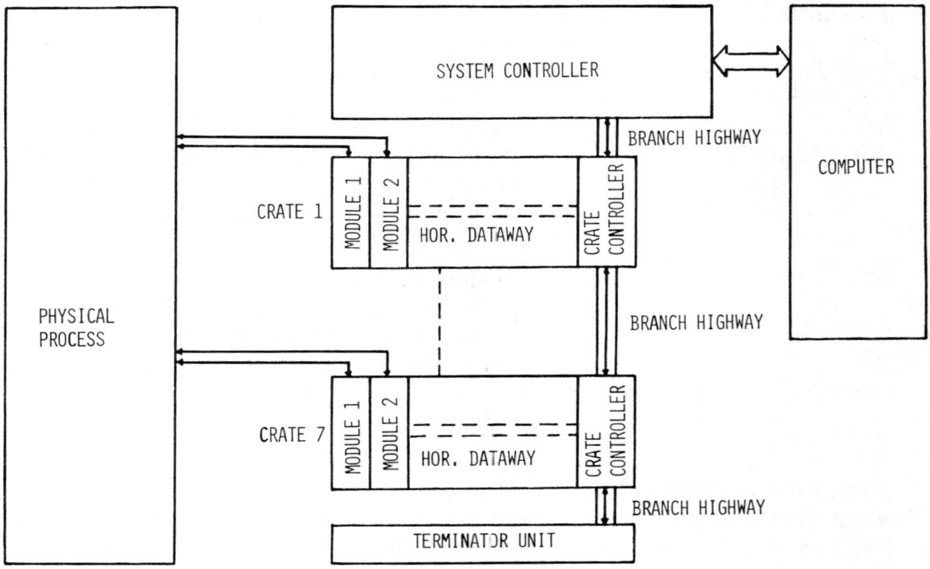

FIG. 9.12 Basic structure of the CAMAC system.

and the computer or another controller. Information can be transferred either bit or byte serially over the branch highway. This highway may contain two or nine twisted pair lines of which one is reserved for the clock signal. For high data rates coaxial cables are used.

The branch highway connects the functional units. These are designed as standardized interface cards to comply with electrical and mechanical specifications for the 19-in. crate. In the crate the slots of 24 cards are connected with each other via a horizontal data path and are supplied with power. The front panel of these modules contains connections to which the process control elements and display units can be plugged in.

In the CAMAC system the crate controller is the link between the horizontal data way and the branch highway:

1. Single modules, their subgroups, and their desired functions are selected via the horizontal data way (Table 9.3). Each module is addressed directly, by a station number line (N). For this purpose 24 lines are installed. In addition, there are 24 read (R) and 24 write (W) lines for data communication between a module and the control station. Subaddresses (A) and functions (F) are

TABLE 9.3 Signal Assignment in the Dataway

Description	Symbol	Pins	Meaning
Instruction			
Unit number	N	1	Selection of the unit
Subaddress	A1, A2, A4,A8	4	Selection of a subaddress within the addressed unit
Function	F1,F2,F4, F8,F16	5	Specifies operation to be executed
Timing			
Strobe 1	S1	1	Timing of the transmission
Strobe 2	S2	1	
Data			
Read	R1, ...,R24	24	Receiver lines
Write	W1, ...,W24	24	Transmitter lines
Status			
Alarm	L	1	Unit generates alarm
Busy	B	1	Dataways are busy
Answer	Q	1	Unit enabled for instruction
Common control lines			
Initiate status	Z	1	Initialized defined status
Reset	C	1	Reset of all registers
Inhibit	I	1	Execute no operation
Reserve			
Bus line	X	1	For later use
Private lines			
Patch pins	P1, ...,P5	5	For special use
Power supply			
+24 V =	+24	1	6 A
+26 V =	+26	1	25 A
+ 6 V =	+ 6	1	25 A
-24 V =	-24	1	6 A
0 V =	0	2	2 mA
Additional power supply			
+200 V =	+200	1	
+ 12 V =	+ 12	1	
- 12 V =	- 12	1	
117 V live	ACL	1	
117 V neutral	ACN	1	
Clear ground	E	1	
Reserve	Y1,Y2	2	
Total number of lines		86	

Source: J. G. Ottes, CAMAC-Ein System rechnergeführter Elektronik, *Elektronik* (Munich), Nr. 10, 1970.

coded and transmitted over special lines. A CAMAC instruction within a crate therefore is composed of the parts NAF, which means that data communication between single modules is only possible over the crate controller. To be able to spontaneously respond to external events each module possesses an alarm line (L) which is linked to the crate controller.

2. Data and signal communication from and to a higher control level is done over the branch highway. The crate controller executes the instructions coming from the higher control station and passes these to the specified module. Up to seven crates can be addressed via the branch highway. For this purpose the NAF instruction has to be supplemented with the corresponding address of the crate (CNAF). The system controller occupies therefore two slots in the frame. Via slot 25, it is connected to the branch highway, and via slot 24, it is connected to the horizontal data path.

9.4.2 Function of the Line Control

An instruction in NAF format is supplied by the line control. The line control determines the location at which the operation in the F part is to be carried out. Five lines are provided to determine this location. These are the address line N_i ($1 = i \leq 24$) and four subaddress lines A_K ($K = 1, \ldots, 4$). The operation itself is determined by signals sent over the five function lines F_i ($O = i \leq 4$). At the beginning of a dataway activity, the line control issues the busy signal B and maintains this until the operation is completed. A module addressed via the address line N decodes the information transferred over the A and F bus. After decoding of the instruction, it immediately responds with the enable signal Q.

The line procedure for a read instruction is shown in Fig. 9.13. Table 9.3 illustrates the function of the participating lines. The frame control provides the instruction NAF and simultaneously actuates the line B. At the same time the clock generator is triggered. It generates, after a delay of 400 nsec, the signal S1. The addressed module decodes the signal B and F and responds over the Q line and R line with requested data. Before the signal S1 is activated, these two signals must have been settled. After a waiting period of 100 nsec, the operation is terminated by the signal S2, and the information NAF is removed from the line. This resets the signal B. The dashed lines in Fig. 9.13 show how long the input data must wait for the R line. This procedure is used by the function "Read and Erase a Register." The edge of signal S1 is used for the read cycle, and the edge of S2 for the erase cycle.

In addition to this group of read instructions, there are write and decode instructions implemented. Their timing is similar to that of the read instructions. There are five lines provided for the transmission of a possible instruction set of 25 to 32 instructions. However,

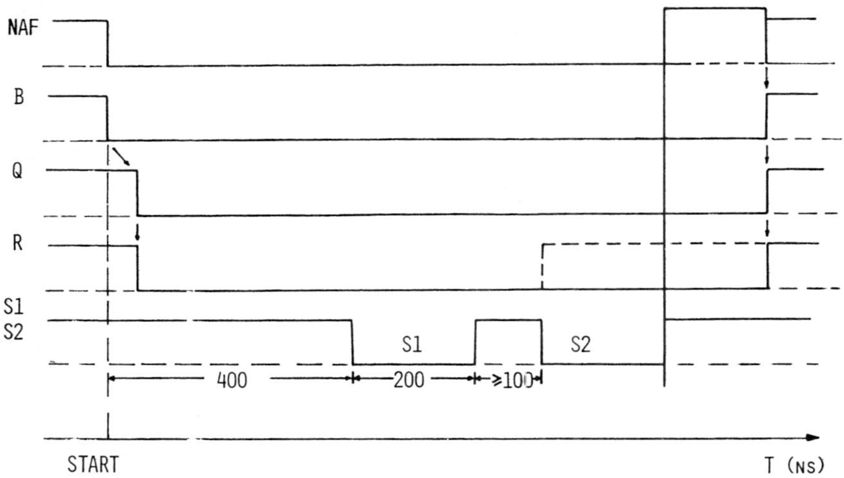

FIG. 9.13 Timing diagram for a read instruction. (From J. G. Ottes, CAMAC-Ein System rechnergeführter Elektronik, *Elektronik* (Munich), Nr. 10, 1970.)

not all instructions are specified, since there is no need to use them all.

Data communication over the branch highway is performed in a similar manner. This communication process obtains its instructions from the higher level digital control via the so-called branch driver. In addition, there are specific procedures for the transmission protocol to be used. These are described in the reports of the European System of Nuclear Electronics (ESONE) Committee.

9.5 ETHERNET

The Ethernet local area network provides a communication system for carrying digital data packet among locally distributed computing stations and other digital communication devices. Its task is similar to that of the PDV bus. This branding broadcast communication system is named for the historical ether through which electromagnetic radiations were formerly assumed to propagate. An Ethernet transmitter broadcasts completely addressed synchronous bit sequences called *packets* onto a logically passive communication medium, called the *Ether*. The packet is heard by all participating stations and is copied from the Ether by the station whose address matches with the leading address bits of the packet. The primary characteristics of this local area network include in its physical layer,

Data rate: 10 Mbits sec^{-1}
Maximum station separation: 2.5 km
Maximum number of stations: 1024
Medium: shielded coaxial cable, twisted pair lines, fiber optics, etc.
Topology: A branching unrooted tree

With this specification the Ethernet is intended primarily for use in automation, distributed data processing, terminal access, and other situations requiring an economical connection to a local communication medium. Originally the Ethernet was not designed for real time computer systems; however, this application was not specifically excluded.

9.5.1 Design Principles

The Ethernet implementation has four major parts (see Fig. 9.14).

Ether
Transceiver
Interface
Controller

The Ether can be constructed from any communication media including coaxial cables, twisted pairs, and optical fibers. It is unrooted because it can be extended from an arbitrary point into any direction. A station wishing to join the Ethernet taps into the channel at the nearest convenient point.

Looking at the problem of interconnection and control, we see that the Ethernet provides interconnection through a passive communication facility. This means that a failure of an active element will affect the communication of only a single station. Unlike the PDV bus (Sec. 9.2) with a central controller for the whole communication system, the Ethernet has distributed control among its stations.

Sharing of the Ether is controlled in such a way that transmission of a data frame, initiated by a station, is deferred to any communication which may already be in progress. If an interference with other packets is detected, the initiated transmission is aborted and rescheduled by its source station. After a randomly determined retransmission interval, the controller tries once more a transmission until a packet is heard by the destination. Because of the probabilistic operation of the Ethernet, its application is limited to process control tasks which do not require fast reaction times.

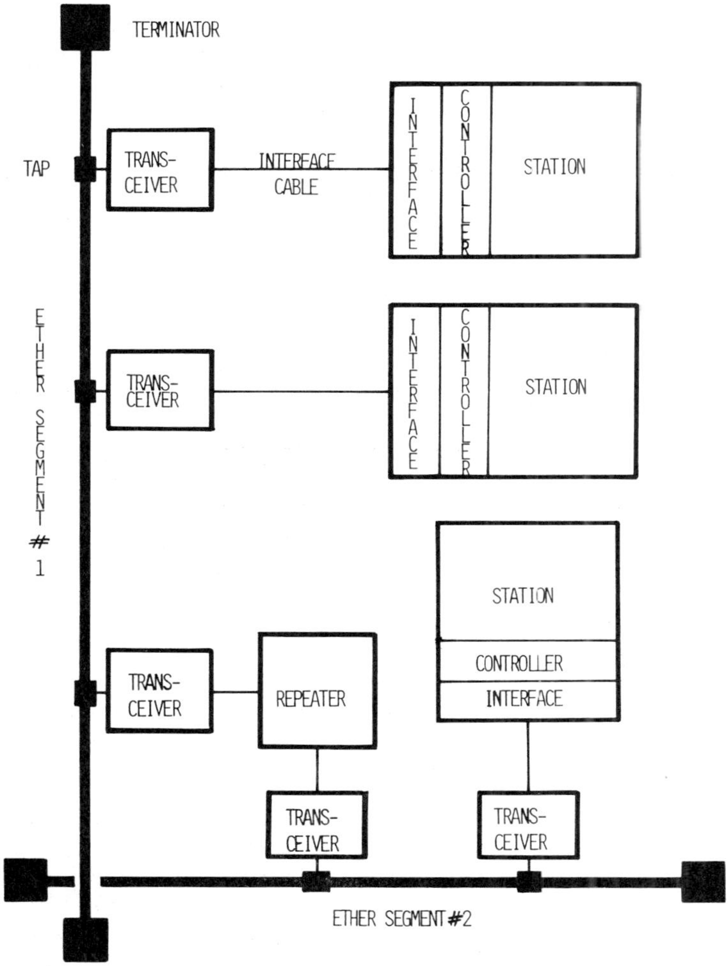

FIG. 9.14 A typical ethernet configuration with two segments.

9.5.2 Protocol

In the Ethernet there are three layers needed for data communication. The specifications [1] for the net only specifies two layers of the network architecture; the physical layer and the data link layer (Fig. 9.15). The highest level, which is the client level, will be defined by

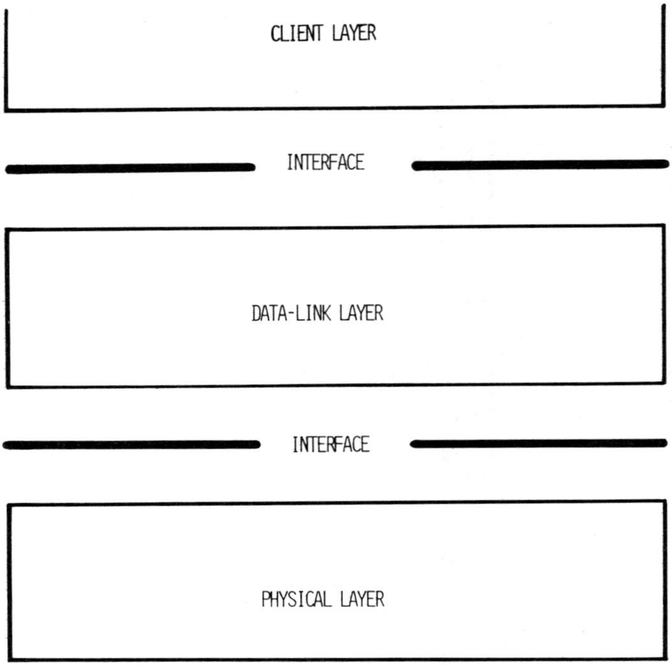

FIG. 9.15 Overall structure of the layered Ethernet architecture.

the user. The user communicates with the data link layer via a specified interface. The physical and data link layers use the same principle as the architecture of an "open systems interconnection." It is documented by the International Organization for Standardization (ISO) [2]. The interface between these two layers includes signals for framing (carrier sense, transmit initiation) and contention resolution (collision detect). In addition there are facilities for passing a pair of serial bit streams (transmit, receive) between the two layers and a wait function for timing.

In Fig. 9.16 the Ethernet frame format is shown, which corresponds with the packet layout for serial transmission over the physical channel. Similar to the PDV bus, only essential characteristics are specified to provide maximum flexibility for the implementer. Five fields are needed for the construction and processing of frames: the source and destination address in the header, a type field for use by higher layers, a data field, and the frame-check sequence field containing a cyclic redundancy check value to detect transmission errors.

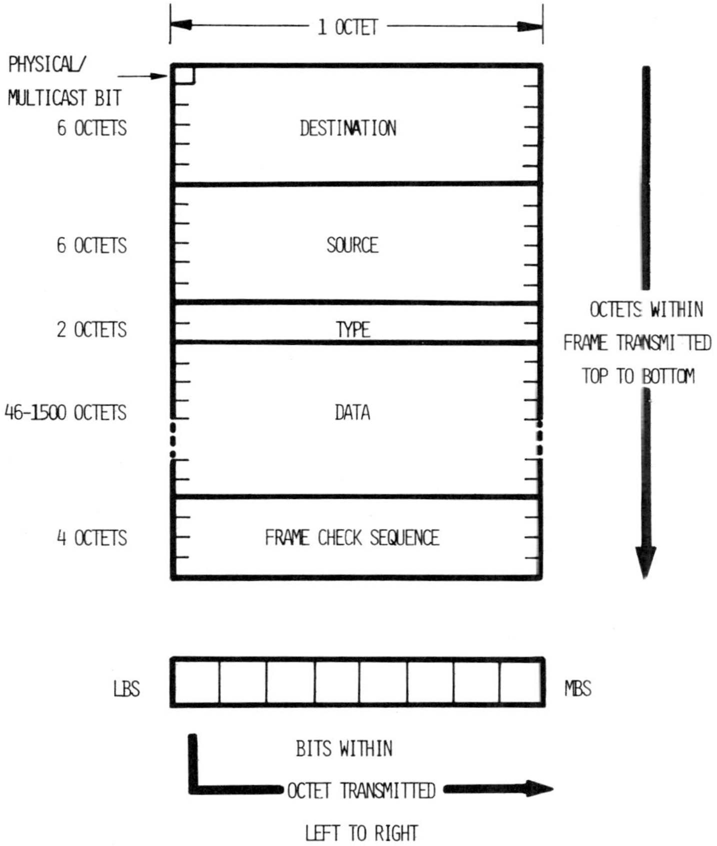

FIG. 9.16 Ethernet frame format.

Of these five fields, all are of fixed length except the date field, which may contain any number of octets within a specified range.

9.6 COMPARISON OF VARIOUS INTERFACE SYSTEMS

The preceding paragraphs discussed examples of bus systems whose configurations and characteristics are internationally standardized. It only portrays a small selection of the multitude of existing interface systems. In addition to the existing or proposed standard systems,

TABLE 9.4 Comparison of Different Bus Systems to Connect Computers with Peripheral Equipment

01	DESIGNATION	PDV - BUS	IEC - BUS	CAMAC - DATAWAY
02	ORIGIN	GFK, TECHNICAL UNIVERSITIES AND INDUSTRY IN GERMANY	IEC/TC66/WG3	EURATOM, ESONE COMM.
03	REFERENCE	DRAFT	IEC 66.22	EUR 4100, TID 25875
04	DATA OF ISSUE	LAST DRAFT 12.75	JAN. 76	1972
11	AREA OF APPLICATION	CONNECTION OF SUBSYSTEMS TO A CENTRAL CONTROL STATION	DEVICE INTERFACE FOR LABORATORIES, PROGRAMMABLE MEASURING APPARATUS	DEVICE INTERFACE FOR LABORATORIES, NUCLEAR PHYSICS, ALSO INDUSTRY
12	EXTENT OF THE SPECIFICATION	PROCEDURES, OPERATIONS, ELECTRICAL SIGNALS, CONNECTORS	COMPLETE, EXCEPT PROCEDURES	COMPLETE, EXCEPT PROCEDURES INCLUDING MECHANICAL BUILD-UP
21	TYPE OF CONNECTION	BUS (DOUPLE LOOP PREFERRED)	BUS	BUS AND BRANCH LINES
22	TYPE OF TRANSFER	SERIAL	BYTE-SERIAL	ADDRESS, COMMAND AND DATA PARALLEL
23	DATA FORMAT	16BITS(2 BYTES)UP TO 16 BYTES	8 BITS, RECOMMENDED: ISO-7-BIT-CODE	24 BITS
24	ADDRESS FORMAT	8 BITS(OF WHICH 1 BIT FOR SPECIAL CHARACTERS)	5 BITS VIA DATA LINES	23 LINES FOR 23 MODULE ADDRESSES, 4 BITS FOR SUB-ADDRESSES
25	COMMAND FORMAT	8 BITS	7 BITS VIA DATA LINES	5 BITS FOR FUNCTIONS
26	DATA RATE	500 kBAUD	10^6 BYTES/SEC IN SPECIAL CONFIGURATIONS $2{,}5\cdot 10^5$ BYTES/SEC	APPROX. 10^6 WORDS/SEC (UPPER LIMIT WITH ONE CRATE)
27	DISTANCE	UP TO 5 km	20 m	19" CRATE
28	NUMBER OF DEVICES	125	15	23 MODULES
31	MODES OF OPERATION	BLOCK MULTIPLEX	BLOCK MULTIPLEX	WORD-SERIAL
32	SERVICE REQUEST	a.) CYCLIC POLLING b.) 24 REQUEST BITS AND CYCLIC POLLING	GENERAL DEMAND+SERIAL POLL; PARALLEL POLL FOR 8 DEVICES POSSIBLE	1 REQUEST LINE PER MODULE
33	SYNCHRONISATION	SYNCHRONISING CHAR.	3-WIRE HANDSHAKE	2 CENTRAL CLOCK PULSES
34	ERROR DETECTION	CRC AT AND OF EACH 16 BITS	8th BIT WITH ASCII POSSIBLE FOR PARITY	
41	UNIDIRECTIONAL SIGNALS	BUS ONE OR TWO (41–42); COUPLER TO 6 DEVICE	2	65 BUS LINES AND 46 BRANCH LINES
42	BIDIRECTIONAL SIGNALS			
43	CONNECTOR PINS	NOT SPECIFIED (43–44)	25	86 AT THE MODULE
44	TECHNOLOGY	14 OUT OF 25, SP 1126A	SINGLE-ENDED TRANSFER, TTL- COMPATIBLE	SINGLE-ENDED TRANSFER TTL-COMPATIBLE
50	SPECIAL FEATURES	DEVICES GALVANICALLY ISOLATED	STATION-TO-STATION TRANSFER; SEVERAL LISTENERS FOR ONE TALKER POSSIBLE	LIMITED TO THE DEVICES WITHIN ONE FRAME

Source: Ref. 7.

TABLE 9.4 (Continued)

CAMAC BRANCH HIGHWAY	PDP 11, UNIBUS	SYNCHRONOUS DATA LINK CONTROL (SDLC)	FUNCTIONS AND ELECTRICAL CHARACTERISTICS
EURATOM, ESONE COMM.	DIGITAL EQUIPMENT CORP.	IBM	CCITT
EUR 4600, TID 25876	PDP 11, PERIPHERALS'HANDBOOK	GA 27-3093-C	V.24 AND V.28, GREEN BOOK
1972	1974	3.74	1973
CHANNEL TYPE INTERFACE FOR LABORATORIES, NUCLEAR PHYSICS	UNIFIED INTERFACE FOR PROCESSOR, MEMORY, PERIPHERALS	TELEPROCESSING IN EXTENDED COMPUTER NETWORKS	INTERFACE BETWEEN COMPUTERS OR TERMINAL UNITS ON THE ONE HAND AND MODEMS ON THE OTHER
COMPLETE, EXCEPT PROCEDURES	COMPLETE	PROCEDURE FOR APPLICATION OF BIT-SERIAL INTERFACES (e.g., V 24)	OPERATIONS AND ELECTRICAL SIGNALS, CONNECTOR
BUS AND BRANCH LINES	BUS	POINT-TO-POINT, MULTIPOINT, LOOP, ALL OF THEM DUPLEX OR SEMI-DUPLEX	FROM ONE POINT TO ANOTHER (THE LINE PROCEEDING FROM THE MODEM MAY HAVE A BUS CONFIGUR.)
ADDRESS, COMMAND AND DATA PARALLEL	PARALLEL, ADDRESS+DATA OR ADDRESS + COMMAND	BIT - SERIAL	BIT - SERIAL, DUPLEX
24 BITS, BIDIRECTIONAL	16 BITS	VARIABLE	PROCEDURE DEPENDENT
1 ADDRESS LINE EACH FOR 7 CRATES, 5 BITS MODULE ADDRESSES, 4 BITS FOR SUBADDRESSES	18 BITS	8 BITS	PROCEDURE DEPENDENT
5 BITS FOR FUNCTIONS, 12 CONTROL SIGNALS	16 BITS	8 BITS	PROCEDURE DEPENDENT
APPROX. $6 \cdot 10^5$ WORDS/SEC	UP TO 500 kWORDS/SEC IN BLOCK MULTIPLEX	NOT SPECIFIED; OVERHEAD PER FRAME: 40 BITS	UP TO 20 kBAUD
APPROX. 20m	15 m	NOT LIMITED	TO THE MODEM: 15 m; FROM THE MODEM: NOT LIMITED
7 CRATES, WITH 23 MODULES EACH	20 UNITS TOTALING UP TO 4096 DEVICES	1 PRIMARY STATION 256 SECONDARY STATIONS	PROCEDURE DEPENDENT
WORD - SERIAL	BLOCK MULTIPLEX CHARACTER MULTIPLEX	BLOCK MULTIPLEX	PROCEDURE DEPENDENT, IN GENERAL BLOCK TRANSFER
COLLECTIVE REQUEST AND POLLING	4 LEVELS; DAISY-CHAIN; GENERATION OF ORIGIN ADDRESS	POLLING	PROCEDURE DEPENDENT
HANDSHAKE	HANDSHAKE	FRAME SYNCHRONISATION BY FLAG BYTE 01111110	BIT SYNCHRONISATION BY CLOCK PULSE LINE: BYTE AND MESSAGE SYNCHRONISATION PROCEDURE DEPENDENT
	NONE	CRC OF 16 BITS PER BLOCK	PROCEDURE DEPENDENT
34	5		MIN. 4; MAX. 48
24	51		
65	72		MAX. 50
SIMILAR TO TTL	TTL, SINGLE-ENDED TRANFER		V.28 SIMILAR TO SP1163A THRESHOLD ± 3V
	DMA	SDLC IS AN IMPLEMENTATION OF HDLC BUT USES MORE SPEC'S	THESE STANDARDS CONSTITUTE A LIST OF ALTERNATIVES FOR VARIOUS APPLICATIONS. NEITHER PROCEDURES NOR SYNCHRONISATION SPECIFIED.

there are many very specific buses tailored to specific process control computers. These were developed independently by computer manufacturers.

Table 9.4, which was taken from Ref. 7, shows the specification of several process computer interfaces of various manufacturers. The details, such as communication protocols, levels of electrical signal, dimensions, and costs, are not taken into consideration in this list. In general, terms for the selection of interfaces can be as follows:

Area of application
Type of data communication
Technology

The criteria shown in Table 9.4 allow the user of process control computers to evaluate the performance of various interface systems.

REFERENCES

1. H. Kürner, Prozessrechner-Schnittstellen, Regelungstechnische Praxis, 1975, Heft 11.
2. H. Klassmann, Die serielle Datenübertragung im CAMAC-System zur dezentralen Datenerfassung und Prozeßsteuerung. GFK-GI-GMR Fachtagung Prozessrechner 1974.
3. Buxmeyer, E., Serielles Bussystem für industrielle Anwendungen unter Echtzeitbedingungen (PDV-Bus), KfK-PDV-Bericht Nr. 70, Mai 1976.
4. IEEE Standard 488-1975, Digital Interface for Programmable Instrumentation, New York, 1975; Entwurf DIN IEC 66.22, Elektrische Meßtechnik, Byteserielles, bitparalleles Schnittstellensystem für programmierbare Meßgeräte, Beuth-Verlag Berlin, 1976.
5. The Ethernet. A local area network, Data Link Layer and Physical Layer Specification, Version 1.0, Xerox Corporation, Stamford, Conn., September 1980.
6. International Organization for Standardization (ISO), Reference Model of Open Systems Interconnection Document Nr. ISO/Tc97/SC16 N227 (June 1979).
7. G. Haußmann and P. Mielentz, Interfaces Between Computers and Their Peripherals, Second regional meeting of Purdue Europe, Zürich, March 1976.

10

INTERFACE PROBLEMS WITH MICROPROCESSORS

10.1 INTRODUCTION

The microprocessor was made available to industry approximately 10 years ago. In that time large-scale integration (LSI) of semiconductor components matured to a point where it became economical to implement an entire CPU, a memory module, or an I/O circuit on an individual silicon chip. As a matter of fact, today it is possible to build a complete simple computer on one chip. In principle a microcomputer system can be conceived similar to that of a control computer system. The function of a control computer was explained in Chap. 2. The majority of all microcomputers are bus oriented. Therefore such a computer can be designed from individual LSI modules and can be very closely tailored for a special control application. The microcomputer communicates with the process with the help of similar input and output components and circuits as they are discussed in this book. However, in many applications interfacing of a microcomputer with a process may be much simpler because of the availability of very many specially designed I/O modules. These circuits include parallel I/O modules and serial-to-parallel conversion modules for digital signals, multiplexers, A/D and D/A converters, and entire disk controllers.

In general microcomputers are used for less complex process control tasks. Typically they are employed in programmable controllers, as a functional entity in an analog controller, as a machine tool controller, or as a satellite computer at the lowest level of a hierarchical computer system (Secs. 4.4 and 4.5).

There is a large spectrum of different microprocessors available, and it can be a very difficult task to select the right processor and I/O components for a particular application. Whereas with control computers the 16-bit processor has become an almost universal industry standard, microprocessors are designed as 4-, 8-, and 16-bit machines. A 4-bit microprocessor can, for example, be a useful device to control a small machine tool which only has digital control elements, and a 16-bit microprocessor would be a logical selection for an control application where many analog controllers are supervised and where an optimization task has to be performed.

The reader may visualize that it is impossible to describe within the framework of this book all different microprocessor and microcomputer concepts. In this chapter the operation of different I/O devices for typical 8-bit processors will be discussed, and an example will be given on how to select a microcomputer system for a model process. The architecture of a central processing unit was described in Chap. 2, and the functions of an universal bus system were discussed in Chaps. 2, 4, and 6. The reader should know the contents of these chapters when trying to familiarize oneself with this chapter.

In Sec. 4.5 future trends in interface technology were discussed. It was shown how a conventional computer control system built from discrete circuits could be substantially improved by the use of LSI circuit modules (Figs. 4.38 and 4.41). This improvement was realized by substituting different components of the control system such as the multiplexer, A/D and D/A converters, and communication system with newly developed integrated circuits which were capable of performing entire subfunctions of the systems. The microprocessor had an important role in this modernization process (Fig. 4.41).

10.2 INPUT/OUTPUT DEVICES FOR A MICROCOMPUTER SYSTEM

10.2.1 A Typical Microcomputer Configuration

A typical microcomputer system for a control application is shown in Fig. 10.1. The heart of the computer is the microprocessor (CPU), the operation of which was described in Chap. 2. There are two memory modules, one read-and-write random access memory (RAM) and one read-only memory (ROM) (Sec. 2.7).

The RAM module is the working memory and serves for storage and manipulation of process operating parameters, such as measurement data, set points, and upper and lower control limits. The ROM module contains the operating system or monitor and the application program. During the execution of the application program, individual instructions will be read out of the memory, brought to the control unit, and executed. In case the process is optimized with an optimization algorithm, this program will also be located in the ROM (Sec. 1.2.7). The parallel digital input/output module allows process communication with digital control elements such as relays, alarm devices, and indicator lights. The serial input/output module serves as an interface for data peripherals such as a CRT or a keyboard. It accepts parallel information from the microcomputer bus and converts it

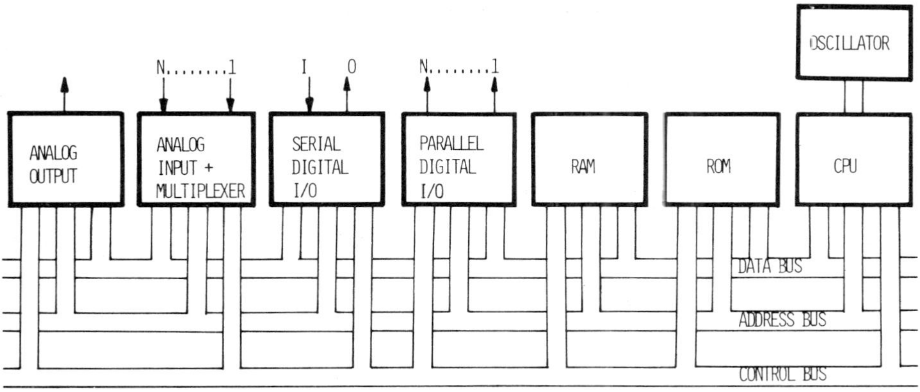

FIG. 10.1 Typical microcomputer configuration. (From Ref. 9.)

to serial information so that it can be sent in serial form over a communication line. This device also can be operated in reverse mode. The A/D and D/A converters process signals coming from or going to analog peripherals such as measuring instruments, sensors, and actuators (Sec. 5.3). The individual modules of this computer system are interconnected with a universal bus system consisting of the data, address, and control buses (Secs. 2.5, 4.3, and 6.5). In the following sections different input/output modules for such a computer system will be described. It is very typical with many of these modules that the I/O control circuit and the interface circuit are incorporated into one module (Fig. 1.20, Sec. 1.3). This concept often makes interfacing tasks with microcomputers much simpler than with control computers (Figs. 4.38 to 4.41).

10.2.2 Parallel Input/Output Module

This module is used for parallel input and output of digital process data. The communication in a microcomputer system is done in a word-serial mode which means that for the example of the 8-bit computer, it is possible to send 8 bits of information in parallel to an output device. Likewise it is also possible to read 8 bits of information simultaneously from an input device (Fig. 10.2). There are several I/O devices available which are based on this concept. One module is the programmable input/output (PIO) interface from Intel, and an other one is the peripheral interface adaptor (PIA) from Motorola. Figure 10.3 shows a schematic diagram of a PIA device [1].

On the computer side, this module is connected to the data bus via eight input/output lines D0 to D7. With the help of the chip select or register select lines it is possible to address internal registers.

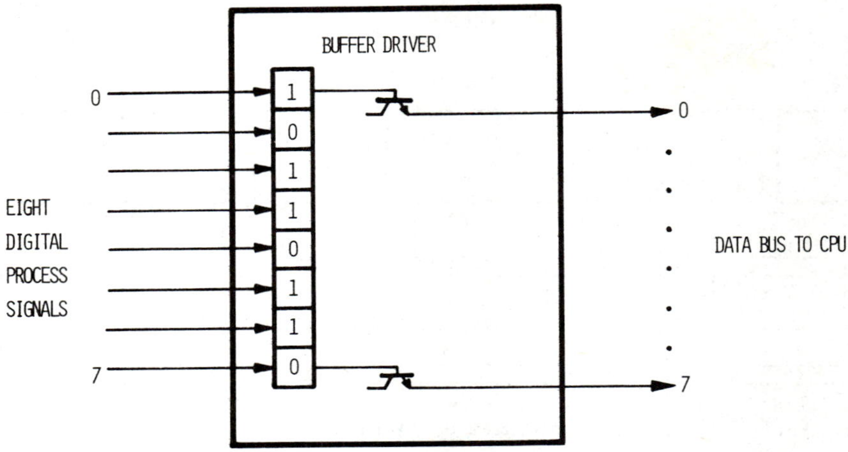

FIG. 10.2 Principle of parallel data input.

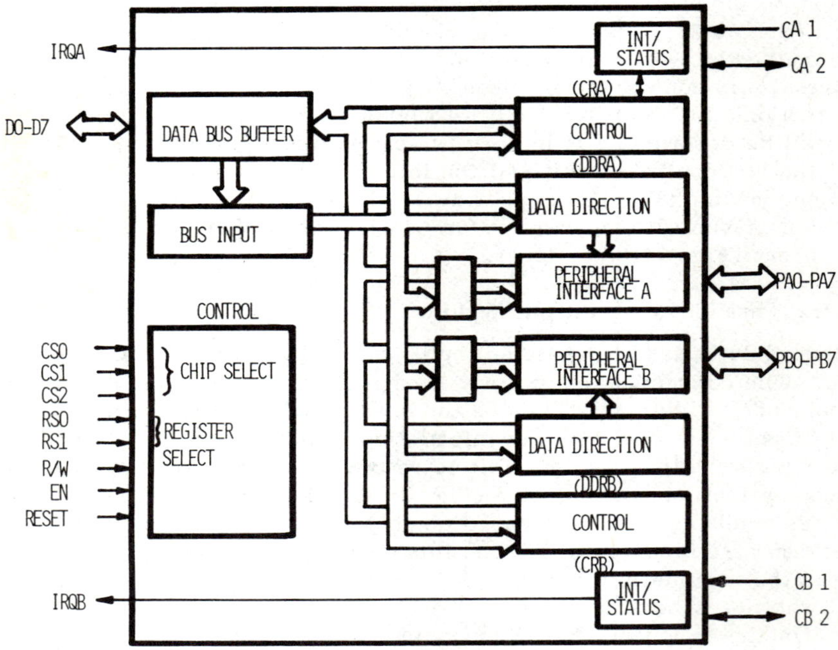

FIG. 10.3 Motorola 6820 PIA. (From Ref. 1.)

The R/W line specifies the read or write mode. The Enable pulse EN is used to condition the PIA's internal interrupt control circuit and also for timing of the peripherals control signals. A reset line allows the device to be initialized. There are two interrupt line IRQA and IRQB one for each output channel.

On the process side there are two individual data channels. Each of which can exchange 8 bits of input or output information with the peripheral. Each channel has two control lines, for example, channel A has the lines CA1 and CA2. Over these lines status information, such as buffer full, transmission acknowledged, or send the next byte of information, may be exchanged between the interface and the device. Some parallel input/output modules contain three output channels. In order to minimize interfacing efforts for many applications, it would even be desirable to have more output channels. Their number, however, is limited by the number of connector pins available at the module.

The module has several registers and one buffer. With their help it is possible to program the direction of data transfer over the indiv-

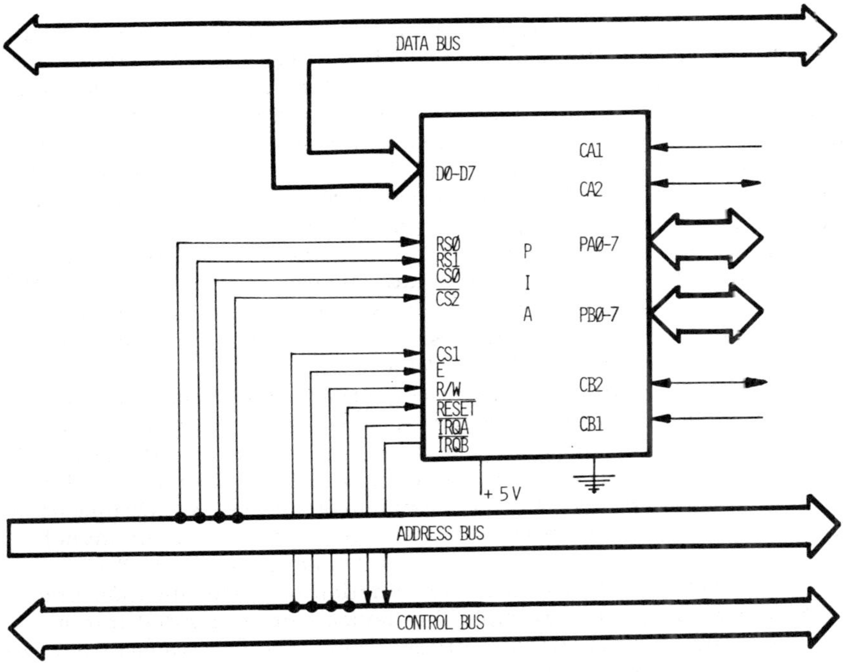

FIG. 10.4 Motorola 6820 and 6800 interface. (From Ref. 1.)

idual lines of the two input/output channels. The data bus buffer serves as temporary buffer for information which comes from or is sent to the computer. The data direction register is used to store direction instructions. The data transfer direction for each individual line can be programmed by placing a HIGH or LOW signal into the bit position which controls an individual input/output line. The control register stores command information which were issued by the program. It will determine for each channel whether or not an interrupt signal is to be issued or when a buffer is full or empty. It also generates status information for a device by setting a corresponding bit (Sec. 7.2.1).

Figure 10.4 shows how the device will be connected to the different bus lines of the computer [2]. The device will be addressed by the memory mapping method (Sec. 6.5). It is not necessary that the module occupies all lines of the address bus since there are usually by far fewer input/output devices connected to the CPU than there are memory locations to be addressed. If, for example, with an 8-bit computer there are more than eight peripherals to be connected to it than an address decoder has to be provided to address the interface chip via the chip select lines CS.

10.2.3 Parallel to Serial Input and Serial to Parallel Output

When digital data have to be sent over a long distance, serial communication lines are in general used (Sec. 4.2). They offer a significant cost advantage over parallel communication lines. Also many data peripherals such as teletypewriters and CRTs are designed for serial input and output of data. The microcomputer of Fig. 10.1 has a serial input/output module connected to it. When the module is addressed, it is capable of accepting an 8-bit word from the data bus. Internally this 8-bit word will be converted to an 8-bit serial message and then sent over the communication line to the peripheral (Fig. 10.5). In a similar fashion 8 bits of serial information can be strobed into the module and converted to parallel information and placed on the data bus of the computer for further processing. The principle of this conversion technique is shown in Figs. 5.74 and 5.75 (Sec. 5.6.2).

A module to perform this data conversion and transmission for an asynchronous input/output is called an *universal asynchronous receiver transmitter* (UART) (Sec. 4.2.1). There are also devices available which handle synchronous data transmission or both. The INTEL 8251 universal synchronous asynchronous receiver transmitter (USART), which can be used as both a synchronous and an asynchronous communication device, will be explained in more detail (Fig. 11.6) [1]. Table 10.1 describes the function of the connection lines of the device [3,4]. In principle the device can be divided into a unit consisting of 4 functional entities. They are the transmitter, receiver, data buffer, and control unit.

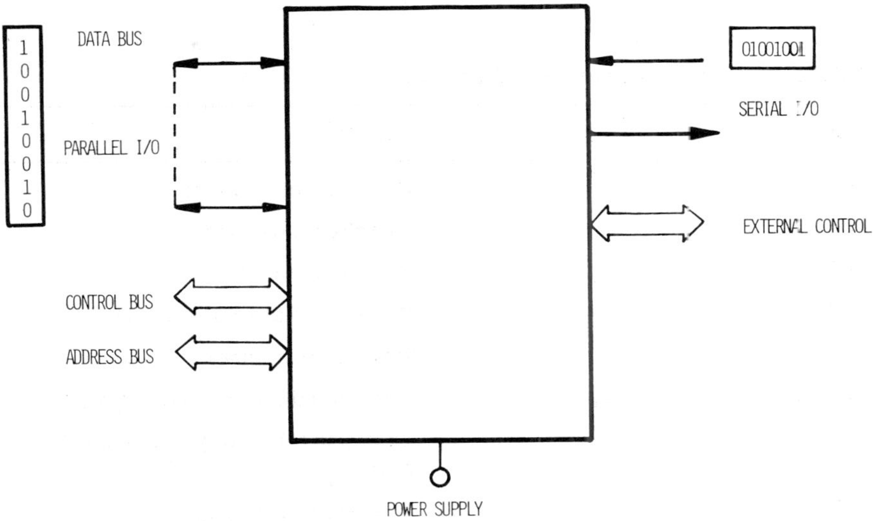

FIG. 10.5 Serial digital I/O. (From Ref. 9.)

During an output operation by the computer, the module serializes the word to be transmitted and automatically attaches a start bit and two stop bits to it (Sec. 4.2.1). In addition, a parity bit is inserted which is used by the receiver to verify the correct transmission. The parity can be odd or even. The length of the word to be transmitted by the USART can be 5 to 8 bits. The operation of the USART is initiated by a control word which will set its mode to synchronous or asynchronous transmission. The next byte will be an instruction to set the word length and additional transmission control bits (Table 10.2) [1].

Data transfer between transmitter and receiver will be initiated by a handshake procedure. The receiver needs timing signals which are synchronized with the incoming bit stream. With the help of these timing signals, which are provided by an external clock, the individual bits can be recognized. The receiver strips the control bits off the incoming word and performs a parity check and a digital-to-parallel data conversion. The parallel data word will be stored in the data register of the receiver for further processing by the peripheral. Figure 10.7 shows how a USART device is connected to the bus system of the computer [1]. The external clock signal is provided by the baud-rate generator (Sec. 4.1.1). Its clock rate must be adjusted to the operation mode of the peripheral. For example, a teletype operates at 110 Bd, and a CRT typically at 9600 Bd.

TABLE 10.1 Line Functions of USART

Line	Function
CS	Chip select (device)
SYNDET	Synchronous mode
$\overline{\text{RxC}}$	Receiver clock
RxD	Receiver data (serial in)
TxC	Transmitter clock
TxD	Transmitter data (serial out)
TxDRY	Transmitter ready (to accept data)
TxE	Transmitter empty
RxRDY	Receiver ready (character ready)
D0-D7	Data bus
RESET	Resets device to idle mode
CLK	Internal clock
C/D	During read selects status or data to be input to CPU
	During write interprets data as command or data
RD	UART gates status or data on data bus
$\overline{\text{WR}}$	UART accept command or data from data bus
Modem control	
$\overline{\text{DSR}}$	Data set ready
$\overline{\text{DRT}}$	Data terminal ready
$\overline{\text{CTS}}$	Clear to send
$\overline{\text{RTS}}$	Request to send

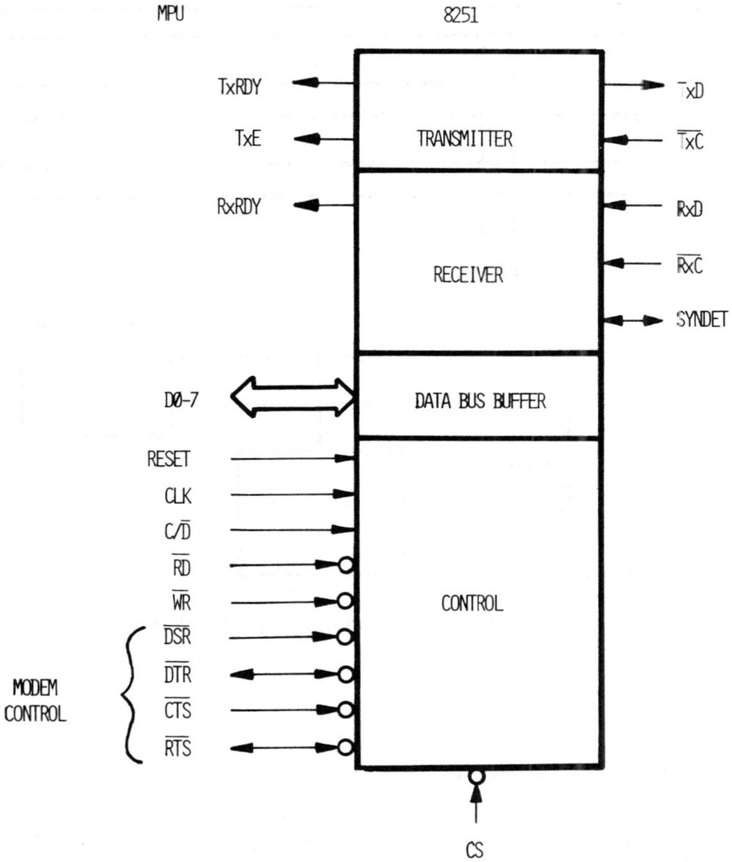

FIG. 10.6 INTEL 8251 USART. (Courtesy of SYBEX, Berkeley, California.)

TABLE 10.2 USART Commands

C/D	RD	WR	CS	Operation
0	0	1	0	8251 to Data Bus (Read)
0	1	0	0	Data Bus to 8251 (Write)
1	0	1	0	Status to Data Bus
1	1	0	0	Data Bus to Control
–	–	–	1	Data Bus to 3-State

Source: Ref. 1.

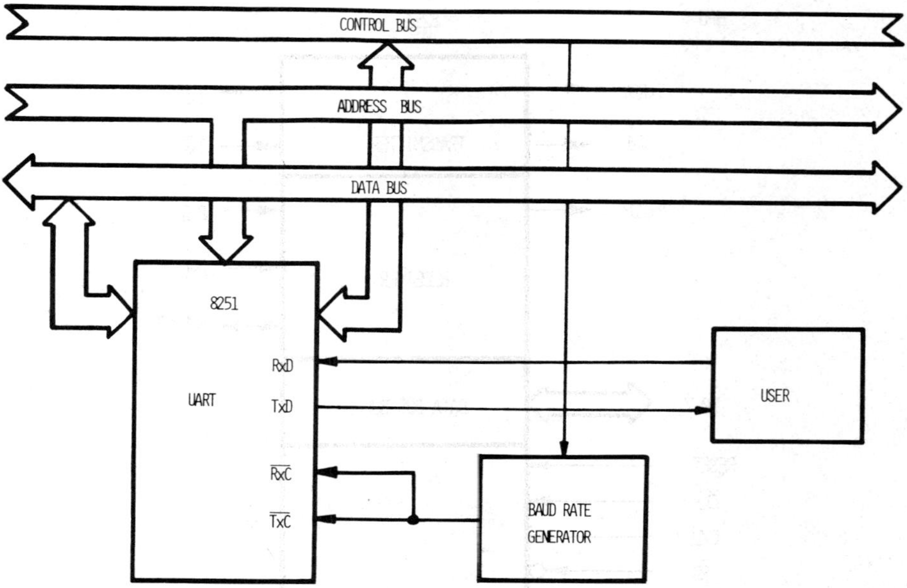

FIG. 10.7 Typical USART application. (Courtesy of SYBEX, Berkeley, California.)

10.2.4 Analog-to-Digital Converter

The microcomputer system shown on Fig. 10.1 is capable of reading analog data from a measuring device with the help of an A/D converter. The principle of A/D conversion was discussed in Sec. 5.3. There are several monolithic interface modules available which contain a converter, sample-and-hold unit, output register, and a control logic. A multiplexer may also be integrated (Sec. 5.6) in this device.

Depending on the design of the module or the application, there are several methods available to integrate the device with the bus system of the microcomputer. The simplest method is to connect the A/D converter directly to the three bus lines (Fig. 10.8) [5]. Here the three A/D converters are addressed via the chip select lines by the processor using the address line. The number of A/D converters which can be addressed in this design depends on the number of available address lines of the bus and the number of chip select lines needed to activate the chip. The system shown in Fig. 10.8 can only operate in a polling mode. Whenever a reading is to be taken, a module will be addressed by the program, and the analog signal will be strobed in, converted, and brought to the CPU. After completion of

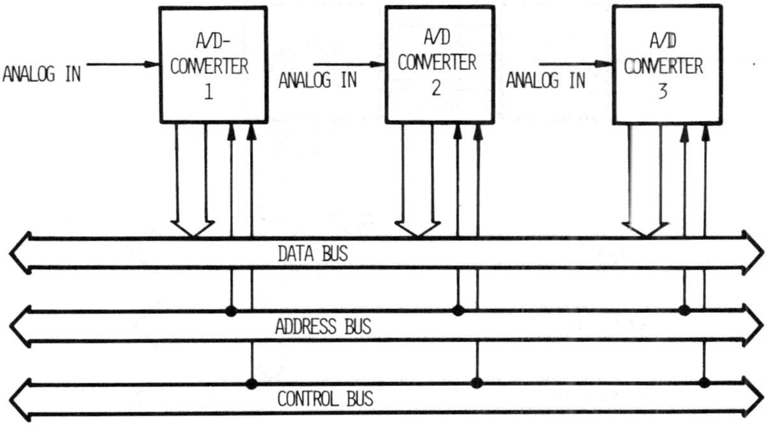

FIG. 10.8 Direct connection of A/D converter to a bus system. (Courtesy of Intersil Inc., Cupertino, Calif.)

this entire operation, the program is continued to perform another task. There are no provisions for an interrupt. If data acquisition under interrupt control is required, a different circuit has to be conceived (Secs. 6.3.3 and 7.2.1).

Figure 10.9 shows an application where the A/D converter can be active at all times. This interface can be operated in read-anytime mode. An parallel I/O module (Figs. 10.3 and 10.4) serves as an interface between the bus and the A/D converter. For the operation of the A/D converter the number of available I/O channels of the parallel I/O module is of importance. A two-channel module can accomodate two 8-bit or one 12-bit A/D converters, whereas a three-channel module will be able to operate three 8-bit or two 12-bit converters. Addressing of the A/D converter is done by addressing the parallel I/O module. This module operates the A/D converter. With the help of control information in its buffers it is capable of keeping the A/D converter active at all times. The interrupt line of the converter may or may not be used depending on the application. There are several A/D converters on the market which have to be connected to the microcomputer via such a parallel I/O module. The possible elimination of this module offers a cost advantage. In this section the function of an 8-bit A/D converter (ADC0808/ADC0809) of the National Semiconductor Corporation will be described [6]. This device is equipped with an 8-bit successive approximation A/D converter, an eight-channel multiplexer, an address input latch, a data output latch, and a control logic (Fig. 10.10 and Table 10.3). Eight analog signal lines can be connected to the multiplexer.

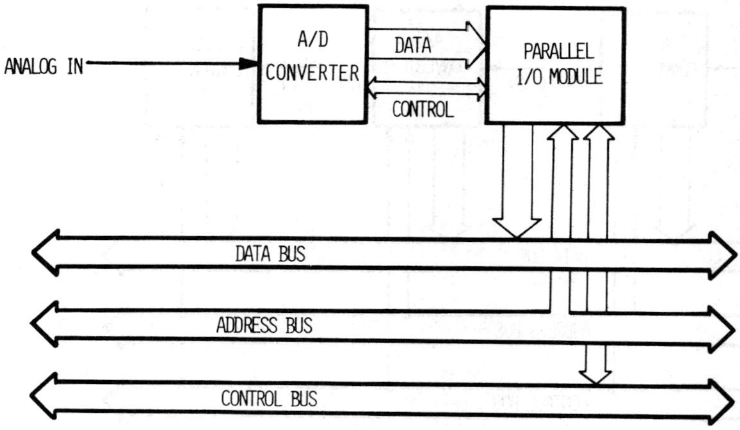

FIG. 10.9 Data conversion using a parallel I/O module. (Courtesy of Intersil, Inc., Cupertino, Calif.)

TABLE 10.3 Line Functions of ADC0808/ADC0809 A/D Converter

Line	Function
IN0-IN7	Analog input lines
A B C	Address of analog input lines IN0-IN7
ALE	Clocks address of IN0-IN7 into multiplexer address register
CLOCK	Clock signal (640-1280 kHz)
START	Start conversion
EOC	Interrupt upon completion of conversion
OE	Output enable (acknowledge)
D0-D7	Data output lines
REF (-), (+)	Reference voltage
V_{cc}	Supply voltage

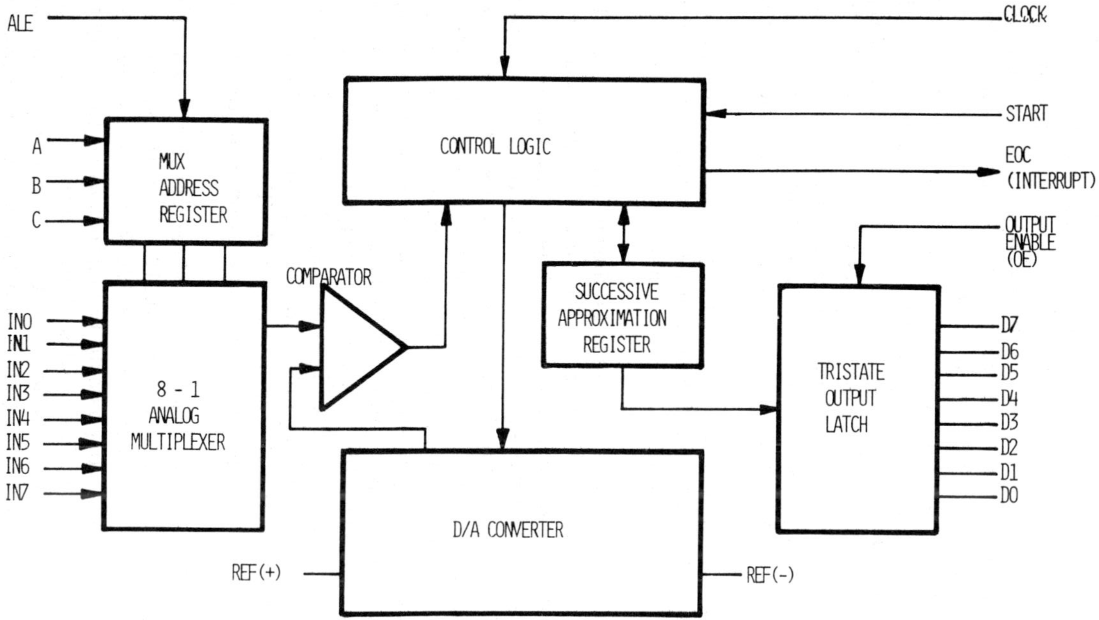

FIG. 10.10 An A/D converter with an integrated analog multiplexer. (Courtesy of National Semiconductor GmbH, Fürstenfeldbruck, Federal Republic of Germany.)

To initiate a conversion, an input channel of the multiplexer is addressed by a software instruction via input A, B, and C. With the help of line ALE the address is clocked into the multiplexer address register. The conversion begins when a signal is placed on the START line. The rising edge of this signal first clears all registers, and the falling edge starts the operation of the successive approximation converter. The function of such a converter was described in Sec. 5.3.3 (Fig. 5.12). The analog voltage to be converted is brought to one leg of a comparator. To the other leg the reference voltage is brought from the D/A converter. The results of the successive comparisons are stored in the successive approximation register. Upon completion of the conversion, the output line EOC goes high to signal to the computer that data are ready to be read (interrupt request) (Secs. 6.3.3 and 7.2.1). The output enable (OE) is raised high by the acknowledge signal of the computer allowing the data to be read via the tristate outputs D0 to D7.

When connecting this A/D converter to a microcomputer, the designer may use different interfacing methods as was discussed above. Two different methods will be described using a ZILOG Z80 computer [6].

An inexpensive simple interface is shown in Fig. 10.11. In order to select the interface the OE port enable line is addressed for this purpose. To it one line of the address bus in this case line A7 is connected. The address lines A, B, and C for the analog input lines are connected to the data bus lines D0 to D2. Information on these lines which has to be placed by the program on the bus will be selecting the analog input lines. The information is removed when the channel has been selected. A signal on line A7 will also reset the interrupt request flip-flop and activate inputs ALE and START. Upon completion of the conversion, the EOC line issues an interrupt request by setting an interrupt flag, and the computer is able to strobe the digitized analog information from the D0 to D7 lines into its register. The two D flip-flops at the bottom of the figure divide the 2-MHz clock signal of the computer to the clock frequency necessary to operate the A/D converter.

An improved (however, more expensive) version of an interface to the Z80 computer is shown in Fig. 10.12. In this design an address decoder is used for the selection of the I/O chip. This increases the address space and makes it possible to connect more than eight peripherals to the CPU. The first 3 bits of the address are used to select the analog input lines, and the last 5 bits to select the chip address ports. The read (RD), write (WR), and I/O request (IOREQ) lines are brought individually to the interface, thus making independent control of all lines possible.

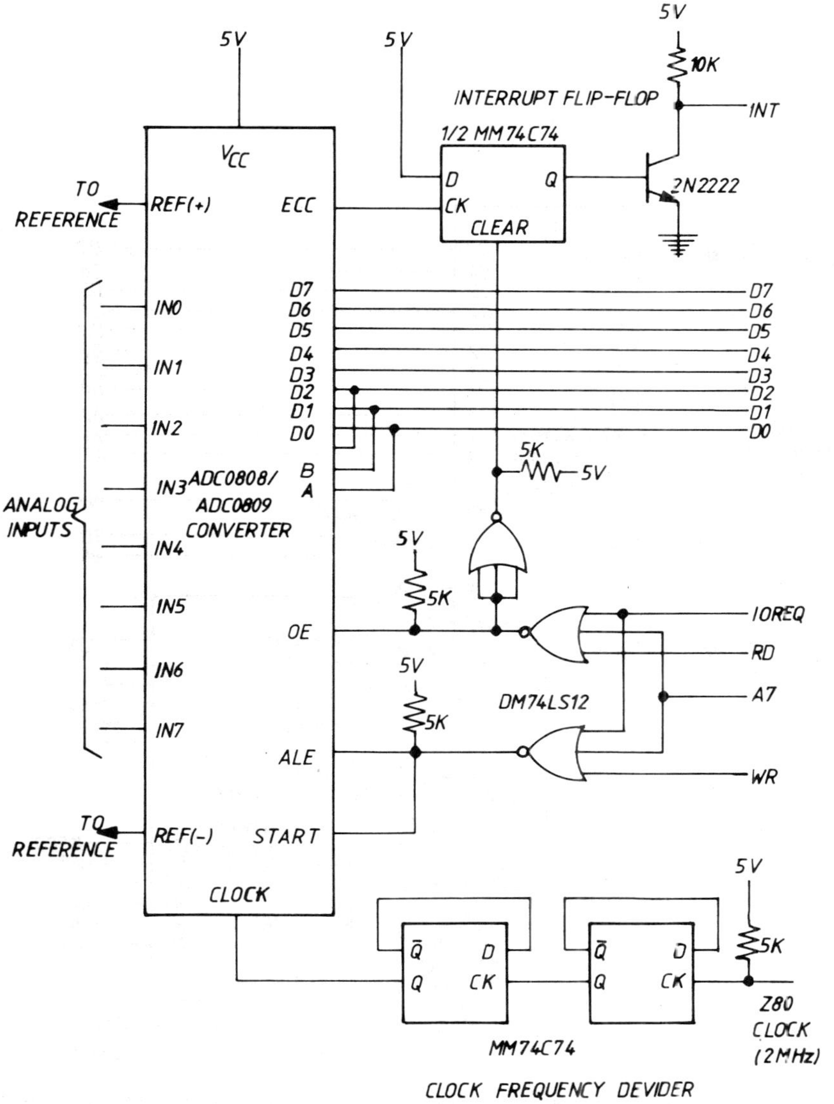

FIG. 10.11 An A/C converter with a simple interface. (Courtesy of National Semiconductor GmbH, Fürstenfeldbruck, Federal Republic of Germany.)

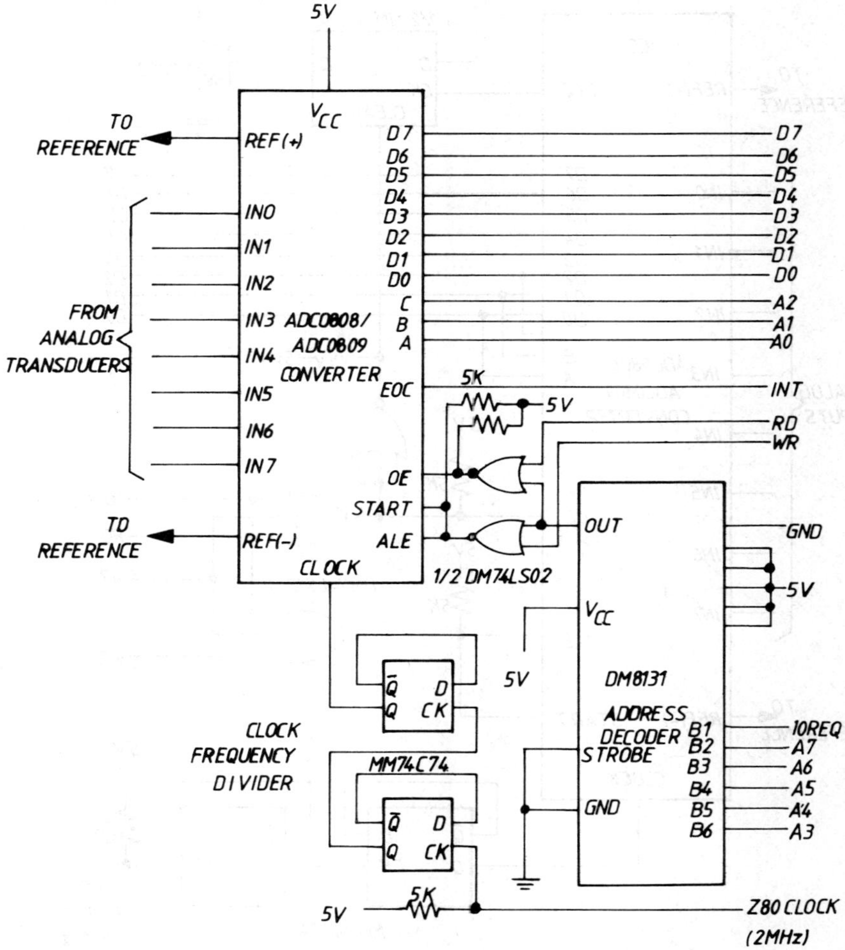

FIG. 10.12 An A/D converter using an address decoder. (Courtesy of National Semiconductor GmbH, Fürstenfeldbruck, Federal Republic of Germany.)

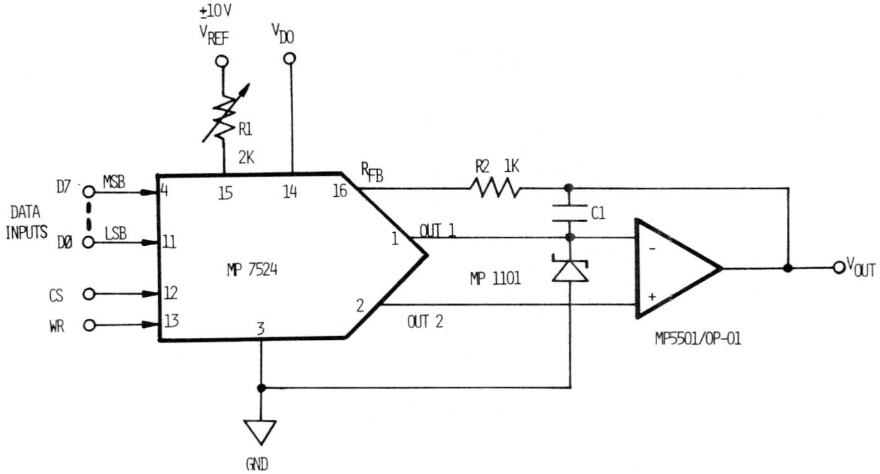

FIG. 10.13 A/D converter MP 7524. (From Ref. 7, courtesy of Micro Power Systems, Inc., Santa Clara, California.)

10.2.5 Digital-to-Analog Converter

When an analog control signal, for example, a new set point for an analog controller has to be transmitted to the process, the microcomputer system has to be provided with a D/A converter (Fig. 10.1). The operating principle of a D/A converter was discussed in Sec. 5.3.3 with the help of Fig. 5.11. A typical D/A converter for microcomputer is shown in Fig. 10.13 [7]. When the microcomputer wants to send an analog signal to the process, it addresses the D/A converter via the chip select input CS and places the information to be converted on the data bus (D0-D7). It strobes the information into the D/A converter by issuing a write command. The data enters the converter when both lines WR and CS are LOW, and it is latched in when WR returns to HIGH. The chip select input CS may either be connected directly to one of the address lines, or if more address space is required, it may get its signal from the output of an address decoder. The converter shown in Fig. 10.13 has an operational amplifier connected to outputs OUT1 and OUT2. The resistors R1 and R2 are used for gain adjustment. The capacitor C1 prevents the amplifier from oscillating or ringing.

10.2.6 Multiplexer

Multiplexers are used when several input or output signal lines are to be connected to one data converter. The A/C converter discussed in

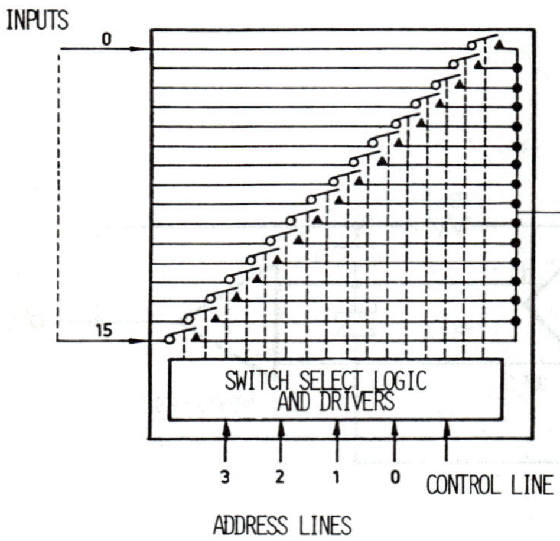

FIG. 10.14 Multiplexer. (From Ref. 9.)

Fig. 10.10 had a multiplexer integrated on the same chip. In many cases, however, it is desired to have an individual multiplexer in the microprocessor system. The concept of such a multiplexer is shown in Fig. 10.14 (compare also Sec. 5.6.2) [2]. The multiplexer is activated with the help of line ENABLE.

The individual signal lines can be selected via the address lines A0 to A3.

10.2.7 The S100 Bus

The concept of different bus systems are discussed in Secs. 2.5, 4.3 and 6.5. Must microcomputer manufacturers use their own bus architecture; however, the principle of operation of most bus systems is similar. The company MITS which produces the Altair microcomputer introduced in 1976 the S100 bus for the hobby market. This well-conceived bus system soon found acceptance by many microprocessor users and computer peripheral manufacturers. It in particular finds its application where microcomputers and data peripherals are to be connected. For this reason numerous manufacturers have designed interface boards and peripherals around the S100 bus system. As its name implies, the S100 bus uses 100 bus lines. They are combined in several parallel subsystems (Fig. 10.15). All lines are operated in a unidirectional mode only. The assignment of these lines is shown in Table 10.4.

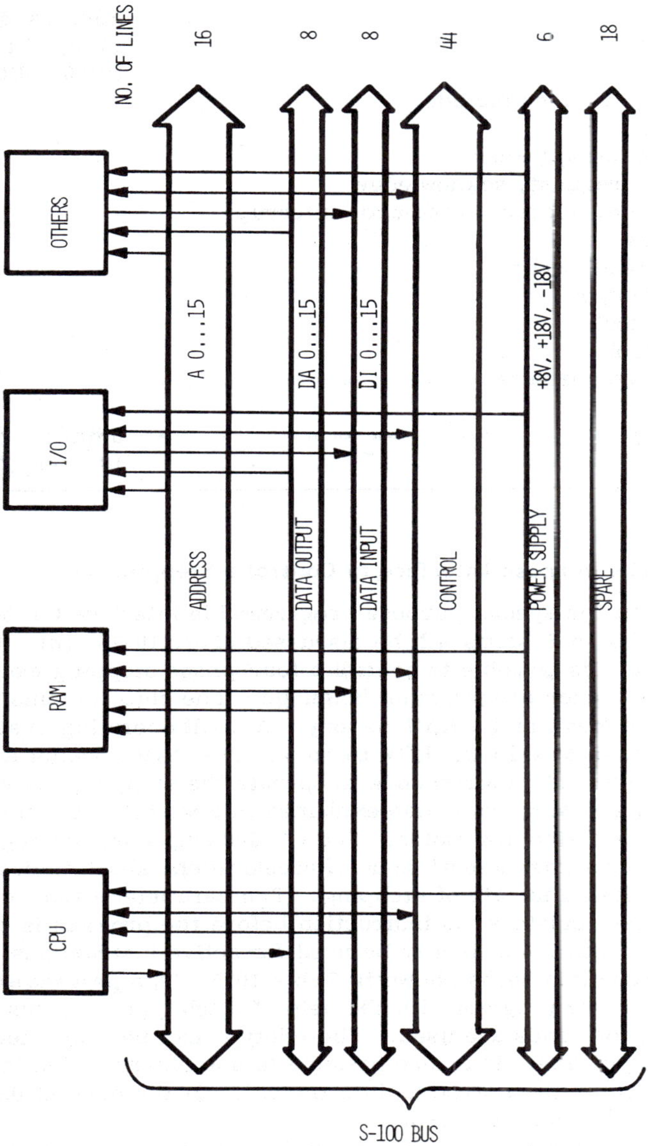

FIG. 10.15 S100 bus.

TABLE 10.4 Line Assignment of the S100 Bus

Address bus	A00...A15
Data input bus	DI00...DI15
Data output bus	DO00...DO00
Control bus: 44 lines include	
Clock lines	
Vectored interrupt lines	
Interrupt: request, acknowledge	
Enable: data, address, command, interrupt	
Status lines	
Memory: read, write	
External control	
Real time clock	
Stack control	
Processor synchronize	
Others	
Power supply	6 lines
Spare	18 lines

10.2.8 Special Purpose Interface to Control a Stepping Motor

In this section the special purpose programmable interface CY 500 of Cybernetics Micro Systems will be discussed (Fig. 10.16) [8]. With this interface it is possible to control a four-phase stepping motor with the help of a higher programming language. The chip contains a dedicated CPU, a ROM, and a RAM memory. A small operating system and a compiler are located in the ROM memory. The RAM working memory can hold 18 bytes of program code to operate the stepping motor. The function of a stepping motor was explained in Sec. 8.3 with an example of a three-phase stepping motor. The CY 500 chip contains registers for the operating parameters' slope of acceleration, absolute position, relative position, and rate of stepping. The parameter values can be stored in these registers via instruction before the program is executed. Thus all memory cells may be used for active instructions. The available instruction set is shown in Table 10.5. The programmer may develop his or her program with the help of a teletype. In this case ASCII coded commands are used. The teletype can be connected to a serial data input line. It is also possible to use binary-coded instructions and to enter the commands into the chip via the parallel data bus lines of the main microprocessor.

Program loading and execution will be explained with the help of a small example. It is desired to turn a stepping motor for 513 steps at a rate of 180 steps/per second. After completion of the programmed action, a control line (trigger) has to be checked, and if it is set, the

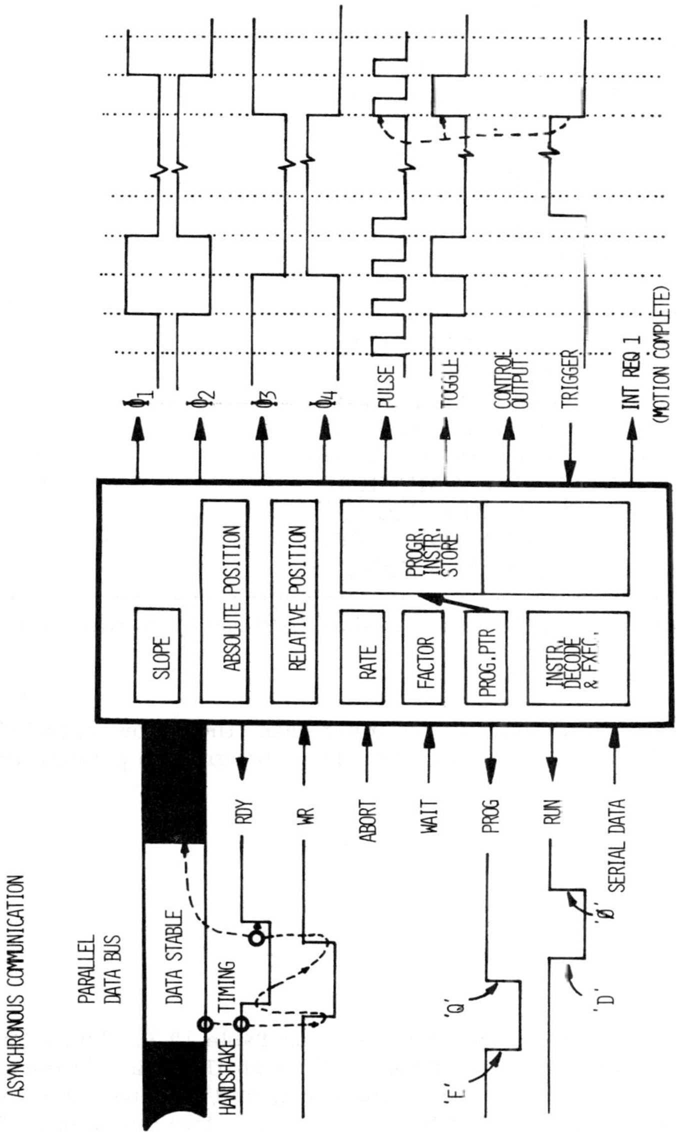

FIG. 10.16 The CY500 stepping motor controller.

TABLE 10.5 Commands of the CY500 Interface Cybernetics

A	ATHOME (DECLARE 0 POSITION)
B	BITSET (CONTROL OUTPUT=1)
C	CLEARBIT (SET CONTROL=0)
D	DO IT NOW (EXECUTE PROGRAM)
E	ENTER (PROGRAM INTO CY500)
F_f	FACTOR (DIVIDES RATE BY f)
G	GO (BEGIN STEPPING)
H	HALFSTEP MODE
I	INITIALIZE CY500
J	JOG (EXTERNAL START/STOP)
L	LEFT/RIGHT (EXT.DIR.CONTROL)
N_n	NUMBER OF STEPS n
O	ONESTEP (IMMEDIATELY)
P_p	POSITION p IS DESTINATION
Q	QUIT PROGRAM MODE
Rr	RATE OF STEPPING SET TO r
Ss	SLOPE OF ACCELERATION (±s)
T	TIL PIN 28 HIGH, REPEAT PROGRAM
U	UNTIL 'WAIT' LOW, WAIT HERE
+	SET CLOCKWISE DIRECTION
-	SET COUNTERCLOCKWISE DIR.
0	RETURN TO COMMAND MODE

Source: Courtesy of Cybernetic Micro Systems, Los Altos, California.

action will be completed. In the other case, the motor control should return to the command mode. The simple program to perform this action is shown below.

```
E
R180
N513
G
T
Ø
```

For the explanation of the commands, see Table 10.5. The instruction E will initiate loading of the program through the main microprocessor. R180 and N513 will input the stepping rate and the number of steps respectively. G signals the beginning of the stepping action. T is the command to check the status of the trigger line. Depending on the state of this line the program will either repeat itself or go back to the command mode.

When the stepping motor is to be actuated, the main microprocessor will load the individual instructions into the interface with the help of a handshaking procedure (Fig. 10.16). The instructions will be interpreted by the compiler of the chip. Because of the limited memory space available, only 18 instructions can be stored at one time. If there are more than 18 instructions, the program has to be loaded and executed in sections.

This intelligent stepping motor interface will relieve the main microcomputer from performing the repetitious task of sending individual control signals to the stepping motor and also from supervising the different modes of operation.

10.3 AN EXAMPLE OF A MICROCOMPUTER SYSTEM

In this section a microcomputer system for the control of the climate of a manufacturing facility for delicate parts will be discussed [9]. This facility consists of several rooms in which assembly operations are to be performed.

The comfort of a human being working in a work space depends on the combination of temperature, humidity, and air velocity to which one is exposed. It is possible to define an effective temperature at which different combinations of these variables will cause the same feeling of comfort. A typical range for the effective temperatures is shown in Fig. 10.17. In this example several parameters would have

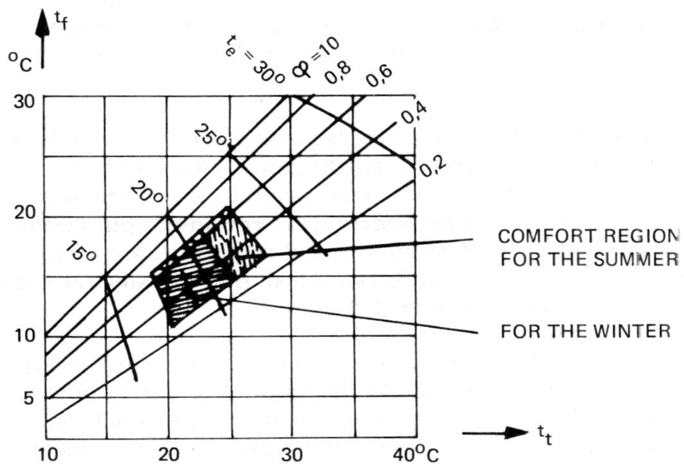

FIG. 10.17 Comfort chart: ϕ = relative humidity; t_t = dry bulb air temperature; t_f = wet bulb air temperature; t_e = effective temperature.

to be controlled. They are the dry-bulb and wet-bulb temperatures and the air movement. It would be possible to store the contents of Fig. 10.17 in a look-up table of a ROM memory. With the help of this data the computer could minimize the power consumption of the air-conditioning equipment by operating it at a point of maximum efficiency, at which the workers in the working environment still feel comfortable. For this purpose the power consumption curves of the cooling equipment would also have to be placed in the ROM memory.

In our example, only the control of the dry-bulb temperature will be discussed. No attempt will be made to explain the possible optimization procedure. The layout of the control system is shown in Fig. 10.18. Each assembly room to be controlled has its own cooling coil, temperature sensor, control valve, and alarm indicators. The microprocessor will operate in a polling mode. At predefined time intervals temperature reading will be taken from the thermocouples and interpreted (Fig. 10.19). The operating and set points will be compared. If the temperature is below 25°C, no control adjustment will be made. If the temperature is above 25°C, the control motor has to be actuated to increase the flow of coolant through the heat exchanger to reduce the air temperature of the room. If at this operating condition the temperature stays below 27°C, a green indicator light has to be set in the room. In case the temperature rises above 27°C, the green light will be turned off, and a red light and an alarm bell will be turned on.

The entire control system consists of the following components:

Sensor, Controller, and Alarm Devices. The temperature sensor (thermocouple) transmits its signal to the computer (Sec. 3.4.1). The motor will be actuated by an analog signal from the computer (Sec. 8.3). Note that the value position will not be fed back to the computer, to simplify the problem. The alarm devices will be actuated by digital signals.

Signal Transmission System. The transmission system consists of the signal lines, amplifiers, and level transformers. Because of the use of amplifiers, low-cost, shielded, twisted pair of cables are selected for the signal lines (Chap. 4 and Sec. 8.3). For the amplification of the low-level thermocouple signals operational amplifiers are used at the signal source (Secs. 3.4.1 and 5.2). All digital signals are converted from TTL levels to 48 V (Sec. 8.1.2).

Input/Output Modules for the Microcomputer. The input/output signals for process communication have to be adapted to the requirements of the computer. The individual tasks of the modules will be described later. In order for the operator to be able to communicate with the computer and the process a data peripheral is needed. It

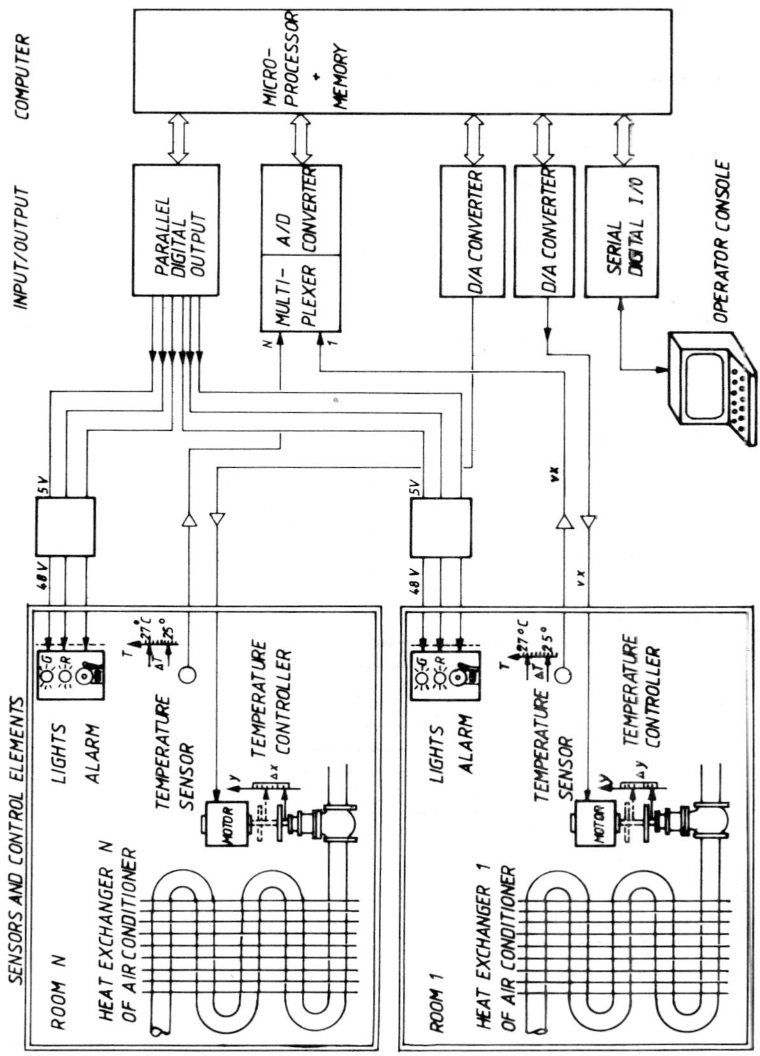

FIG. 10.18 Climate control by microcomputer. (From Ref. 9.)

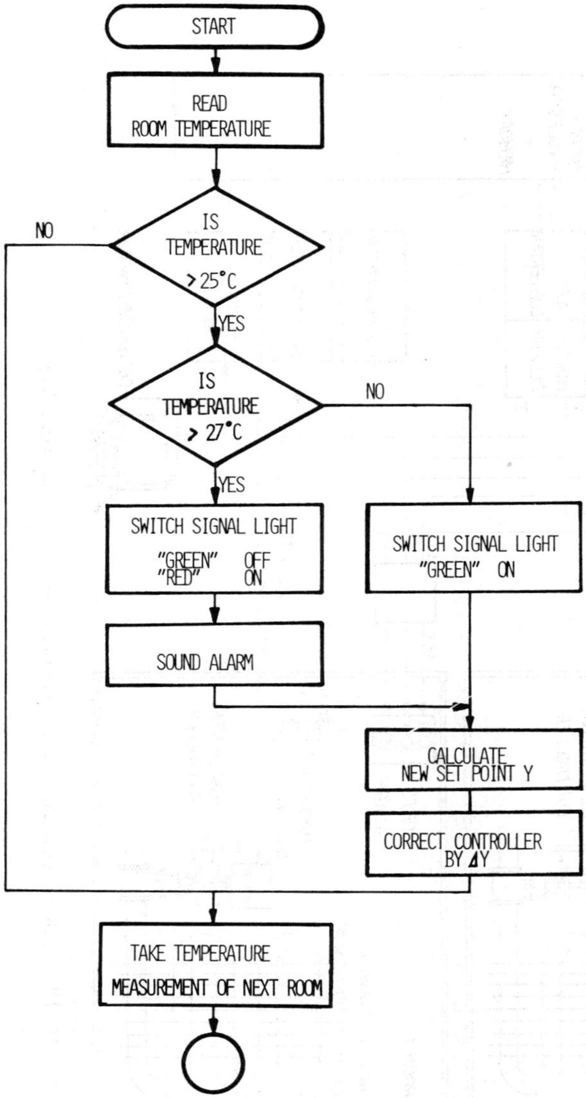

FIG. 10.19 Program module "temperature measurement". (From Ref. 9.)

serves for the input and output of process parameters and for the initiation of service and diagnostics routines.

Microcomputer. A typical computer architecture was discussed in Chap. 2. The microprocessor selected is bus oriented, and the input/output peripherals can be connected to the bus via interface modules. The computer system contains a ROM memory in which the application and optimization program and diagnostics and service routines are permanently stored. A RAM memory serves as temporary data buffer for process data, set-point variables, upper and lower process limits, and temporary results obtained from calculations with the control algorithms.

In the context of this example the selected interface modules to drive the data and process peripherals are of particular interest. Figure 10.1 shows a microcomputer system which may have been conceived for this application. The interface modules are as follows:

1. Parallel Input/Output Module: With the help of this module the signals for the indicator lights and alarm devices in the room are controlled (Sec. 10.2.2). Each room needs three digital signal lines. Thus it is possible to serve five rooms with a two-channel module. For additional signals another module would be necessary.
2. Serial Input/Output Module: This module is required to connect a data peripheral such as a teletype or CRT to the computer system for operator communication (Sec. 10.2.3). There is one module required.
3. Analog/Digital Converter: The analog signals from the thermocouples are converted to digital information by the A/D converter. Several different methods of connecting A/D converters to a microcomputer were discussed in Sec. 10.2.4. It was decided to use a A/D converter in connection with a multiplexer. Because of the slow polling process in this application analog data only arrive at long intervals. Thus a slow A/D converter in connection with a multiplexer will be able to handle all analog data. With this scheme the cost for the A/D converter will be kept at a minimum.
4. Digital/Analog Converter: The analog signals to adjust the set points for the control valves are generated by the D/A converter (Sec. 10.2.5). An individual D/A converter was provided for each control motor. This made it possible to keep the control information for a longer hold period on the line, and thus the controller had enough time to complete the necessary valve adjustment.

REFERENCES

1. Rodnay Zacks, *Microprocessors*, Sybex, Berkeley, Calif., 1977.
2. Austin Lesea and Rodnay Zacks, *Microprocessors Interfacing Techniques*, Sybex, Berkeley, Calif., 1977.
3. A. Ebright, 8255 Programmable Peripheral Interface Applications, Intel Corporation, Santa Clara, Application Notes AP-15, 1976.
4. L. Smith, Using the 9251 Universal Synchronous/Asynchronous Receiver/Transmitter, Intel Corporation, Santa Clara, Calif., Application Notes AP-16, 1976.
5. INTERSIL ICL 7109 CMOS 12 Bit Binary A/D Converters for Microcomputer Systems, INTERSIL Data Sheet, 1978, Intersil Corp., Cupertino, Calif.
6. Using the ADC0808/ADC0809 8-Bit μP Compatible A/D Converters with 8-Channel Analog Multiplexer, National Semiconductor Corporation Data Sheet, Santa Clara, Calif., 1980.
7. 8-Bit Buffered Multiplying DAC, Micro Power Systems, Inc., Santa Clara, Calif., Data Sheet, 1980.
8. CY500 Stored Program Stepper Motor Controller, Cybernetic Micro Systems, Los Altos, Calif., Data Sheet, 1980.
9. U. Rembold, Wie funktioniert ein Mikrorechner und was kann er? VDI-Bericht 348, VDI-Verlag GmbH, Düsseldorf, 1979.

INDEX